Studies in Computational Intelligence 422

Editor-in-Chief

Prof. Janusz Kacprzyk
Systems Research Institute
Polish Academy of Sciences
ul. Newelska 6
01-447 Warsaw
Poland
E-mail: kacprzyk@ibspan.waw.pl

For further volumes:
http://www.springer.com/series/7092

Joanna Kołodziej, Samee Ullah Khan,
and Tadeusz Burczyński (Eds.)

Advances in Intelligent Modelling and Simulation

Artificial Intelligence-Based Models and Techniques in Scalable Computing

Springer

Editors
Prof. Joanna Kołodziej
Cracow University of Technology
Cracow
Poland

Prof. Tadeusz Burczyński
Cracow University of Technology
Cracow
Poland

Prof. Samee Ullah Khan
North Dakota State University
Fargo, ND
USA

ISSN 1860-949X e-ISSN 1860-9503
ISBN 978-3-642-43261-3 ISBN 978-3-642-30154-4 (eBook)
DOI 10.1007/978-3-642-30154-4
Springer Heidelberg New York Dordrecht London

Softcover reprint of the hardcover 1st edition 2012

Printed on acid-free paper

Springer is part of Springer Science+Business Media (www.springer.com)

To our Families and Friends with Love and Gratitude

Preface

In recent years, there has been a growing interest in the need for designing intelligent models and methodologies to address and solve complex problems and models within the realm of large-scale distributed computing systems. One of the most challenging issues for such methods and models is to effectively manage the whole computational infrastructure of the distributed systems, such as Grids, Clouds, Clusters, and ad hoc and P2P networks that are operated by various types of users with varying relations, which change under unpredicatele enviornments. In complex large-scale distributed systems, such as Computational Grids (CGs) the IT resources usually belong to different owners (institutions, enterprises or individuals) and are managed by various administrators. Resource administrators conform to diverse sets of rules and configuration directives, and can impose conflicting usage policies on the system users. Moreover, in such complex systems there are some uncertainties, such as of the available information that is incomplete, imprecise, fragmentary, and overloading, which results in a lack of full and precise determining and selection of the evaluation criteria, the assignment scores, and the final integrated decision results.

Solving complex global optimization problems in large-scale distributed systems requires huge computations for intricate objective functions. Conventional global optimization based on stochastic and probability algorithms cannot guarantee a global optimal within a finite search iteration. Scalability is a desirable feature for the optimization techniques in highly distributed and dynamic environments, where the storage and computing capabilities can be spread over a wide geographical area. Intelligent scalable systems enable the flexible routing and charging, advanced user interactions, and the aggregation and sharing of geographically-distributed resources from different organization with distinct owners, administrators, and policies.

Artificial Intelligence-based techniques, including fuzzy logic, neural networks, evolutionary, and memetic algorithms, which are complimentary to the existing traditional techniques, have shown great potential to solve these demanding, real-world decision problems that exist in uncertain large-scale environments. These technologies have formed the foundation for the intelligent scalable computing that we know of today.

With the advent of intelligent large-scale systems, in which the efficient inter-domain operation is one of the most important features, it is arguably required to investigate novel methods and techniques to enable secure access to data and resources, flexible communication, efficient scheduling, self-adaptation, decentralization, and self-organization. The book discusses the advanced topics of theory, models, technologies, system architectures, and implementation of applications in intelligent scalable computing and systems.

The fifteen chapters of the book are structured into the following four main parts:

I. Future Generation Fuzzy Systems

This section comprises of two chapters. In the first chapter, Pedrycz details the basics of fuzzy modeling to describe large-scale distributed systems. The salient feature of such descriptive model is that it can simultaneously capture the symbolic and numerical perceptions of system modeling. Kosinski and Wegrzn-Woska's work composes the seond chapter of the book on defuzzification function approximation. The work describes an artificial neural network tool to facilitate in obtaining step ordered fuzzy numbers.

II. Parallel and Multiobjective Evo-Based Techniques and Architectures in Large-Scale Global Optimization

This sectioning is composed of four chapters. Mora et al., propose two parallelization procedure applied to three different multiobjective ACO. The salient feature of their parallization approach is to improve both solution quality and also to reduce the overall turnaround time. The benchmark their approach on the bi-criteria TCP problem. Miranda et al., tackle an important problem in bioinformatics, the multi sequence alignment. The approach that the team undertakes is significantly different from current practices by utilizing a parallel island injection genetic algorithm, which results in considerable improvement in execution times without losing solution quality. Castillo et al., benchmark SOAP and REST Web services to perform evolutionary computations. They draw some excellent conclusions and deduce that it is possible to use Web services for evolutionary computations in a traditional master-slave setting; however, SOAP should be the preffered medium of Web services. Arenas et al., complete this sectioning with a thorough review of bioinsprired methodologies on GPUs. The study provides qualitative background information on GPUs and their relations or analogous concepts with respect to CPU. The study then performs classification of bioinspired algorithms with some insightful conclusions.

III. Nature-Inspired Solutions for Intelligent Networking

Five chapters are classified in this section. Ludwig introduces a new methodology based on particle swarm optimization for designing overlay networks with data mirroring as the primary aspect to measure the quality of the overlay network. The proposed method is compared with a traditional genetic algorithm and an artificial immune system based procedure. The author shows a significant improvement in terms of system scalability, which often is the bottleneck of overlay networks. Gundry et al., develop a genetic algorithm based MANET topology control mechanism by allowing mobile nodes to adapt their respective speed and directions from a small neighboordhood of nodes within a confined space fraught with unertanities. The proposed technique show superior performance that were benchmarked with realistic case studies. Ohnishi et al., present an evolutionary algorithm that can evolve an unstructured P2P network into a highperformance structured P2P network. The proposed technique exhibits great promises to realize quick, realiable, and accurate P2P resource search and discovery. Kolodziej et al., contribute by providing a thorough overview and taxonomy of evolutionary algorithms for green computing. The work classifies existing problems and corresponding solutions, which may enable future generation researchers to tackle outstanding issues within the general domain of green computing. Dobre et al., deliver the final chapter in this sectioning by developing a simulation model within the domain of P2P systems when critical decisions must be taken to predict the performance of the underlying system based on some predefined criteria. Such simulations can improve system performance and bring realibility to the system design in general.

IV. Economical and Biological Approaches

The final sectioning is a collection of four chapters. Gairola and Chong model and simulate phychic noise in cooperative games. The deduce some very interesting conclusions that may entail further investigation of cooperative games within the context of large-scale distributed systems that are fraught with uncertanities, which may trigger noise that in turn affect the overall system performance. Warden et al., present a decentralized, flexible, and scalble control for logistical networks. The work is implressibe because it can ensure knowledge management with machine learning insprired approach for adaptability in planning and management. Jacob et al., describe a virtual human, LINDSAY to aid in providing affordable healthcare to the masses. The system can be used as an instructional tool that can simulate physiological processes at all levels (organ to cell) of the human body. LINDSAY also provides a graphical programming environment to compose complex but interactive modules for human anatomy study. Dossou et al., contribute the closing chapter in this book. The work compares enterprise modeling concepts and develop a toolkit to compare various methodologies that can aid in future research in enterprise integration.

Acknowledgements

We are grateful to all of the contributors of this book, for their willingness to work on this interdisciplinary project. We thank all of the authors for their interesting book chapter proposals, their time, efforts, support, and their excellent research results, which makes this volume an interesting and complete monograph of the latest research advances and technology development for the next generation distributed and complex emergent intelligent decision systems. We also would like to express our sincere gratitude to the reviewers, who have helped us to ensure the highest quality of work.

Our special appreciation goes to Prof. Janusz Kacprzyk, editor-in-chief of the Springer's Studies in Computational Intelligence Series, Dr. Thomas Ditzinger, Holger Schaepe, and all of the Springer editorial team for their editorial assistance and excellent cooperative collaboration in this project.

Finally, we would like to send our warmest gratitude message to our families and friends for their patience, love, and support during the course of this project.

We strongly believe that this book ought to serve as a reference for students, researchers, and industry practitioners interested or currently working in the evolving interdisciplinary area of green computing using emergent natural heuristics paradigm.

Cracow and Fargo
January 2012

Joanna Kołodziej
Samee Ullah Khan
Tadeusz Burczyński

Contents

Part III: Nature-Inspired Solutions for Intelligent Networking

List of Contributors

M.G. Arenas
Departamento de Arquitectura y Tecnologia de Computadores, Universidad de Granada, Spain; GeNeura. Department of Architecture and Computer Technology, CITIC, University of Granada, Spain;
e-mail: maribel@geneura.ugr.es

J.L. Bernier
GeNeura Department of Architecture and Computer Technology, CITIC, University of Granada, Spain;
e-mail: jbernier@geneura.ugr.es

J. L. Bordim
Department of Computer Science, University of Brasilia (UnB), Brazil;
e-mail: bordim@unb.br

M. F. Caetano
Department of Computer Science, University of Brasilia (UnB), Brazil;
e-mail: caetano@cic.unb.br

P.A. Castillo
GeNeura, Department of Architecture and Computer Technology, CITIC, University of Granada, Spain;
e-mail: pedro@geneura.ugr.es

S. Y. Chong
University of Nottingham, Malaysia Campus, Jalan Broga, 43500 Semenyih, Selangor Darul Ehsan, Malaysia;
e-mail: Siang-Yew.Chong@nottingham.edu.my

J. M. Correa
Department of Computer Science, University of Brasilia (UnB), Brazil;
e-mail: jan@cic.unb.br

V. Cristea
Computer Science Department, Faculty of Automatic Control and Computers, University POLITEHNICA of Bucharest, Splaiul Independentei 313, Sector 6, 060042 Bucharest, Romania;
e-mail: valentin.cristea@cs.pub.ro

T. Davison
Department of Computer Science, University of Calgary, 2500 University Dr., NW Calgary T2N 1N4, Canada;
e-mail: tbdaviso@ucalgary.ca

C. Dobre
Computer Science Department, Faculty of Automatic Control and Computers, University POLITEHNICA of Bucharest, Splaiul Independentei 313, Sector 6, 060042 Bucharest, Romania
e-mail: ciprian.dobre@cs.pub.ro

P. E. Dossou
Icam Group, Icam Vendee, 28 Boulevard d'Angleterre, 85000 La Roche-Sur-Yon, France;
e-mail: Paul-eric.dossou@icam.fr

A. Esmaeili
Department of Computer Science, University of Calgary, 2500 University Dr., NW Calgary T2N 1N4, Canada;
e-mail: a.esmaeili@ucalgary.ca

D. Gairola
University of Nottingham, Malaysia Campus, Jalan Broga, 43500 Semenyih, Selangor Darul Ehsan, Malaysia;
e-mail: dhruv.gairola.89@gmail.com

P. Garcia-Sanchez
GeNeura, Department of Architecture and Computer Technology, CITIC, University of Granada, Spain;
e-mail: pgarcia@geneura.ugr.es

C. Gingras
Undergraduate Medical Education, University of Calgary, 2500 University Dr., NW Calgary T2N 1N4, Canada;
e-mail: crgingra@ucalgary.ca

S. Gundry
Department of Electrical Engineering, The City College of New York, NY 10031, USA;
e-mail: sgundry00@ccny.cuny.edu

B. Hallgrimsson
Department of Cell Biology & Anatomy, University of Calgary, 2500 University Dr., NW Calgary T2N 1N4, Canada;
e-mail: bhallgri@ucalgary.ca

O. Herzog
Center for Computing and Communication Technologies, Universitat Bremen, Bremen, Germany;
e-mail: herzog@tzi.de

C. Jacob
Department of Computer Science, Department of Biochemistry & Molecular Biology,
University of Calgary, 2500 University Dr., NW Calgary T2N 1N4, Canada;
e-mail: cjacob@ucalgary.ca

H. Jamniczky
Department of Cell Biology & Anatomy, University of Calgary, 2500 University Dr., NW Calgary T2N 1N4, Canada;
e-mail: hajamnic@ucalgary.ca

L. S. Jaques
Department of Computer Science, University of Brasilia (UnB), Brazil;
e-mail: luizajaques@gmail.com

S. U. Khan
North Dakota State University, ND 58108, USA;
e-mail: samee.khan@ndsu.edu

J. Kołodziej
Department of Mathematics and Computer Science, University of Bielsko-Biała , 43-309 Bielsko-Biała, Poland;
e-mail: jkolodziej@ath.bielsko.pl

M. Koeppen
Kyushu Institute of Technology, 680-4 Kawazu, Iizuka, Fukuoka 820-8502, Japan;
e-mail: mario@ndrc.kyutech.ac.jp

J. Kusyk
The Graduate Center of the City University of New York, New York, NY 10016 USA;
e-mail: jkusyk@gc.cuny.edu

J.L.J. Laredo
Departamento de Arquitectura y Tecnologia de Computadores, Universidad de Granada, Spain;
e-mail: juanlu@geneura.ugr.es

S. A. Ludwig
Department of Computer Science, North Dakota State University, Fargo, ND, USA;
e-mail: Simone.Ludwig@ndsu.edu

S. von Mammen
Department of Computer Science, University of Calgary, 2500 University Dr., NW Calgary T2N 1N4, Canada;
e-mail: s.vonmammen@ucalgary.ca

A. C. M. A. de Melo
Department of Computer Science, University of Brasilia (UnB), Brazil ;
e-mail: albamm@cic.unb.br

J.J. Merelo
GeNeura. Department of Architecture and Computer Technology, CITIC, University of Granada, Spain;
e-mail: jmerelo@geneura.ugr.es

L. A. Miranda
Department of Computer Science, University of Brasilia (UnB), Brazil;
e-mail: lidiamir@gmail.com

P. Mitchell
Icam Group, Icam Vendée, 28 Boulevard d'Angleterre, 85000 La Roche-Sur-Yon, France;
e-mail: philip.mitchell@icam.fr

A.M. Mora
Department of Architecture and Computer Technology, CITIC, University of Granada, Spain;
e-mail: amorag@geneura.ugr.es

S. Novakowski
Undergraduate Medical Education, University of Calgary, 2500 University Dr., NW Calgary T2N 1N4, Canada;
e-mail: snovakow@ucalgary.ca

K. Ohnishi
Kyushu Institute of Technology, 680-4 Kawazu, Iizuka, Fukuoka 820-8502, Japan;
e-mail: ohnishi@cse.kyutech.ac.jp

Y. Oie
Kyushu Institute of Technology, 680-4 Kawazu, Iizuka, Fukuoka 820-8502, Japan;
e-mail: oie@cse.kyutech.ac.jp

P. Pawlewski
Poznań University of Technology, ul.Strzelecka 11 60-965 Poznań, Poland;
e-mail: pawel.pawlewski@put.poznan.pl

D. Phillips
Department of Computer Science, University of Calgary, 2500 University Dr., NW Calgary T2N 1N4, Canada;
e-mail: dbphilli@ucalgary.ca

F. Pop
Computer Science Department, Faculty of Automatic Control and Computers, University POLITEHNICA of Bucharest, Splaiul Independentei 313, Sector 6, 060042 Bucharest, Romania
e-mail: florin.pop@cs.pub.ro

G. Romero
Department of Architecture and Computer Technology, CITIC, University of Granada, Spain;
e-mail: gustavo@geneura.ugr.es

C. Safak-Sahin
BAE Systems - AIT, Burlington, MA 01803 USA;
e-mail: csafaksahin@gmail.com

V. Sarpe
Department of Computer Science, University of Calgary, 2500 University Dr., NW Calgary T2N 1N4, Canada;
e-mail: vsarpe@ucalgary.ca

A. Sarraf-Shirazi
Department of Computer Science, University of Calgary, 2500 University Dr., NW Calgary T2N 1N4, Canada;
e-mail: asarrafs@ucalgary.ca

S. Steil
Undergraduate Medical Education, University of Calgary, 2500 University Dr., NW Calgary T2N 1N4, Canada;
e-mail: sasteil@ucalgary.ca

E. Urrea
Lehman College of the City University of New York, Bronx, NY 10468 USA;
e-mail: eurrea@gc.cuny.edu

M. U. Uyar
Department of Electrical Engineering, The City College of New York, NY 10031 USA;
and
The Graduate Center of the City University of New York, New York, NY 10016 USA;
e-mail: uyar@ccny.cuny.edu

T. Warden
Center for Computing and Communication Technologies (tzi), Universitat Bremen Bremen, Germany
e-mail: warden@tzi.de

J. Wojtusiak
Machine Learning and Inference Laboratory (mli), George Mason University Fairfax, VA, USA
e-mail: jwojtusi@gmu.edu

B. Wright
Undergraduate Medical Education, University of Calgary, 2500 University Dr., NW Calgary T2N 1N4, Canada;
e-mail: wrightb@ucalgary.ca

I. Yazdanbod
Department of Computer Science, University of Calgary, 2500 University Dr., NW Calgary T2N 1N4, Canada;
e-mail: iyazdanb@ucalgary.ca

K. Yoshida
Kyushu Institute of Technology, 680-4 Kawazu, Iizuka, Fukuoka 820-8502, Japan
e-mail: kaori@ai.kyutech.ac.jp

A. Y. Zomaya
School of Information Technologies, University of Sydney, Sydney, NSW 2006, Australia;
e-mail: albert.zomaya@sydney.edu.au

J. Zou
Department of Electrical Engineering, The City College of New York, NY 10031 USA;
e-mail: jzou@ccny.cuny.edu

Part I
Future Generation Fuzzy Systems

Chapter 1
Towards Designing Human Centric Systems: A New View at Fuzzy Modeling with Granular Membership Functions

Witold Pedrycz

Abstract. A comprehensive modeling realized with the aid of fuzzy sets (referred to as fuzzy modeling) has emerged as an interesting alternative in system modeling. Fuzzy models are numeric constructs. Likewise fuzzy sets being used as integral building blocks are inherently *numeric* constructs (by being commonly described by their numeric membership functions). The objective of this study is to explore a new avenue of fuzzy modeling, referred to as *granular* fuzzy modeling where fuzzy sets could be represented by some granular counterparts such as e.g., interval fuzzy sets, shadowed sets, or fuzzy fuzzy (fuzzy^2) sets. Several ways of constructing (transforming) numeric membership functions into granular constructs (granular mappings) are conceptualized along with a discussion on selected algorithmic aspects of fuzzy models, in particular those emerging in the area of fuzzy decision-making.

1.1 Introductory Notes

From their very inception, fuzzy sets with their commonly used characterization provided in terms of membership functions are numeric constructs. In spite of the notion of information granularity, which is inherently associated with fuzzy sets, we encounter numeric membership grades which are subject to the detailed, numerically - cast processing subsequently producing numeric results. We witness membership grades reported up to six decimal digits,

Witold Pedrycz
Department of Electrical & Computer Engineering University of Alberta, Edmonton, Canada
and
Systems Research Institute, Polish Academy of Sciences Warsaw, Poland
e-mail: wpedrycz@ualberta.ca

J. Kołodziej et al. (Eds.): Advances in Intelligent Modelling and Simulation, SCI 422, pp. 3–24.
springerlink.com

which is somewhat surprising and counterintuitive. Interestingly, this trend of numerically inclined constructs of fuzzy models has been overwhelming in the recent years. This stands in a sharp contrast to fuzzy models encountered in the 70ies where the linguistic facet (manifesting in so-called linguistic models) was vividly present. Overall, there is an interesting and a quite broad spectrum of views at information granules and their processing. We can envision two well-positioned extremes at the representation and processing fuzzy sets:

symbolic perspective A concept - information granule is viewed as a single symbol (entity). This position is very much present in the AI community, where computing revolves around symbolic processing. Symbols are subject to processing rules giving rise to results, which are again symbols coming from the same vocabulary one has started with [14]. One may also refer here to [15], [22].

numeric perspective Here information granules are associated with a detailed numeric characterization. Fuzzy sets are profound examples here. We start with numeric membership functions. All ensuing processing involves numeric membership grades so in essence it focuses on number crunching. The results are inherently numeric. The progress present here has resulted in a diversity of numeric constructs. Because of the commonly encountered numeric treatment of fuzzy sets, the same applies to logic operators (connectives) encountered in fuzzy sets. There has been a plethora of developments, refinements and an emergence of new categories of logic connectives, say uninorms and nullnorms [1], [2], [3], [4], [5], [7], [8], [9], [11]. The reader may refer to a long, impressive list of accomplishments and enhancements including theoretical investigations [23] [25] and their usage in system modeling aggregation [5], and decision-making.

Interestingly, alternatives positioned in-between symbolic and numeric treatment of information granules have not been fully exploited however they may offer several interesting variants and bring some conceptual and computational flexibility. One may refer to the discussion realized in the context of symbolic - numeric processing covered in [15]. Fuzzy sets (membership functions) are captured as information granules with intent of departing from handling numeric membership grades and elevating the processing to the more abstract and less detailed level.

The granular representation of fuzzy sets gives rise to two modes of processing (and modeling, in particular) in which the granularity of information manifests in various ways. The first one follows the general scheme arising in the form

$$\begin{aligned}&\text{Fuzzy sets} A, B, C \ldots \rightarrow \text{granular representation} G(A), G(B), G(C) \ldots \rightarrow \\ &\quad \rightarrow \text{processing at the level of information granules}\end{aligned} \tag{1.1}$$

where $G(.)$ stands for the granular representation (characterization) of a given fuzzy set. The processing completed at the level of information granules invokes the use of the tabular representation of operations on fuzzy sets. The computing is realized at the granular (symbolic, non-numeric) level of information granules meaning that an amount of computing could be quite reduced. This aspect of processing could be of interest when dealing with neurocomputing, including fuzzy neural networks [18] with a tangible benefit of reducing an overall computing overhead.

The second mode of processing adheres to the processing of fuzzy sets at the level of numeric membership values while the final result is represented as a granular construct. In other words, we have

$$\begin{array}{l}\text{Fuzzy sets}\, A, B, C \ldots \rightarrow \text{ processing} \rightarrow \\ \text{fuzzy set}\, W \rightarrow \text{ granular representation}\, G(W)\end{array} \tag{1.2}$$

Fig. 1.1 highlights these two modes of processing, which relate directly with the essence of system modeling. The first mode, Fig. 1.1(a), we are concerned with a granular representation of membership functions (which supports a formation of the output interface) while the second one, Fig. 1.1(b), is linked with the input interface where we arrive with granular format of fuzzy sets to facilitate all ensuing processing.

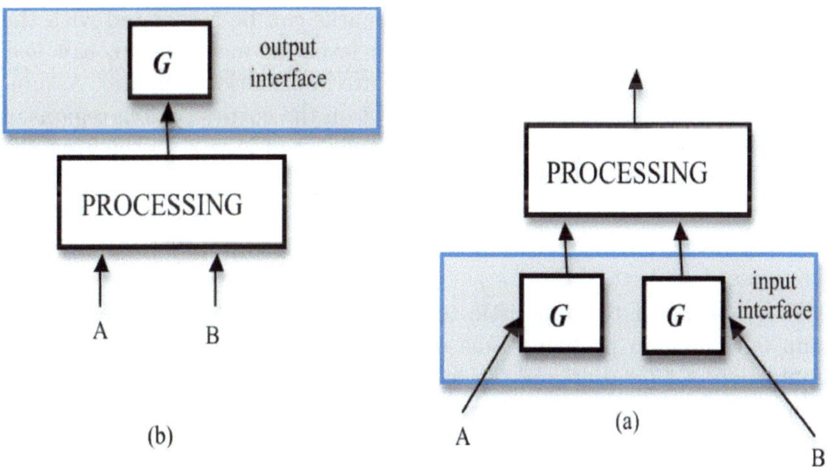

Fig. 1.1. From numeric membership functions to their granular counterparts: (a) formation of input interfaces preceding processing of information granules; (b) in the realization of output interfaces

There are some interesting and compelling arguments behind forming information granules of membership grades; see e.g., [17] [26]. We witness interesting pursuits along the line of generalizations of fuzzy sets by bringing linguistic aspects as discussed in [10]. The non-numeric quantification of membership grades comes with type-2 fuzzy sets [16] and shadowed sets [19] [21].

The study presented here is structured into 5 sections. Considering the taxonomy presented above, we discuss a way of representing membership functions by patches induced by the granular linguistic terms expressed in the membership space (Section 1.2). This offers one of the representative alternatives of the realization of the output interface. The diversity of input interfaces is visible and specific constructs depend upon the nature of processing. By taking this into consideration, we look at logic-consistent granular representations of fuzzy sets, viz. the constructs which are implied by the logic fabric of computing (Section 1.3) as well introduce granular fuzzy sets in the setting of decision-making, Section 1.4, followed by the optimization considerations.

1.2 Granular Representation of Membership Functions- A Design of Output Interfaces

The objective is to represent (capture) a numeric membership function by a collection of information granules - linguistic landmarks defined over the universe of membership grades. Such landmarks can be associated with the semantics of terms describing nonnumeric levels of membership, say *low*, *medium*, *high*, etc. These levels themselves are information granules. Assuming such granules have been specified, we obtain the corresponding regions of the universe of discourse the fuzzy set is defined, see Fig. 1.2. In this way, the granular interpretation of the fuzzy set is a collection of the pairs (Ω_i, L_i) as illustrated in this figure. In a nutshell, the fuzzy set A can be represented as a union of Cartesian products (patches) formed over the membership function and expressed as $A = \bigcup_j \left(\Omega_j \times L_{i.j}\right)$.

The formation of the linguistic terms (landmarks) can be a result of a certain optimization problem. One of possible formulations can concern a minimization of a total area of the regions formed by the Cartesian products presented above. In other words, we require that the sum of the cardinalities of the individual patches is minimized with respect to $L_1, L_2, \ldots, L_c$ with "c" being a collection of linguistic landmarks.

As an illustrative example, let us consider two landmarks L_1 and L_2 (treated as intervals) and the membership function of the form

$$A(x) = \begin{cases} 1 - \left(\frac{x}{m}\right)^{\gamma}, & \text{if} x \in [0, m] \\ 0 & \text{otherwise} \end{cases} \tag{1.3}$$

where the parameter $\gamma (> 1)$ controls its shape. The minimized performance index (the area of the Cartesian products) reads as

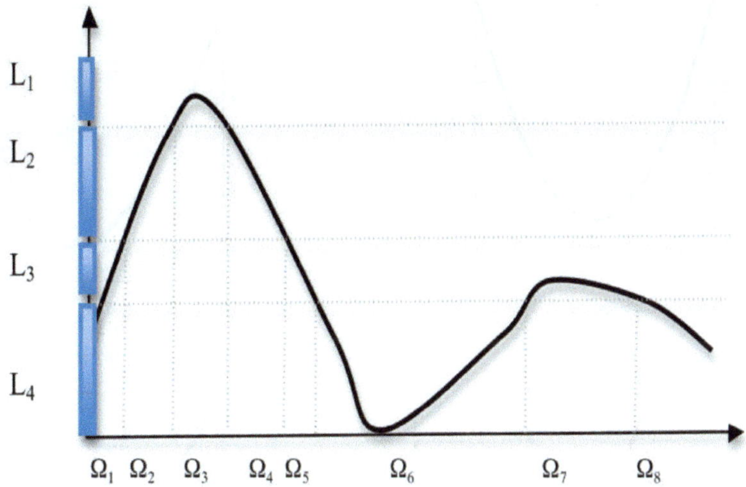

Fig. 1.2. Fuzzy set and its granular representation realized in terms of the linguistic descriptors of membership, $A = (\Omega_1 \times L_3) \cup (\Omega_2 \times L_2) \cup (\Omega_3 \times L_1) \cup (\Omega_4 \times L_2)\ (\Omega_5 \times L_3) \cup (\Omega_6 \times L_4) \cup (\Omega_7 \times L_3)(\Omega_8 \times L_4)$

$$Q = (1-\alpha)\cdot m \cdot \sqrt[\gamma]{1-\alpha} + \alpha \cdot (m - m \cdot \sqrt[\gamma]{1-\alpha}) \tag{1.4}$$

Here α is a cutoff point of the information granules (intervals) L_1 and L_2. The plots of the index expressed as a function of α are displayed in Fig. 1.3; note that in all cases there is a clearly delineated value of the cutoff (threshold) whose location depends on the value of "γ"; in general with the increase of his parameter, the optimal cutoff assumes higher values (meaning that the linguistic term L_1 occupies a broader range of the unit interval).

1.3 The Design of Input Interface-Construction of Logic-Consistent Granular Representations of Fuzzy Sets

Our interest is to develop a granular representation of membership functions in such a manner that the granular membership functions (granular manifestations of numeric membership functions) coincide (viz. approximate or represent) the original membership function. More formally, the problem is formulated as follows. Consider fuzzy sets A and B defined in the same space (universe of discourse) X. Given is a certain numeric transformation of A and B, denoted here by ϕ, which produces $C = \phi(A,B)$. In particular, ϕ can be a

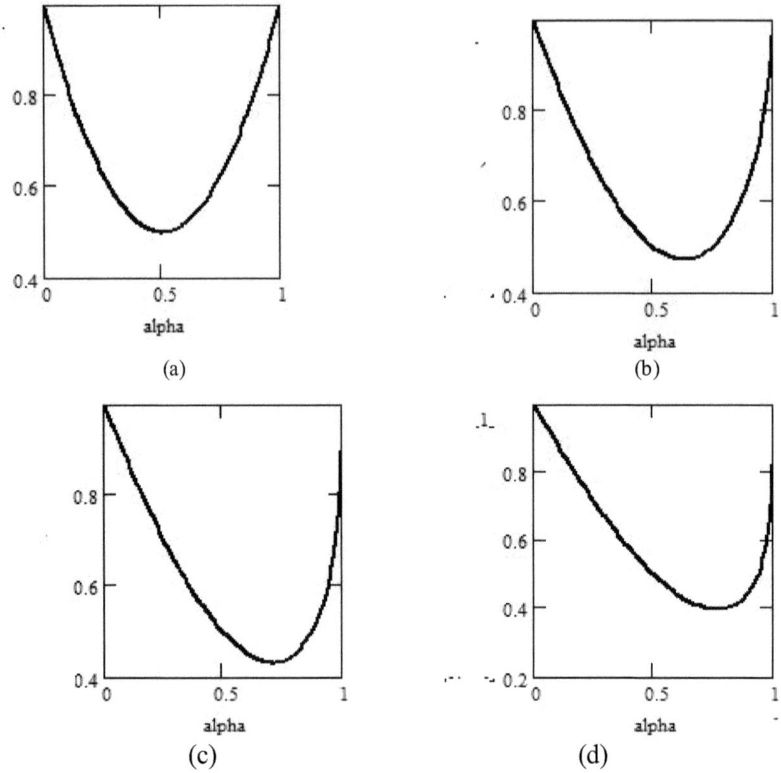

Fig. 1.3. Plots of Q as functions of a for selected values of "γ": (a) $\gamma = 1$, (b) $\gamma = 2$, (c) $\gamma = 3$ and (d) $\gamma = 4$

logic operator. A granulation of membership functions G is a way of representing the unit interval of membership values as a finite and small collection of information granules such as e.g., sets, fuzzy sets, rough sets, shadowed sets, etc. These information granules come with a well-defined semantics, say *Low*, *Medium*, *High*, *Very High* membership, etc. A vocabulary comprising a finite number of information granules coming as a result of $G(A)$ is used as a granular representation of the original numeric membership of A. In other words, instead of talking about membership values, say 0.74, 0.236, etc., we are concerned about a more abstract view at the membership concept. The underlying task is to construct such information granules - components of the vocabulary $\{L_1, L_2, \ldots, L_c\}$ so that the following relationship:

$$G(\phi(A,B)) \approx \phi(G(A), G(B)) \tag{1.5}$$

becomes satisfied to the highest extent. The detailed meaning of the "approximately equal" relationship ($\approx$) shown above will be clarified later on.

We show it entails a certain optimization problem. As the transformation (mapping) ϕ appears explicitly in the construction of the information granules - granular descriptors of fuzzy sets, we will be referring to such descriptors as ϕ-consistent. For instance, we can talk about *and, or, complement, match* - consistent granular descriptors of fuzzy sets, which relate with the corresponding logic operators (*and, or*) or some matching operations. Likewise, in case of several operators being considered at the same time (en block), say ϕ, γ, η... we arrive at (ϕ, γ, η...)-consistent granular description of fuzzy sets. For instance, in this way we can talk about (*and, or*) - consistent granular description (representation) of fuzzy sets. It is worth noting that logic operators could be constructed by taking into account of some background knowledge of statistical character, refer to [20].

As before, for the sake of clarity, the problem of a granular representation or description of fuzzy sets is concerned with a formation of a family of information granules- intervals formed over the unit interval. These interval information granules come in the form of intervals $[a_i,\ a_{i+1}]$ $[0,1]$ viz. information granules $L_1, L_2, \ldots, L_c$ where $L_1 = [0, a_1), L_2 = [a_1, a_2) \ldots L_i = [a_{i-1}, a_i) \ldots L_c = [a_{c-1}, 1]$. The above intervals form a partition of the unit interval where $0 < a_1 < \ldots < a_{c-1} < 1$. The interval format of granulation of the unit interval is fully characterized by the vector of the parameters of the granular transformation of the unit interval, $a = [a_1\, a_2 \ldots a_{c-1}]^T$.

Before moving on to the general concept and ensuing optimization problems, let us start with an illustrative example highlighting the essence of the construct. When applying the granulation process (process) G with a vocabulary of information granules $L_1, L_2, \ldots, L_c$ to the membership grade A, we obtain a vector with the following entries:

$$G(A) : A \rightarrow [L_1(A) L_2(A) \ldots L_c(A)] \tag{1.6}$$

where $G_1(A), G_2(A), \ldots, G_c(A)$ are the levels of matching of the numeric membership grade A with the information granules developed during the granulation process for a fixed value of the argument.

While the granular transformation is of a general character, there could be various formal frameworks in which information granules are specified as sets, fuzzy sets, rough sets, etc. For instance, if we consider a granulation mechanism G producing interval granulation of the space of membership values [0,1], such as *L (low), M (medium), H (high)* membership, the above granular representation produces a 3-dimensional Boolean vector with the entries 0 or 1 as illustrated in Fig. 1.4.

There is also a two-argument operation ϕ, for which we form the ϕ - consistent granular representation of the fuzzy sets. As noted earlier, the objective is to attain the equality $G(\phi(A,B)) = \phi(G(A), G(B))$ with the granular realization of A and B, $G(A)$, $G(B)$ being a subject of the optimization. Formally, we express the optimization task in the integral or discrete format

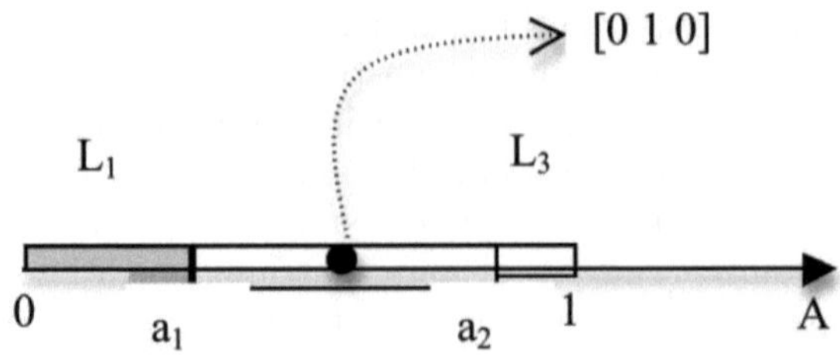

Fig. 1.4. Interval granulation mechanism applied to membership grade A; L_1 = *low*, L_2 = *medium*, L_3 = *high* and the resulting 3-dimensional Boolean representation vector (with 2 cutoff points a_1 and a_2)

$$V(a) = \int_0^1 \int_0^1 \|G(\phi(A,B)) - \phi(G(A),G(B))\| dA\, dB \tag{1.7}$$

where the integral (sum) is taken over all membership grades of A and B taking values in the unit interval. The distance function |.| is typically specified to be the Euclidean one. The solution to (1.7) arises in the form

$$a_{opt} = arg\, Min_a V(a) \tag{1.8}$$

where the entries of a_{opt} are arranged in an increasing order. The discrete version of the optimization problem (which is more suitable from the practical perspective) is expressed in the form

$$V(a) = \sum_{i=1}^{n} \sum_{j=1}^{n} \|G(\phi(A_i,B_j) - \phi(G(A_i),G(B_j))\| \tag{1.9}$$

where $A_1,A_2,\ldots,A_n$ and $B_1,B_2,\ldots,B_n$ are the membership grades used in the optimization process. For instance, one can consider these grades to be uniformly distributed in the [0,1] interval. In the analogous fashion, we formulate an optimization problem for (ϕ ,δ)- consistent granular representation of fuzzy sets. Say, if ϕ and δ are the two-argument operators (such as logic *and* and *or* operators realized by some *t*-norms and *t*-conorms), the optimization problem in its continuous version along with its solution reads as follows

$$\begin{aligned} V(a) = & \int_0^1 \int_0^1 \|G(\phi(A,B)) - \phi(G(A),G(B))\| dA\, dB + \\ & + \int_0^1 \int_0^1 \|G(\delta(A,B)) - \delta(G(A),G(B))\| dA\, dB \end{aligned} \tag{1.10}$$

1.3.1 *The Components of the Optimization Problem*

The formulation of the optimization problem needs to be now specified so that all technical details are addressed. In what follows, we consider a discrete version of the ϕ -consistent granulation problem, which is relevant from a perspective of further applications. We are concerned with a finite number of membership grades of A and B, say $A_1, A_2, \ldots, A_n$ and $B_1, B_2, \ldots, B_n$.

The Optimization Criterion

The essence of the objective function guiding the optimization problem is to form a vocabulary of information granules so that the results $\phi(G(A), G(B))$ for all membership grades A_k and B_i falling under some fixed elements of the vocabulary are the same granular "meaning", viz. invoking the same information granule. Of course, more dispersion is encountered and through an optimization of the information granules, an intent is to minimize it. Take all pairs possible pairs A_k and B_l, $k, l = 1, 2, \ldots, n$. Subsequently we produce the membership grade $\phi(A_k, B_l)$ After the granulation process, see Eq. (1.2), the corresponding membership grades A_k and B_l and $\phi(A_k, B_l)$, give rise to a triple of integers corresponding to the elements in the vocabulary of information granules, say i_1, i_2, and i_3. The results are arranged in the tabular format, namely a contingency matrix T. The rows and columns of T are labeled by the elements of the vocabulary of information granules used in the realization of G. With the (i_1, i_2) coordinate of the matrix T, we associate a vector with the following entries $[00\ldots010\ldots0]$ where the 1 is located at the ith entry of this vector. The results are successively accumulated in the contingency matrix by updating (incrementing) the content of the vectors associated with the corresponding locations of the table. Thus the binary vector resulting at the end of the counting process consists of the counts of occurrence of the elements of the vocabulary used in the granulation process G. For the $(i, j)^{th}$ coordinate of T we obtain

$$n_{ij} - \left[n_{ij}(1)\, n_{ij}(2) \ldots n_{ij}(c)\right] \tag{1.11}$$

$i, j = 1, 2, \ldots, c$. Denote by k_{ij}^* the largest entry of the vector (1.11), that is

$$k_{ij}^* = arg\,max_{(k=1,2,\ldots,c)} n_{ij}(k) \tag{1.12}$$

The essence of the ϕ-consistent granular representation of fuzzy sets is to choose the information granules in the granulation scheme G in such a way so that the following performance index becomes minimized

$$V(a) = \sum_{i=1}^{c} \sum_{j=1}^{c} \sum_{\substack{k=1 \\ k=k^*_{ij}}}^{c} n_{ij}(k) \tag{1.13}$$

In essence, we select the elements of the vocabulary in such a way that the dispersion of entries of n_{ij} is made as low as possible. The minimization is carried out through the adjustments of the cutoff points (bounds of the intervals) a. Note that the optimization is associated with the use of a certain logic operation. This stipulates that the intervals may be different when dealing with a different realization of the logic operator (say, a given t-norm).

One can look at the aggregate process of granulation of [0,1] by forming an overall performance index, which combines those used in the formation of the (*or*, *and*) – consistent granulation of membership functions, namely

$$V(a) = V(and, a) + V(or, a) \tag{1.14}$$

which, as before, is optimized with regard to the coordinates of the vector a. This formulation is legitimate considering that a great deal of processing in fuzzy sets is realized by means of *and* and *or* logic operators. We emphasize the existence of the logic operators used in the consistency optimization. By choosing some other t-norm or t-conorm, we arrive at the optimized information granules.

1.4 Decision-Making with a Granular Representation of Fuzzy Sets: AHP Modeling Revisited

In decision-making problems, we are commonly faced with information provided by humans, which is inherently non-numeric. Partial evaluations, preferences, weights are expressed linguistically. The evident role of fuzzy sets in decision-making and associated important processes such as e.g., consensus building is well documented in the literature.

While fuzzy sets raise awareness about the non-numeric nature of information, its importance, a need for its handling and provided a great deal of techniques of processing fuzzy sets, the fundamental issue is about a transformation of available pieces of linguistic information into formal constructs of information granules. The resulting information granules are afterwards effectively processed within the computing setting pertinent to the assumed framework of information granulation. The linguistic terms quantifying preferences such as *high, medium,* etc. are in common use. There is some linear order, as we can say that *high, medium* etc. however the numeric details are usually not present. It is not clear, however, how they have to be translated into such numeric entities, which can be further seamlessly processed using the formalisms of sets, fuzzy sets, rough sets and alike. Likewise, it is not straightforward what optimization criterion can be envisioned when arriving at the formalization of the linguistic terms through information granules.

Given the diversity of decision-making problems and being alerted to the fact that each of their categories could come with some underlying specificity and particular requirements, in this study we concentrate on the Analytic Hierarchy Process (AHP) model, which addresses a large and important category of decision-making schemes.

The pairwise comparisons of alternatives are articulated in terms of linguistic quantifications, say *highly* preferred, *moderately* preferred, etc. Each term is associated with some numeric values. It has been identified quite early in the development of the AHP-like models that the single numeric values taken from the 1-9 scale does not necessarily fully reflect the complexity of the non-numeric nature of the pairwise comparisons. The first approach presented along this line involved the use of triangular fuzzy numbers defined in the scale. There have been a significant number of pursuits along this line. The granular nature of the pairwise assessments was discussed in context of a group decision-making where reaching consensus calls for some flexibility of evaluations individual assessments have to be endowed with to facilitate processes of consensus building.

1.4.1 The AHP Method-A Brief Review

Given a finite number of alternatives, say various options, solutions, etc. $a_1, a_2, \ldots, a_n$, etc. to be considered in a certain investment scenario, the objective is to order them by associating with them some degrees of preference expressed in the [0,1] interval. The essence of the method introduced by Saaty is to determine such preference values through running a series of pairwise comparisons of the alternatives. The results are organized in an $n \times n$ reciprocal matrix $R = [r_{ij}]$, $i, j = 1, 2, \ldots, n$. The matrix exhibits two important features. The diagonal values of the matrix are equal to 1. The entries that are symmetrically positioned with respect to the diagonal satisfy the condition of reciprocality that is $r_{ij} = 1/r_{ji}$.

The starting point of the estimation process of the fuzzy set of preferences are entries of the reciprocal matrix which are obtained through collecting results of pairwise evaluations offered by an expert, designer or user (depending on the character of the task at hand). Prior to making any assessment, the expert is provided with a finite scale with values in-between 1 to 7. Some other alternatives of the scales such as those involving 5 or 9 levels could be sought as well. If a_i is strongly preferred over a_j when being considered in the context of the fuzzy set whose membership function we would like to estimate, then this judgment is expressed by assigning high values of the available scale, say 6 or 7. If we still sense that a_i is preferred over a_j yet the strength of this preference is lower in comparison with the previous case, then this is quantified using some intermediate values of the scale, say 3 or 4. If no difference is considered, the values close to 1 are the preferred choice, say 2 or 1. The value of 1 indicates that a_i and a_j are equally preferred. The general quantification of preferences positioned on the scale of 1-7 can be described as follows:

- equal importance – 1
- moderate importance of one element over another – 3
- strong importance – 5
- demonstrated importance – 7
- extreme importance – 9

There are also some intermediate values, which could be used to further quantify the relative dominance. On the other hand, if a_j is preferred over a_i, the corresponding entry assumes values below one. Given the reciprocal nature of the assessment, once the preference of a_i over a_j has been quantified, the inverse of this number is inserted into the entry of the matrix that is located at the $(j,i)^{th}$ coordinate. Next the maximal eigenvalue is computed along with its corresponding eigenvector. The normalized version of the eigenvector is then the membership function of the fuzzy set we considered when doing all pairwise assessments of the elements of its universe of discourse. The effort to complete pairwise evaluations is far more manageable in comparison to any experimental overhead we need when assigning membership grades to all elements (alternatives) of the universe in a single step. Practically, the pairwise comparison helps the expert focus only on two elements once at a time thus reducing uncertainty and hesitation while leading to the higher level of consistency. The assessments are not free of bias and could exhibit some inconsistency. In particular, we cannot expect that the transitivity requirement could be fully satisfied. Fortunately, the lack of consistency could be quantified and monitored. The largest eigenvalue computed for R is always greater than the dimensionality of the reciprocal matrix (recall that in reciprocal matrices the elements positioned symmetrically along the main diagonal are inverse of each other), $\lambda_{max} > n$ where the equality $\lambda_{max} = n$ occurs only if the results are fully consistent. The ratio

$$\nu = (\lambda_{max} - n)/(n-1) \tag{1.15}$$

can be regarded as an inconsistency index of the data; the higher its value, the less consistent are the collected experimental results. This expression can be sought as the indicator of the quality of the pairwise assessments provided by the expert. If the value of ν is too high exceeding a certain superimposed threshold, the experiment may need to be repeated. Typically if ν is less than 0.1 the assessment is sought to be consistent while higher values of ν call for the re-examination of the experimental data and a re-run of the experiment.

1.4.2 A Quantification (Granulation) of Linguistic Terms as Their Operational Realization

The linguistic terms used in a pairwise comparison of alternatives are expressed linguistically by admitting qualitative terms. They can be organized in a linear fashion, as there is some apparent linear order among them. The terms themselves are not operational meaning that no further processing can be realized, which involves a quantification of the linguistic terms. Schematically, we can portray the process of arriving at the operational representation of linguistic terms as illustrated in Fig. 1.5.

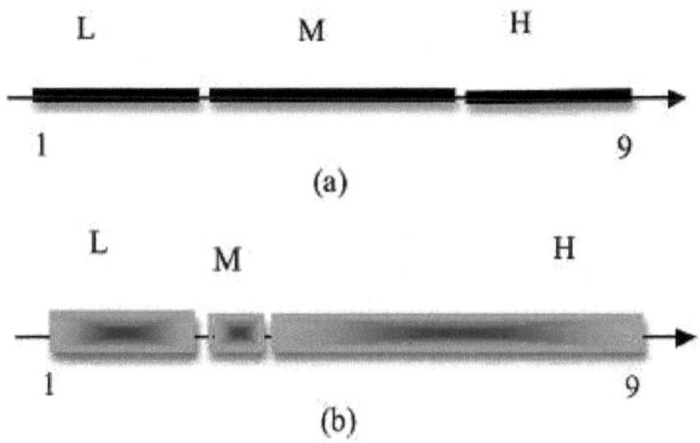

Fig. 1.5. Towards the operational realization of linguistic terms: (a) realization with the aid of intervals, and (b) fuzzy set-based implementation with the membership functions expressed over the [0,1] interval of membership grades

The two important features of such granulation mechanisms are worth noting here: (a) the mapping is by no means linear that is a localization of the associated information granules on the scale is not uniform, (b) the semantics of the terms allocated in the process of granulation is retained. Various information granulation formalisms can be contemplated. The question on how to arrive at the operational version of the information granules can be reformulated as a certain optimization problem.

1.4.3 The Optimization of the Granulation Problem

The construction of information granules is realized as a certain optimization problem. In this section, we elaborate on the fitness function, its realization, and the particle swarm optimization (PSO) [6] [12] [24] along with the corresponding formation of the components of the swarm. Given the granular nature of the reciprocal matrix, a Monte Carlo scheme is used to express the corresponding values of the fitness function.

Evaluation of the Mapping from Linguistic Terms to Information Granules

The objective of the fitness function is to provide a quantification of the information granules on which information granules are to be mapped. Considering the nature of the AHP model, the quality of the solution (preference vector) is expressed in terms of the inconsistency index. For the given vector of cutoff points, their quality associates with the corresponding value of the inconsistency index. The minimization of the values of the index by adjusting the positions of the cutoff points in the 1 ... 9 scale is realized by the PSO. When it comes to the formation of the fitness function, its determination has to take into account a fact that interval-valued entries of the reciprocal matrix have to return numeric values of the fitness function. This is realized as follows. As we encounter information granules in the form of intervals, we consider a series of their realizations. Let us consider that an individual generated by the PSO has produced a collection of cutoff points specified by the individual in the swarm on which PSO operate, which is located on the [1, 9] scale. For instance, if there are three linguistic terms in the reciprocal matrix, say *L (Low)*, *M (Medium)*, and *H (High)*, and the corresponding cutoff points are a_1 and a_2, respectively, we arrive at the mapping (representation) of the terms as follows $L:[1,a_1]\quad M:[a_1,a_2]\quad H:[a_2,9]$. If we consider *"m"* linguistic values, this results in *"m-1"* cutoff points. Being arranged in a single vector, they constitute an individual in the swarm of the PSO.

A finite series of realizations of the information granules is formed by randomly generating entries from the above intervals, plugging them into the reciprocal matrix, and computing the largest eigenvalue and the corresponding value of the inconsistency index. The average of the values of the inconsistency index is the fitness function associated with the particle formed by the cutoff points a_1 and a_2. A way of the formation of the fitness function is in line with the standard practices encountered in Monte Carlo simulations.

As an example, we consider a 5 by 5 reciprocal matrix with the three linguistic entries

$$R = \begin{bmatrix} 1 & H & L & 1/L & 1/L \\ 1/H & 1 & 1/L & 1/M & 1/M \\ 1/L & L & 1 & 1/M & 1/M \\ L & M & M & 1 & 1/L \\ L & M & M & L & 1 \end{bmatrix} \quad (1.16)$$

The granular matrix R is sampled 500 time (the numbers drawn from the uniform distribution defined over the corresponding subintervals of the [1,9] scale). Recall that the fitness function is the average of the inconsistency index computed over each collection of 500 reciprocal matrices. The process of learning realized by the PSO is illustrated in Fig. 1.6 where we show the values of the fitness function obtained in successive generations. In the PSO we use the generic form of the algorithm where the updates of the velocity are realized in the form

$$v(t+1) = \xi(t) \times v(t) + c_1 a \bullet (x_p - x) + c_2 b \bullet (x_g - x) \quad (1.17)$$

where "t" is an index of the generation and $\bullet$ denotes a vector multiplication realized coordinatewise, a and b are vectors of random numbers generated from a uniform distribution expressed over the unit interval, x_p represents the local best solution and x_g represents the global best solution.

The next position of the particle is computed in a straightforward manner:

$$x(t+1) = x(t) + v(t+1) \quad (1.18)$$

The constriction coefficient ξ decreases linearly over successive generations, i.e.:

$$\xi(t) = (num\_iter - t) * (\xi_{max} - \xi_{min}) / num\_iter + \xi_{min}, \quad (1.19)$$

where ξ_{max} =0.9, ξ_{min} =0.4 and "num_iter" is the total number of generations the PSO and "t" denotes the index of the current generation. The parameters of the PSO were set up as follows: the number of particles is 100, the number of iterations is set to 500, while $c_1 = c_2$ =2.

To put the obtained optimization results in a certain context, we report the performance obtained when considering a uniform distribution of the cutoff points over the scale, which are equal to 3.67 and 6.34, respectively. The average inconsistency index assumes the value of 0.1159 with a standard deviation of 0.0332. The histogram of the inconsistency rates provides a more comprehensive view at the results: there is a visible presence of a longer tail of the distribution spread towards higher values of the inconsistency index, Fig. 1.7.

The PSO returns the optimal cutoff points of 2.2 and 2.4, which are evidently shifted towards the lower end of the scale. The inconstancy index

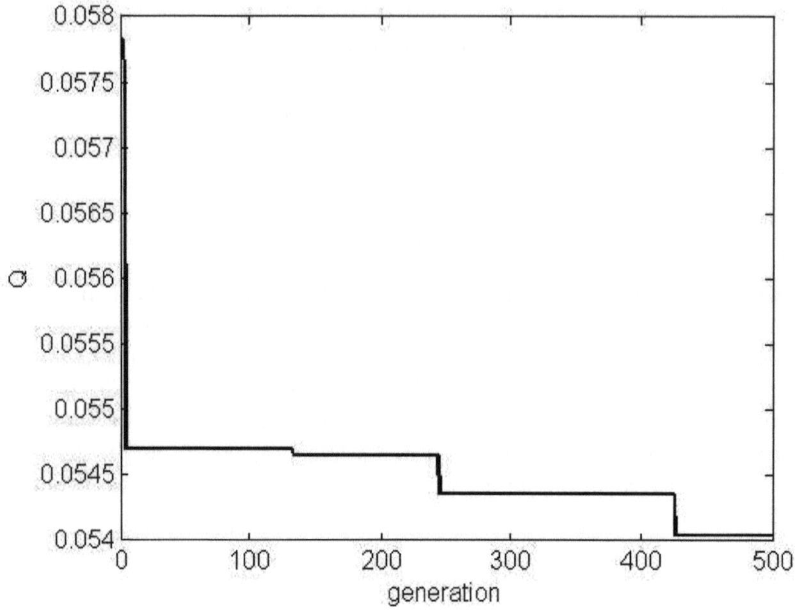

Fig. 1.6. The values of the fitness versus generations of the PSO

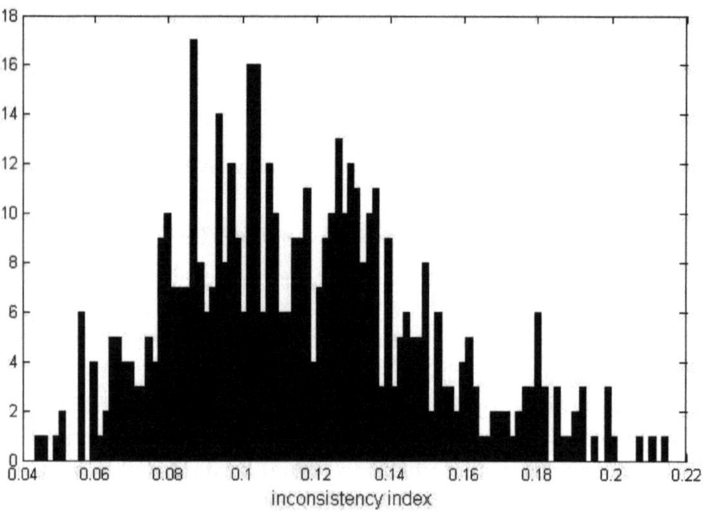

Fig. 1.7. The distribution of the values of the inconsistency index – a uniform distribution of the cutoff points

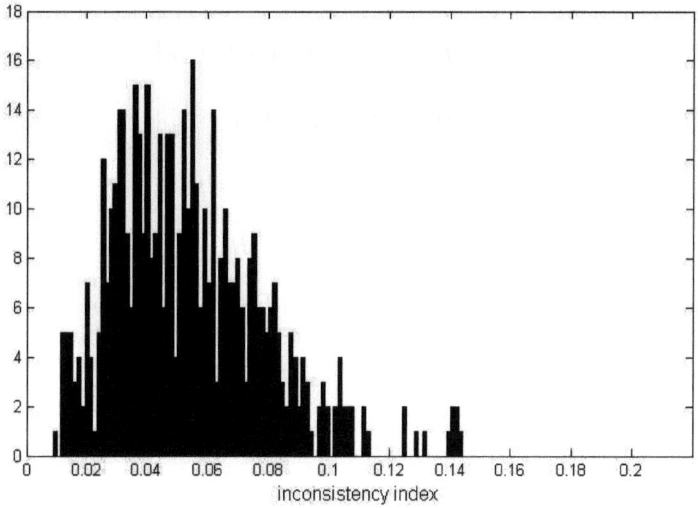

Fig. 1.8. The distribution of the values of the inconsistency index- a PSO-optimized distribution of the cutoff points

takes on now lower values and is equal to 0.054 with the standard deviation of 0.0252. The corresponding histogram is shown in Fig. 1.8.

For the optimal splits of the scale, a reciprocal matrix with the lowest inconsistency index is given below

$$R = \begin{bmatrix} 1 & 2.2 & 1/1.61 & 1/2.35 & 1/2.4 \\ 1/2.2 & 1 & 1/2.1 & 1/4.05 & 1/6.15 \\ 1.65 & 2.1 & 1 & 1/1.45 & 1/2.45 \\ 2.35 & 4.05 & 1.45 & 1 & 1/1.25 \\ 2.4 & 6.15 & 2.45 & 1.25 & 1 \end{bmatrix} \tag{1.20}$$

with the normalized eigenvector corresponding to the largest eigenvalue of this reciprocal matrix equal to $e = [0.79 \quad 1.00 \quad 0.65 \quad 0.30 \quad 0.00]^T$, which identifies the second alternative as an optimal one.

As another example, let us consider another 5×5 reciprocal matrix with 5 linguistic terms, VL, L, M, H, and VH

$$R = \begin{bmatrix} 1 & M & L & VL & H \\ 1/L & 1 & M & 1/VL & VH \\ 1/M & 1/M & 1 & 1/H & L \\ 1/VL & VL & H & 1 & H \\ 1/H & 1/VH & 1/L & 1/H & 1 \end{bmatrix} \quad (1.21)$$

The results of the optimization are shown in Fig. 1.9, here the parameters of the PSO were set up as follows: number of particles: 100, number of iterations: 500, c_1=c_2=2.

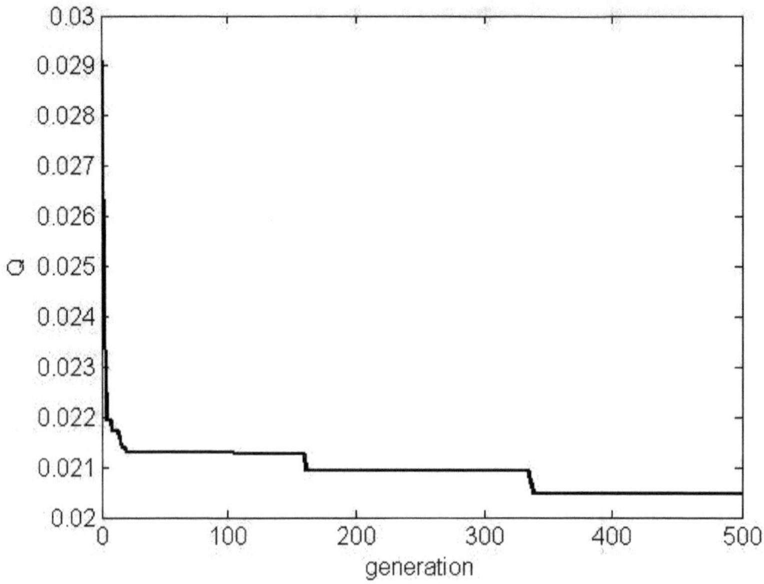

Fig. 1.9. The values of the fitness function obtained in successive generations of the PSO

The results corresponding with the uniform distribution of the cutoff points (that is 2.6, 4.2, 5.8, and 7.4) come with the average inconsistency index of 0.0888 with a standard deviation of 0.0162. The PSO produces the cutoff points of 1.1, 1.8, 5.3, and 5.6. The value of the inconsistency index is now lower and equal to 0.0205 with the standard deviation of 0.0102. The corresponding histograms both for the uniform and PSO-optimized cutoff points are shown in Fig. 1.10.

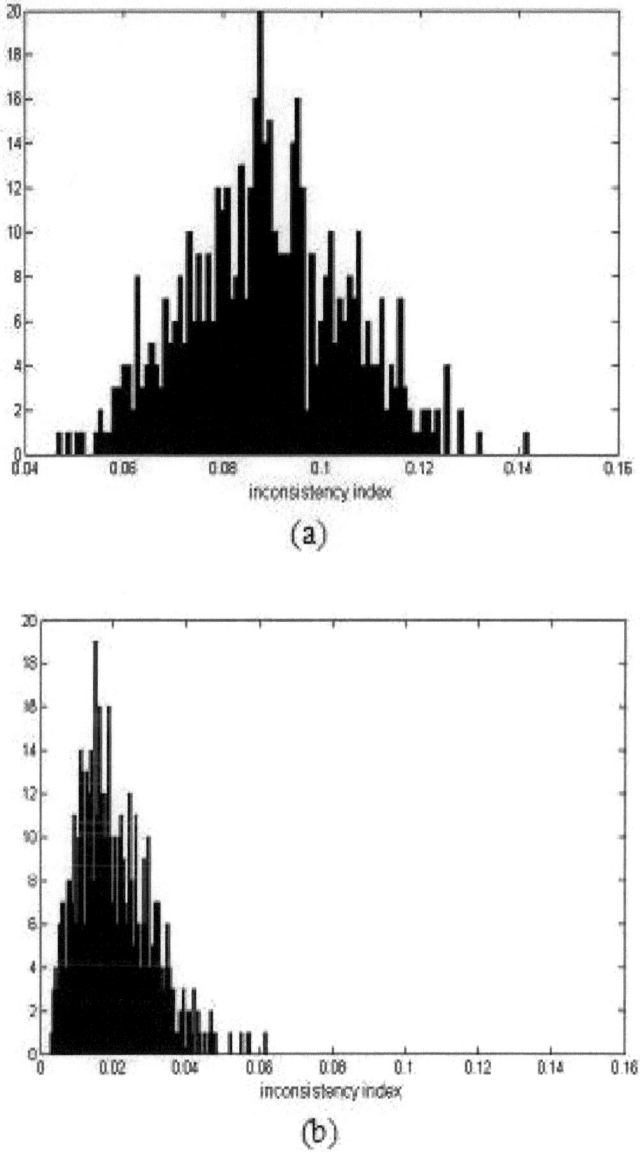

Fig. 1.10. The distribution of the inconsistency index: (a) uniform distribution of cutoff points, (b) - PSO-optimized distribution of the cutoff points

For the optimal split of the scale, a reciprocal matrix with the lowest inconsistency index is given below

$$R = \begin{bmatrix} 1 & 1.1 & 4.55 & 1.05 & 5.5 \\ 1/1.11 & 1 & 3.85 & 1/1.05 & 6.05 \\ 1/4.55 & 1/3.85 & 1 & 1/5.3 & 1.3 \\ 1/1.05 & 1.05 & 5.3 & 1 & 5.3 \\ 1/5.5 & 1/6.05 & 1/1.3 & 1/5.3 & 1 \end{bmatrix} \quad (1.22)$$

with the corresponding eigenvalue equal to e=[1.00 0.92 0.06 0.97 0.00]^T.

1.5 Conclusions

The study offered a concept of a granular view at membership functions, which positions all processing pursuits in-between symbolic and numeric perspective (and ensuing computing processing) of fuzzy sets. The presented study brings forward an alternative approach to the numerically-oriented constructs of membership functions, where one looks at them from the perspective of information granules and delivers their granular interpretation. The in-between alternatives have not been not fully exploited however they offer an interesting option and bring some conceptual and computational flexibility. The adjustable number of information granules realizes this flexibility feature. Both the input and output interfaces are critical in building an effective communication vehicle in which numeric details are hidden thereby providing more flexibility as well as reducing computing overhead. This approach provides estimation mechanisms, which endow qualitative evaluations with the numeric details whose determination is a result of a certain optimization problem.

The logic operators play a fundamental role in processing fuzzy sets. In this way, an amazing diversity of numeric realizations of logic operators can be analyzed at the higher non-numeric level and some relationships among logic operators that manifests at the predetermined level of granularity can be revealed. The or (and)-based consistency of granular descriptors of fuzzy sets is useful in forming a granular description of membership functions being subsequently viewed as a series of Cartesian products of information granules $\Omega_i \times L_i$ expressed in the universe of discourse and the [0,1] space of membership values. The logic operator(-s) -consistent granular description of membership functions can be used in non-numeric processing of fuzzy sets and an interpretation of results of fuzzy modeling. While we have elaborated on the granular realization of fuzzy sets that are *or* and *and* consistent, in the same way one can analyze the granular realization and interpretation of other logic operators such as nullnorms, uninorms, aggregation operators and others. The main difference in comparison with the interval-valued split of the unit interval, lies in the formulation of the fitness function. Here we used the counts of occurrences of combinations of intervals. In case of fuzzy sets, the counts are replaced by σ-counts because of the matching of numeric membership degrees with the fuzzy sets of information granules. By making use of multilevel truth tables produced through the PSO optimization, we realize calculations such as max-t composition, min-s compositions and alike not engaging the minute numeric calculations involving membership values thus covering the spectrum of processing spread in-between numeric and symbolic calculations.

References

1. Akella, P.: Structure of n-uninorms. Fuzzy Sets and Systems 158, 1631–1651 (2007)
2. De Baets, B.: Idempotent uninorms. European Journal of Operational Research 118, 631–642 (1999)
3. De Baets, B., De Meyer, H., Mesiar, R.: Piecewise linear aggregation functions based on triangulation. Information Sciences 181, 466–478 (2011)
4. Calvo, T., De Baets, B., Fodor, J.: The functional equations of Frank and Alsina for uninorms and nullnorms. Fuzzy Sets and Systems 120, 385–394 (2001)
5. Calvo, T., Mayor, G., Mesiar, R.: Aggregation Operators: New Trends and Applications. Physica-Verlag, New York (2002)
6. Coello Coello, C.A., Pulido, G.T., Lechuga, M.S.: Handling multiple objectives with particle swarm optimization. IEEE Trans. on Evolutionary Computation 8(3), 256–279 (2004)
7. Dombi, J.: DeMorgan systems with an infinitely many negations in the strict monotone operator case. Information Sciences 181, 1440–1453 (2011)
8. Drewniak, J., Drygas, P.: On a class of uninorms. Int. J. of Uncertainty Fuzziness Knowledge-Based Systems 10, 5–10 (2002)
9. Drygas, P.: On properties of uninorms with underlying *t*-norm and *t*-conorm given as ordinal sums. Fuzzy Sets and Systems 161, 149–157 (2010)
10. García-Honrado, I., Trillas, E.: An essay on the linguistic roots of fuzzy sets. Information Sciences (in press) (available online June 6, 2011)
11. Fodor, J., Yager, R., Rybalov, A.: Structure of uninorms. Int. J. of Uncertainty, Fuzziness and Knowledge-Based Systems 5(4), 113–126 (1997)
12. Kennedy, J., Eberhart, R.C.: Particle swarm optimization. In: Proc. IEEE Int. Conf. on Neural Networks, vol. 4, pp. 1942–1948. IEEE Press, NJ (1948)
13. Klement, E., Mesiar, R., Pap, E.: Triangular Norms. Kluwer Academic Publishers, Dordrecht (2000)
14. Kuipers, B.: Qualitative Reasoning. MIT Press (1994)
15. Mira, J.: Symbols versus connections: 50 years of artificial intelligence. Neurocomputing 71(4-6), 671–680 (2008)

16. Mizumoto, M., Tanaka, K.: Fuzzy sets of type-2 under algebraic product and algebraic sum. Fuzzy Sets and Systems 5, 277–290 (1981)
17. Pedrycz, A.: Finite cut-based approximation of fuzzy sets and its evolutionary optimization. Fuzzy Sets and Systems 160(24), 3550–3564 (2009)
18. Pedrycz, W.: Fuzzy neural networks and neurocomputations. Fuzzy Sets & Systems 56, 1–28 (1993)
19. Pedrycz, W.: Shadowed sets: representing and processing fuzzy sets. IEEE Trans. on Systems, Man, and Cybernetics 28, pt. B, 103–109 (1998)
20. Pedrycz, W.: Statistically grounded logic operators in fuzzy sets. European Journal of Operational Research 193(2), 520–529 (2009)
21. Pedrycz, W.: From fuzzy sets to shadowed sets: interpretation and computing. Int. J. of Intelligent Systems 24(1), 48–61 (2009)
22. Shapiro, R.R. (ed.): Encyclopedia of Artificial Intelligence, vol. 1 & 2. J. Wiley, New York (1990)
23. Yager, R.R., Rybalov, A.: Uninorm aggregation operators. Fuzzy Sets and Systems 80, 111–120 (1996)
24. Van den Bergh, F., Engelbrecht, A.P.: A study of particle swarm optimization particle trajectories. Information Sciences 176(8), 937–971 (2006)
25. Wu, J., Luo, M.: Some properties of weak uninorms. Information Sciences 181(18), 3917–3924 (2011)
26. Zadeh, L.A.: A note on Z-numbers. Information Sciences 181, 2923–2932 (2011)

Chapter 2
Step Fuzzy Numbers and Neural Networks in Defuzzification Functional Approximation

Witold Kosiński and Katarzyna Węgrzyn-Wolska

Abstract. Ordered fuzzy numbers as generalization of convex fuzzy numbers are defined together with four algebraic operations. For defuzzification operators, that play the main role when dealing with fuzzy controllers and fuzzy inference systems, new representation formulae are given. Step ordered fuzzy numbers are considered. Approximation method based on forward neural networks is presented for determining defuzzification functionals when training sets of data are given. Results of approximation are given.

2.1 Introduction

Fuzzy numbers [41] are very special fuzzy sets defined on the universe of all real numbers **R**. There are two commonly accepted methods of dealing with fuzzy numbers, both basing on the classical concept of fuzzy sets, namely on the membership functions. The first, more general approach deals with the so-called convex fuzzy numbers (CFN) of Nguyen [33], while the second one deals with shape functions and $L-R$ numbers, set up by Dubois and Prade [7]. In applications the $L-R$ numbers as a restricted class of membership functions, are often in use. When operating on convex fuzzy numbers we have the interval arithmetic for our disposal. However, the approximations of shape functions and operations are needed, if one wants to remain within the $L-R$ numbers while following the Zadeh's extension principle [42]. Then in

Witold Kosiński
Polish-Japanese Institute of Information Technology, Warsaw, Poland and
Kazimierz-Wielki University, Bydgoszcz, Poland
e-mail: wkos@pjwstk.edu.pl

Katarzyna Węgrzyn-Wolska
École Supérieure d'Ingénieurs en Informatique et Génie de Télécommunication (ESIGETEL), France
e-mail: katarzyna.wegrzyn@esigetel.fr

J. Kołodziej et al. (Eds.): Advances in Intelligent Modelling and Simulation, SCI 422, pp. 25–41.
springerlink.com

most cases calculation results are not exact and are questionable since unexpected and uncontrollable results of repeatedly applied operations, caused by the need of making intermediate approximations (remarked in [38, 39]), can appear. However, even in the case of convex fuzzy numbers (CFN) multiply operations are leading to the large grow of the fuzziness, and depend on the order of operations.

This rises the heavy argument for those who still criticize the fuzzy number calculus. Fortunately, it was already noticed by both Dubois and Prade in their recent publication [4] that something is missing in the definition of the fuzzy numbers and the operations on them. This as well as other drawbacks have forced us to think about a generalization. Number of attempts to introduce non-standard operations on fuzzy numbers has been made [6, 11, 36, 38]. Our main observation made in [23] was: a kind of quasi-invertibility (or quasi-convexity [32]) of membership functions is crucial. Invertibility of membership functions of convex fuzzy number A makes it possible to define two functions a_1, a_2 on $[0,1]$ that give lower and upper bounds of each α-cut of the membership function μ_A of the number A defined by

$$A[\alpha] = \{x : \mu_A(x) \geq \alpha\} = [a_1(\alpha), a_2(\alpha)]$$

with $a_1(\alpha) = \mu_A|_{incr}^{-1}(\alpha)$ and $a_2(\alpha) = \mu_A|_{decr}^{-1}(\alpha)$, where $|_{incr}$ and $|_{decr}$ denote the restrictions of the function μ_A to its sub-domains on which is increasing or decreasing, respectively. Both functions $a_1(\alpha), a_2(\alpha)$ were used for the first time by the authors of [8] in their parametric representation of fuzzy numbers; they also introduced a linear structure to convex fuzzy numbers.

As long as one works with fuzzy numbers that possess continuous membership functions, the two procedures: the extension principle and the α-cut and the interval arithmetic method, give the same results (cf. [1]). Since results of multiple operations on convex fuzzy numbers are leading to a large growth of the fuzziness, and depend on the order of the operations due to the lack of the distributive law, new approaches could be required as well as more general definitions.

Additional to the growth of fuzziness another drawback is the lack of a solution X to the most simple fuzzy arithmetic equation

$$A + X = C, \tag{2.1}$$

with known fuzzy numbers A and C in which the support of A is greater than that of C. Another drawback of CFN is related to the fact that $A + B - A$ is not equal to B, in general.

2.2 Ordered Fuzzy Numbers

In the series of papers [23, 24], [25] we have introduced and then developed main concepts of the space of ordered fuzzy numbers (OFNs). In our approach the concept of membership functions has been weakened by requiring a mere *membership relation.*

Definition 2.1. Pair (f,g) of continuous functions such that $f,g:[0,1]\to\mathbf{R}$ is called ordered fuzzy number A.

Notice that f and g need not be inverse functions of some membership function. If, however, f is increasing and g – decreasing, both on the unit interval I, and $f\le g$, then one can attach to this pair a continuous function μ and regard it as a membership function a convex fuzzy number with an extra feature, namely the orientation of the number. This attachment can be done by the formula $f^{-1}=\mu|_{incr}$ and $g^{-1}=\mu|_{decr}$. Notice that pairs (f,g) and (g,f) represent two different ordered fuzzy numbers, unless $f=g$. They differ by their orientations.

Definition 2.2. Let $A=(f_A,g_A), B=(f_B,g_B)$ and $C=(f_C,g_C)$ are mathematical objects called ordered fuzzy numbers. The sum $C=A+B$, subtraction $C=A-B$, product $C=A\cdot B$, and division $C=A\div B$ ca be defined by the unified formula, in which the symbol "$\star$" works for those operations, i.e. for "+", "−", "·", and "÷", respectively,

$$f_C(y)=f_A(y)\star f_B(y), g_C(y)=g_A(y)\star g_B(y), \tag{2.2}$$

and where $A\div B$ is defined, if the functions $|f_B|$ and $|g_B|$ are bigger than zero.

Scalar multiplication by real $r\in\mathbf{R}$ is defined as $r\cdot A=(rf_A,rg_A)$. The subtraction of B is the same as the addition of the opposite of B, and consequently $B-B=0$, where $0\in\mathbf{R}$ is the crisp zero. It means that subtraction is not compatible with the extension principle of Zadeh, if we confine OFNs to CFN. However, the addition operation is compatible, if its components have the same orientations. Notice, however, that addition, as well as the subtraction, of two OFNs that are represented by affine functions and possess classical membership functions, may lead to result which may not possess its membership functions (in general $f(1)$ needs not be less than $g(1)$). Notice that to any convex and continuous[1] fuzzy number correspond two OFNs, they differ by the orientation: one has positive, say (f,g) , another (g,f) has negative.

Relation of *partial ordering* in the space $\mathcal{R}$ of all OFN, can be introduced by defining the subset of *positive* ordered fuzzy numbers: a number $A=(f,g)$ is not less than zero, and write

$$A\ge 0 \text{ if } f\ge 0, g\ge 0, \tag{2.3}$$

[1] However, the recent extension presented in [15] includes all convex fuzzy numbers.

and consequently, we say that A is greater of equal to B, and write

$$A \geq B \text{ if } A - B \geq 0. \tag{2.4}$$

In this way the space $\mathcal{R}$ becomes *a partially ordered ring*, since in the space two operations: addition and mutiplication have been introduced, and they satisfy the distributivity law. Neutral element of addition in $\mathcal{R}$ is a pair of constant functions $(0^\dagger, 0^\dagger) =: 0^\ddagger$ equal to crisp zero, where $0^\dagger(s) = s$, with $s \in [0,1]$.

Operations introduced in the space $\mathcal{R}$ of all ordered fuzzy numbers (OFN) make it an algebra. Moreover the space can be equipped with a sup norm (the Ĉebyshev norm)

$$||A|| = \max(\sup_{s\in I}|f_A(s)|, \sup_{s\in I}|g_A(s)|), \text{ if } A = (f_A, g_A). \tag{2.5}$$

In this norm $\mathcal{R}$ is not only unormed vector space but also complete[2] It means that $\mathcal{R}$ becomes a Banach space, and it is isomorphic to a Cartesian product of $C(0,1)$ - the space of continuous functions on $[0,1]$. It is also a Banach algebra with unity, i.e. results of the multiplication of its elements stay in $\mathcal{R}$. The multiplication has a neutral element - the pair of two constant functions equal to one, $(1^\dagger, 1^\dagger) =: 1^\ddagger$) which represents the crisp one from **R**.

What is more important in $\mathcal{R}$ any algebraic equation $A + X = C$ for X, with arbitrarily given fuzzy numbers A and C, can be solved.

Notice, that for each two fuzzy numbers $A = (f_A, g_A), B = (f_B, g_B)$ as above, we may define $\inf(A,B) =: F$ and $\sup(A,B) =: G$, both from $\mathcal{R}$, by the relations:

$$F = (f_F, g_F), \text{ if } f_A = \inf\{f_A, f_B\}, g_A = \inf\{g_A, g_B\}. \tag{2.6}$$

Similarly, we define $G = \sup(A,B)$.

Some interpretations of the concepts of OFN are given in [22]. Fuzzy implications within OFN are discussed in [12],
[18, 20].

2.3 Defuzzification Functionals

In dealing with applications of fuzzy numbers we need a set of functionals that map each fuzzy number into real number and in such a way that this map is consistent with operations on real numbers. Those functionals are called defuzzifications. To be more strict we introduce

Definition 2.3. A map ϕ from the space $\mathcal{R}$ of all OFN's to reals is called a defuzzification functional if is satisfies:

[2] It means that any Cauchy sequence of its elements is converegent to an element of this space.

1. $\phi(c^{\ddagger}) = c$,
2. $\phi(A + c^{\ddagger}) = \phi(A) + c$,
3. $\phi(cA) = c\phi(A)$, for any $c \in \mathbf{R}$ and $A \in \mathcal{R}$.

where $c^{\ddagger}(s) = (c, c), s \in [0,1]$, represents crisp number (a real) $c \in \mathbf{R}$.

Here by writing $\phi(c^{\ddagger})$ we understand the action of the functional ϕ on the crisp number $c^{\ddagger}$ from $\mathbf{R}$, which is represented by the pair of constant functions $(c^{\dagger}, c^{\dagger})$, with $c^{\dagger}(s) = c, s \in [0,1]$. The condition 2. is a *restricted additivity*, since the second component is crisp number. The condition 3. requires from ϕ to be homogeneous of order one, while the condition 1. requires $\int_0^1 dh_1(s) + \int_0^1 dh_2(s) = 1$, in the representation (2.7).

In the course of defuzzification operation in CFN to a membership function a real, crisp number is attached. We know a number of defuzzification procedures from the literature [37]. Continuous, linear functionals on $\mathcal{R}$ give the class of defuzzification functionals. Each of them, say ϕ, has the representation by the sum of two Stieltjes integrals with respect to two functions h_1, h_2 of bounded variation,

$$\phi(f,g) = \int_0^1 f(s)dh_1(s) + \int_0^1 g(s)dh_2(s)\ . \tag{2.7}$$

Notice that if for $h_1(s)$ and $h_2(s)$ we put $\lambda H(s)$ and $(1-\lambda)H(s)$, respectively, with $0 \le \lambda \le 1$ and $H(s)$ as the Heaviside function with the unit jump at $s = 1$, then the defuzzification functional in (2.7) will lead to the classical MOM – middle of maximum, FOM (first of maximum), LOM (last of maximum) and RCOM (random choice of maximum), with an appropriate choice of λ. For example if for $h_1(s)$ and $h_2(s)$ we put $1/2\ H(s)$ then the defuzzification functional in (2.7) will represent the classical MOM – middle of maximum

$$\phi(f,g) = 1/2(f(1) + g(1))\ . \tag{2.8}$$

New model gives the continuum number of defuzzification operators both linear and nonlinear, which map ordered fuzzy numbers into real numbers. Nonlinear functional can be defined, see [29], as an example we have the center of gravity defuzzification functional (COG) calculated at OFN (f,g) as

$$\phi_G(f,g) = \frac{\int_0^1 \frac{f(s)+g(s)}{2} |f(s) - g(s)| ds}{\int_0^1 |f(s) - g(s)| ds}\ . \tag{2.9}$$

If $A = c^{\ddagger}$ then we put $\bar{\phi}_G(c^{\ddagger}) = c$. When $\int_0^1 |f(s) - g(s)| ds = 0$ in (2.9) we write

$$\phi_G(f,f) = \int_0^1 f(s) ds / \int_0^1 ds. \qquad (2.10)$$

Now we pass to next functional. It is nonlinear and originally proposed by [40]. She has done for the case of trapezoidal[3] fuzzy numbers. However, it can be easily extended to the general case.

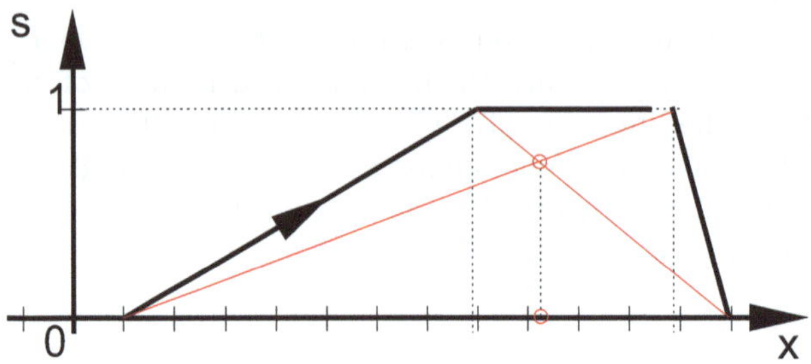

Fig. 2.1. Example of calculation of defuzzification of the geometrical mean of a trapezoidal ordered fuzzy number

Definition 2.4. Method of defuzzification by the geometrical mean is defined by the formula

$$\phi_{GM}(f,g) = \frac{g(1)g(0) - f(0)f(1)}{g(1) + g(0) - (f(0) + f(1))} \qquad (2.11)$$

in which the denotation used are obvious.

The derivation of this formula made for trapezoidal, ordered fuzzy numbers (cf. Fig.1), is rather trivial and gives the horizontal coordinate of the point of intersection of two lines (diagonals): one passing through the points $(0, f(0))$ and $(1, g(1))$, the second line passing through the points $(0, g(0))$ and $(1, f(1))$. In this defuzzification method the generated output for the trapezoidal number always belongs to the interval $[f(1), g(1)$. This means that the value of the functional is a kind of interval defuzzification method and takes into account those arguments x at which the membership function attains its maximum value 1. This method is not a random and for this it can be treated superior with respect to the RCOM method.

[3] An ordered fuzzy number (f,g) is trapezoidal if f and g are affine functions, i.e. their graphs are parts of straight lines.

2.4 Step Numbers

It is worthwhile to point out that the class of ordered fuzzy numbers (OFNs) represents the whole class of convex fuzzy numbers with continuous membership functions. To include all CFN with piecewise continuous membership functions more general class of functions f and g in Def.1 is needed. This has been already done by the first author who in [15] assumed they are functions of bounded variation. The new space is denoted by $\mathcal{R}_{BV}$. Then operations on elements of $\mathcal{R}_{BV}$ are defined in the similar way, the norm, however, will change into the norm of the Cartesian product of the space of functions of bounded variations (BV). Then all convex fuzzy numbers are contained in this new space $\mathcal{R}_{BV}$ of OFN. Notice that functions from BV [31] are continuous except for a countable numbers of points.

Important consequence of this generalization is the possibility of introducing the subspace of OFN composed of pairs of step functions. It will be done as follows. If we fix a natural number K and split $[0,1)$ into $K-1$ subintervals $[a_i,a_{i+1}), i=1,2,...,K$, i.e. $\bigcup_{i-1}^{K-1}[a_i,a_{i+1})=[0,1)$, where $0=a_1<a_2<...<a_K=1$, we may define step function f of resolution K by putting value $f(s)=u_i\in\mathbf{R}$, for $s\in[a_i,a_{i+1})$, then each such function f can be identified with a K-dimensional vector $f\sim\underline{u}=(u_1,u_2...u_K)\in\mathbf{R}^K$, the K-th value u_K corresponds to $s=1$, i.e. $f(1)=u_K$. Taking pair of such functions we have an ordered fuzzy number from $\mathcal{R}_{BV}$. Now we introduce

Definition 2.5. By step ordered fuzzy number A of resolution K we mean ordered pair (f,g) of functions such that $f,g:[0,1]\rightarrow\mathbf{R}$ are K-step function.

We use $\mathcal{R}_K$ for denotation the set of elements satisfying Def.2.5. The set $\mathcal{R}_K\subset\mathcal{R}_{BV}$ has been extensively elaborated by our students in [30] and [9]. We can identify $\mathcal{R}_K$ with the Cartesian product of $\mathbf{R}^K\times\mathbf{R}^K$ since each K-step function is represented by its K values.

It is obvious that each element of the space $\mathcal{R}_K$ may be regarded as approximation of elements from $\mathcal{R}_{BV}$, by increasing the number K of steps we are getting the better approximation of elements from $\mathcal{R}_{BV}$. The norm of $\mathcal{R}_K$ is assumed to be the classical Euclidean one of $\mathbf{R}^{2K}$, then in $\mathcal{R}_{BV}$ we have an inner product (scalar product) structure for our disposal.

On the space $\mathcal{R}_K$ a representation formula for a general non-linear defuzzification functional $H:\mathbf{R}^K\times\mathbf{R}^K\rightarrow\mathbf{R}$ satisfying the conditions 1.– 3. of Def. 3, can be given as a linear composition [29] of arbitrary homogeneous of order one, continuous function G of $2K-1$ variables, with the 1D identity function, i.e.

$$H(\underline{u},\underline{v})=u_1+G(u_2-u_1,u_3-u_1,..., \\ u_K-u_1,v_1-u_1,v_2-u_1,...,v_K-u_1), \tag{2.12}$$

with $\underline{u} = (u_1,...,u_K), \underline{v} = (v_1,...,v_K)$. It is obvious that u_i or v_i, with $1 \le i \le K$, can substitute the variable u_1 in the representation (2.12) to get another defuzzification functional on $\mathcal{R}_K$.

Due to the fact that $\mathcal{R}_K$ is isomorphic to $\mathbf{R}^K \times \mathbf{R}^K$ we conclude, from the Riesz theorem on functional representation in inner-product space and the condition 1. that a general linear defuzzification functional on $\mathcal{R}_K$ has the representation

$$H(\underline{u},\underline{v}) = \underline{u} \cdot \underline{b} + \underline{v} \cdot \underline{d}, \tag{2.13}$$
$$\text{with arbitrary}\, \underline{b}, \underline{d} \in \mathbf{R}^K, \text{such that}\, \underline{1} \cdot \underline{b} + \underline{1} \cdot \underline{d} = 1\ ,$$

where $\cdot$ denotes the inner (scalar) product in $\mathbf{R}^K$ and $\underline{1} = (1,1,...,1) \in \mathbf{R}^K$ is the unit vector in $\mathbf{R}^K$, while the pair $(\underline{1},\underline{1})$ represents a crisp one in $\mathcal{R}_K$. It means that such functional is represented by the vector $(\underline{b},\underline{d}) \in \mathbf{R}^{2K}$. Notice that functionals of the type $\phi_j = \underline{e}_j, j = 1,2,...,2K$, where $\underline{e}_j \in \mathbf{R}^{2K}$ has all zero component except for 1 on the j-th position, form a basis of $\mathcal{R}_K{}^*$ - the space adjoint to $\mathcal{R}_K$, they are called *fundamental functionals.* Here by writing $\phi(c^\ddagger)$ we understand the action of the functional ϕ on the crisp number $c^\ddagger$ from $\mathbf{R}$, which is represented by the pair of constant functions $(c^\dagger, c^\dagger)$, with $c^\dagger(i) = c, i = 1,2,...,K$, in the case of $\mathcal{R}_K$.

Notice that each real-valued function $\psi(z)$ of a real variable $z \in \mathbf{R}$ may be transformed to a fuzzy-valued function on $\mathcal{R}_{BV}$, and even simpler on $\mathcal{R}_K$. Here we have used the representation for $\underline{u} = (u_1,...,u_K)$ and for $\underline{v} = (v_1,...,v_K)$.

2.5 Approximation of Defuzzification Functionals

Ultimate goal of fuzzy logic is to provide foundations for approximate reasoning. It uses imprecise propositions based on a fuzzy set theory developed by L.Zadeh [41], in a way similar to the classical reasoning using precise propositions based on the classical set theory. Defuzzification is the main operation which appears in fuzzy controllers and fuzzy inference systems where fuzzy rules are present. It was extensively discussed by the authors of [37]. They have classified the most widely used defuzzification techniques into different groups, and examined the prototypes of each group with respect to the defuzzification criteria.

The problem arises when membership functions are not continuous or do not exist at all. Here on particular subsets of fuzzy sets, namely *step ordered fuzzy numbers* approximation formula of a defuzzification functionals will be searched based on some number of training data. This is a quite new problem never investigated within step ordered fuzzy numbers.

Problem formulation. Let finite set of training data be given in the form of N pairs: ordered fuzzy number and value (of action) of a defuzzification functional on it, i.e. $\mathsf{TRE} = \{(A_1,r_1),(A_2,r_2),...,(A_N,r_N)\}$. For a given small ϵ find a continuous functional $H : \mathcal{R}_K \to \mathbf{R}$ which approximates the values of the set TRE within the error smaller than ϵ, i.e. $\max_{1\leq p\leq N} |H(A_p)-r_p| \leq \epsilon$, where $(A_p,r_p) \in \mathsf{TRE}$.

Problem may possess several solution methods, e.g. a dedicated evolutionary algorithm [16, 19] or an artificial neural network. We have use the representation (2.12) of the searched defuzzification functional in which a homogeneous, of order one, function Ψ appears. It means that values of this function are determined from its arguments situated on the unit sphere $\mathcal{S}_{2K-1}$ in $2K-1$ D space. Here an artificial neural network will be in use.

Training and test sets used in the further section (from now denoted as TRE and TES, respectively) have the following form. Set of N elements is composed of N pairs of OFN and value of a defuzzification functional on it, i.e.:

$$\{(A_1,r_1),(A_2,r_2),...,(A_N,r_N)\}. \tag{2.14}$$

We are training an artificial feed-forward neural network on this set to find the approximated form of the functional.

2.6 Neural Network Simulations

In order to make approximation of linear and the nonlinear defuzzification functionals on step ordered fuzzy numbers (SOFN) a package of artificial neural networks (ANN) has been used. Since each SOFN is represented by a vector of $2K$ number, each input to artificial neural networks has $2K$ real-valued components. In our case it was:

- MLP neural network with one hidden layer:
  - 20 inputs
  - 5 neurons in hidden layers
  - 1 output neuron
- 500 iterations
- data set:
  - training - TRE_0, TRE_4
  - testing - TES_0, TES_4

Data Generation

The procedure to generate TRE and TES sets was the following.

1. Generate 60 random points on a $2K-1$ dimensional hyper-sphere $\mathcal{S}_{2K-1}$, where $K = 10$. Let $\varphi = (u_2, u_3, ..., u_{K-1}, v_1, v_2, ..., v_K)$ be one of these points. All points fulfill the conditions $u_n < u_{n+1}$ and $v_m > v_{m+1}$. This ensures that the generated fuzzy numbers have a trapezoidal shape. In the further parts this assumption has been omitted.
2. Generate two sets of fuzzy numbers using the following methods of generating a value of u
   - $u = 0$
   - u is a random value from $(-4, 4)$
3. For each fuzzy number find the defuzzified value and split the sets in the ratio 2:1 to form:
   - TRE_0 and TES_0 from fuzzy numbers with $u_1 = 0$
   - TRE_4 and TES_4 from fuzzy numbers with $u_1 \in (-4, 4)$

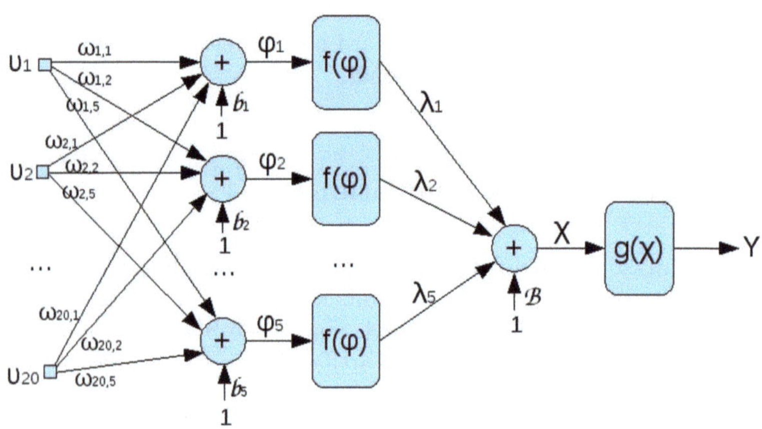

Fig. 2.2. Neural network structure

The general strategy was to train the network with data sets having $2K$ inputs and one output representing the discrete values of fuzzy intput values and the crisp output, repectively, calculated according to selected standard defuzzification algorithms. For the linear defuzzification we have used: MOM (middle of maximum), LOM (last of maximum) and FOM (first of maximum).

Table 2.1 presents the final training Root Mean Square Error (RMSE[%]) for all the used methods. Table 2.2 presents the final training gradient for all the used methods. The RMSE error was calculated as

$$Error^2 = \frac{1}{P}\sum_{i=1}^{i=P}(H(A_i) - r_i)^2 .$$

We have used the Lavenberg-Marquardt adaptation algorithm.

Table 2.1. Final training RMSE

Training Set	MOM	LOM	FOM
TRE_0	1.196156E-11	1.17966E-11	3.2052E-11
TRE_4	8.22773E-10	1.51997E-9	3.1339E-9

Table 2.2. Final training gradient

Training Set	MOM	LOM	FOM
TRE_0	3.57907E-6	1.14851E-6	1.09344E-6
TRE_4	0.001232	0.0001864	0.00311

The validation of our neural network is done by testing the network with TES_0 and TES_4 data sets generated with all of the following defuzzification methods: MOM (middle of maximum), LOM (last of maximum) and FOM (first of maximum). The δ error was calculated as

$$\delta = \frac{\mathsf{V} - \mathsf{V}_{\mathsf{approx}}}{\mathsf{V}}[\%]$$

The validation of data TES_0 and TES_4 defuzzified with MOM strategy converges successfully. The results after learning process are presented at the figures: on TRE_0 the RMSE [%] and the gradient (Figure 2.3), on TRE_4 the RMSE and the gradient (Figure 2.4). The X axis presents the number of iterations.
The results for testing (δ for each test) are presented at the figures: δ_{TES_0} on TES_0 (Figure 2.5), δ_{TES_4} on TES_4 (Figure 2.6). The X axis in histograms presents the number of sample. Similar results have been obtained for other defuzzification methods.

Performed simulation proved that ANN can successfully represent the defuzzification strategies. Linear approximations of defuzzification functionals with MOM, LOM and FOM were correct. The trained ANN approximations for all the methods were successfully tested with TES_0 and TES_4 data sets. Table 2.3 presents the final validation MSE for all the used methods.

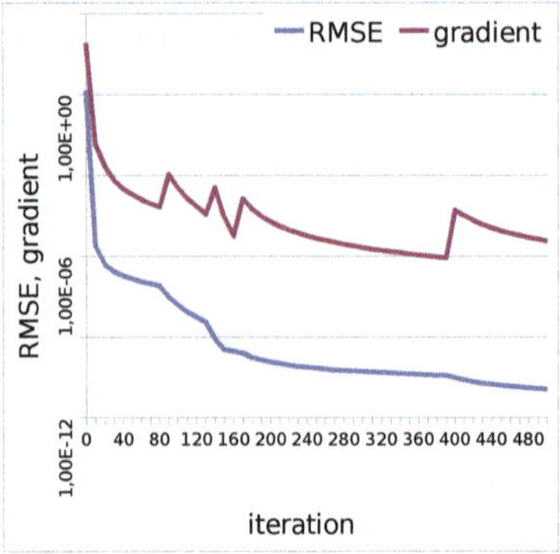

Fig. 2.3. MOM on TRE_0: RMSE (bleu) [%], Gradient (green)

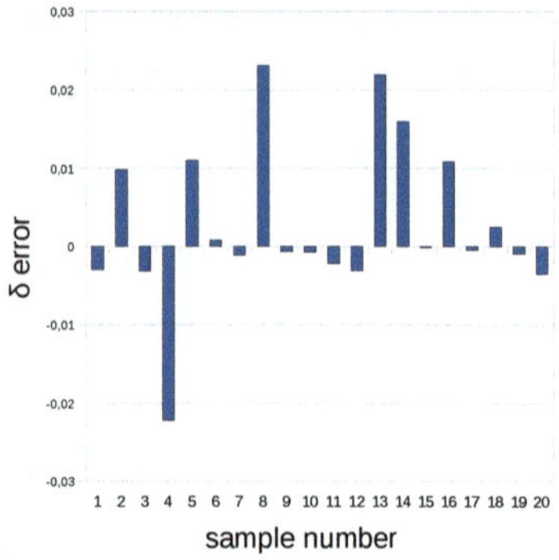

Fig. 2.4. MOM on TES_0: δ_{TES_0} [%]

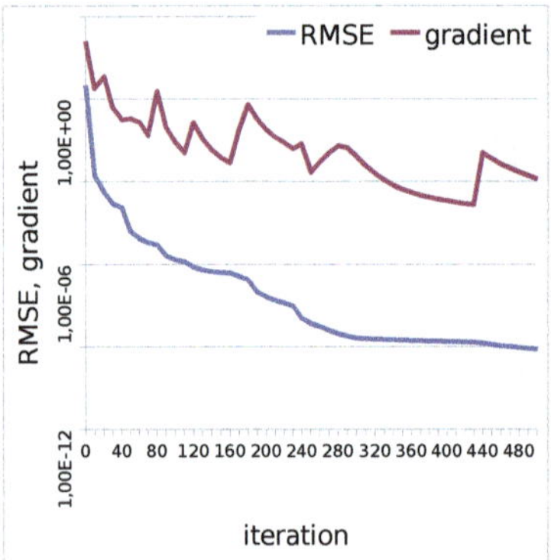

Fig. 2.5. MOM on TRE_4: RMSE (bleu) [%], Gradient (green)

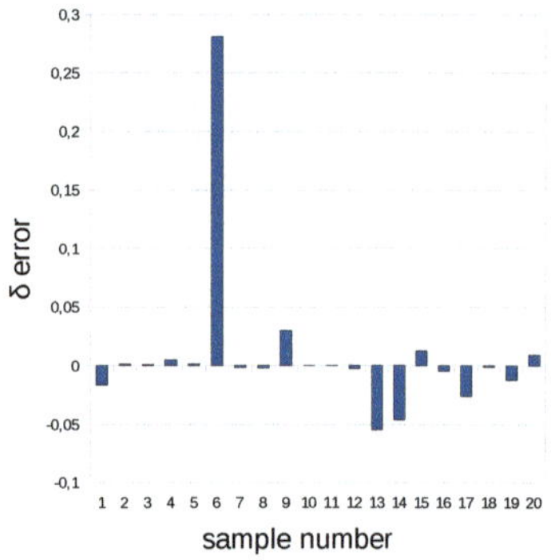

Fig. 2.6. MOM on TES_4: δ_{TES_4} [%]

Table 2.3. Final linear validating δ

Testing Set	MOM	LOM	FOM
δ_{TES_0}	1.781138E-5	3.020065E-5	0.0001056
δ_{TES_4}	4.300E-9	2.02054E-6	0.0006829

2.7 Nonlinear Defuzzification Functional

Similar method has been used for nonlinear defuzzification functional, namely for the center of gravity (COG). The validation of data TES_0 and TES_4 defuzzified with COG strategy converges successfully.

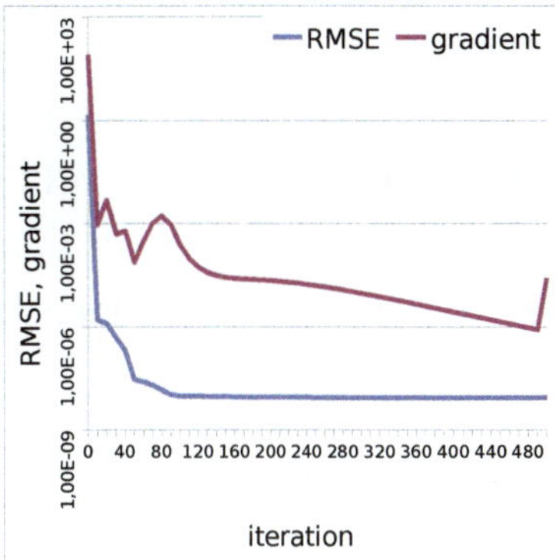

Fig. 2.7. COG on TRE_0: RMSE and gradient

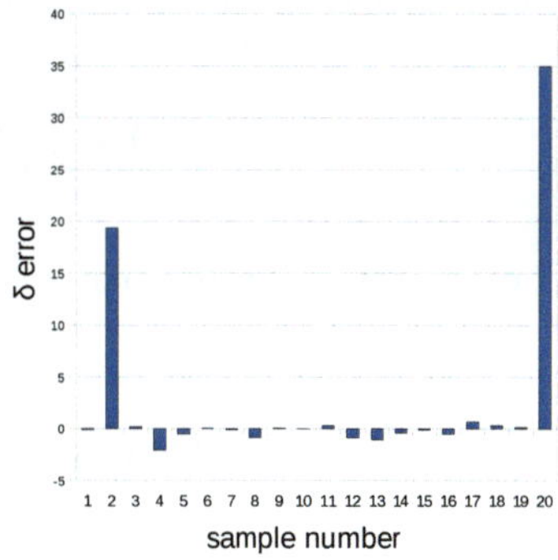

Fig. 2.8. COG on TES_0: RMSE

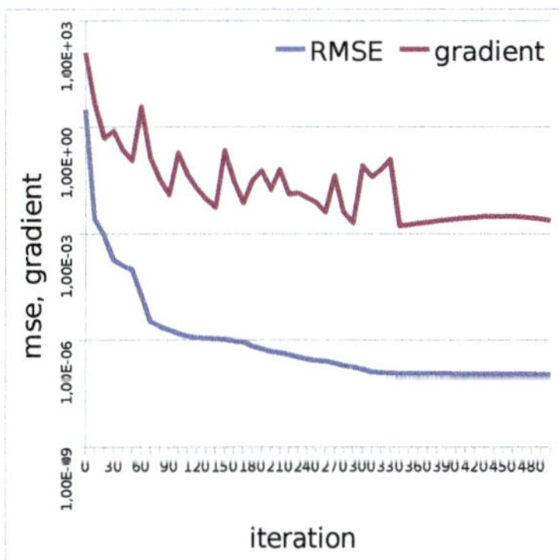

Fig. 2.9. COG on TES_4: RMSE and gradient

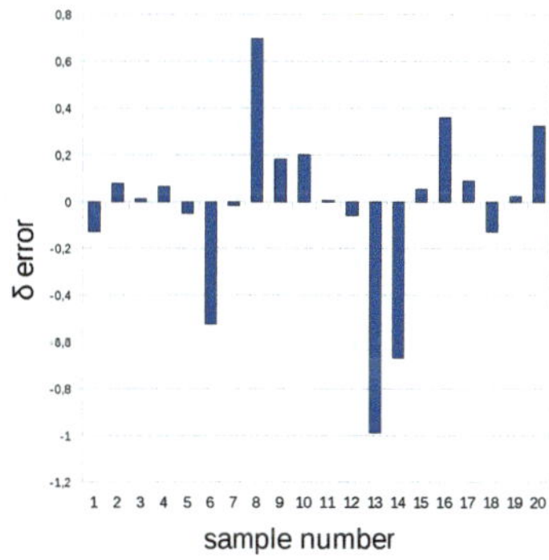

Fig. 2.10. COG on TES_4: RMSE

2.8 Activation Functions

For the first layer we have assumed the bipolar sigmoidal activation function is given by the formula:

$$f(x) = \frac{2}{1+e^{-2x}} - 1\,. \tag{2.15}$$

The hidden layer activation function is given by $g(x) = x$, and the output is given by

$$Y = g(X) = X = \sum_{j=1}^{5} \Phi_j \lambda_j + B\,, \tag{2.16}$$

where $\Phi_j = f(\varphi_j) = f(\sum_{i=1}^{20} u_i * \omega_{i,j} + b_j)$. Hence we have the following functional dependence of the output on the input vector $\underline{u}$

$$Y = \sum_{j=1}^{5} f(\sum_{i=1}^{20} u_i \omega_{i,j} + b_j))\lambda_j + B. \tag{2.17}$$

The weights and other parameters can be listed in the form of tables.

Table 2.4. NN structure after learning

Input	ω_{i1}	ω_{i2}	ω_{i3}	ω_{i4}	ω_{i5}
U_1	-1.500000	1.500000	-0.500000	0.500000	0.100000
U_2	-1.764203	1.721917	-1.163269	0.325977	-0.766464
U_3	-0.909134	0.969637	-1.070874	0.164166	-0.183957
U_4	-1.888874	1.870025	-1.096356	0.652329	-0.950350
U_5	-1.551447	1.618123	-0.028711	-0.164044	-1.205991
U_6	-1.339349	1.297634	0.030845	0.789391	-1.386863
U_7	-0.924551	0.758894	-0.292342	0.406089	0.273688
U_8	-1.444772	1.426924	0.270408	0.433603	-0.521424
U_9	-2.290977	2.078550	0.003299	0.771404	1.510400
U_{10}	-2.035505	1.984894	0.007624	.370735	-0.662857
U_{11}	-1.543991	1.181070	0.511984	0.784363	0.820112
U_{12}	-1.893619	1.545661	0.350312	0.843925	1.318368
U_{13}	-1.690412	1.296557	-0.424741	1.410183	1.098513
U_{14}	-1.583763	1.208434	0.315078	0.782784	1.159102
U_{15}	-1.477475	1.123565	-0.048310	0.786998	0.350696
U_{16}	-1.629397	1.389360	-0.185788	0.619010	0.175965
U_{17}	-1.758531	1.715217	0.851437	0.196586	-0.565551
U_{18}	-1.477277	1.287993	-0.478476	0.377743	1.132364
U_{19}	-0.877315	0.820842	-0.353786	-0.066219	0.204631
U_{20}	-2.090962	1.904204	-0.355554	0.233310	0.775427

Table 2.5. First layer bias

Training Set	b_1	b_2	b_3	b_4	b_5
Nonlinear TRE_0	0.562239483	0.425288679	1.616618821	1.759599096	1.0704185326

Table 2.6. Hidden layer weight

Training Set	$\lambda 1$	λ_2	λ_3	λ_4	λ_5
Nonlinear TRE_0	0.0814377120	0.001417061	-28.63429759	-0.0005601	14.19819490

2.9 Conclusion

The present paper brings an outline of the results of approximation of defuzzification functional of step ordered fuzzy numbers that have been obtained with the help of the tool of the computational intelligence, namely of artificial neural networks. We can conclude that the tool is helpful. It is rather evident that further research in this field should follow.

References

1. Buckley, J.J., Eslami, E.: An Introduction to Fuzzy Logic and Fuzzy Sets. Physica-Verlag, Springer, Heidelberg (2005)
2. Chen, G., Pham, T.T.: Fuzzy Sets, Fuzzy Logic, and Fuzzy Control Systems. CRS Press, Boca Raton (2001)
3. Dubois, D., Fargier, H., Fortin, J.: A generalized vertex method for computing with fuzzy intervals. In: Proc. IEEE Int. Conf. Fuzzy Syst., Budapest, Hungary, pp. 541–546 (2005)
4. Dubois, D., Prade, H.: Gradual elements in a fuzzy set. Soft. Comput. 12, 165–175 (2008), doi:10.1007/s00500-007-0187-6
5. Fortin, J., Dubois, D., Fargier, H.: Gradual numbers and their application to fuzzy interval analysis. IEEE Trans. Fuzzy Syst. 16(2), 388–402 (2008), doi:10.1109/TFUZZ.2006.890680
6. Drewniak, J.: Fuzzy numbers. In: Chojcan, J., Łęski, J. (eds.) Fuzzy Sets and their Applications, pp. 103–129. Wydawnictwo Politechniki Śląskiej, Gliwice (2001) (in Polish)
7. Dubois, D., Prade, H.: Operations on fuzzy numbers. Int. J. System Science 9, 576–578 (1978)
8. Goetschel Jr., R., Voxman, W.: Elementary fuzzy calculus. Fuzzy Sets and Systems 18, 31–43 (1986)
9. Gruszczyńska, A., Krajewska, I.: Fuzzy calculator on step ordered fuzzy numbers, UKW, Bydgoszcz (2008) (in Polish)
10. Kaucher, E.: Interval analysis in the extended interval space IR. Computing, Suppl. 2, 33–49 (1980)
11. Klir, G.J.: Fuzzy arithmetic with requisite constraints. Fuzzy Sets and Systems 91, 165–175 (1997)
12. Kacprzak, M., Kosiński, W.: On lattice structure and implications on ordered fuzzy numbers. In: Proc. EUSFLAT Conference, France (July 2011)
13. Koleśnik, R., Kosiński, W., Prokopowicz, P., Frischmuth, K.: On algebra of ordered fuzzy numbers. In: Atanassov, K.T., Hryniewicz, O., Kacprzyk, J. (eds.) Soft Computing – Foundations and Theoretical Aspects, pp. 291–302. Akademicka Oficyna Wydawnicza EXIT, Warszawa (2004)
14. Kosiński, W.: On Defuzzyfication of Ordered Fuzzy Numbers. In: Rutkowski, L., Siekmann, J.H., Tadeusiewicz, R., Zadeh, L.A. (eds.) ICAISC 2004. LNCS (LNAI), vol. 3070, pp. 326–331. Springer, Heidelberg (2004)
15. Kosiński, W.: On fuzzy number calculus. Int. J. Appl. Math. Comput. Sci. 16(1), 51–57 (2006)

16. Kosiński, W.: Evolutionary algorithm determining defuzzyfication operators. Engineering Applications of Artificial Intelligence 20(5), 619–627 (2007), doi:10.1016/j.engappai.2007.03.003
17. Kosiński, W.: Optimization with fuzzy data via evolutionary algorithms. In: International Conference on Numerical Analysis and Applied Mathematics, ICNAAM 2010, Rhodes Greece, September 19-25, 2010 American Institute of Physics (2010) CD-ROM ISBN978-0-7354-08-31-9
18. Kosiński, W., Kacprzak, M.: Fuzzy implications on lattice of ordered fuzzy numbers. In: Atanssov, K.T., Baczyński, M., Drewniak, J., Kacprzyk, J., Krawczyk, M., Szmidt, E., Wygralek, M., Zadrożny, S. (eds.) Recent Advances in Fuzzy Sets, Intuitionistic Fuzzy sets, Generalized Nets and Related Topics, Volume I: Foundations, pp. 95–110. IBS PAN - SRI PAS, Warsaw (2010)
19. Kosiński, W., Markowska-Kaczmar, U.: An evolutionary algorithm determining a defuzzyfication functional. Task Quarterly 11(1-2), 47–58 (2007)
20. Kosiński, W., Piasecki, W., Wilczyńska-Sztyma, D.: On Fuzzy Rules and Defuzzification Functionals for Ordered Fuzzy Numbers. In: Burczyński, T., Cholewa, W., Moczulski, W. (eds.) Proc. of AI-Meth 2009 Conference, Gliwice. AI-METH Series, pp. 161–178 (November 2009)
21. Kosiński, W., Prokopowicz, P.: Algebra of fuzzy numbers. Matematyka Stosowana. Matematyka dla Społeczeństwa 5/46, 37–63 (2004) (in Polish)
22. Kosiński, W., Prokopowicz, P., Kacprzak, D.: Fuzziness – Representation of Dynamic Changes by Ordered Fuzzy Numbers. In: Seising, R. (ed.) Views on Fuzzy Sets and Systems from Different Perspectives. STUDFUZZ, vol. 243, pp. 485–508. Springer, Heidelberg (2009)
23. Kosiński, W., Prokopowicz, P., Ślęzak, D.: Fuzzy numbers with algebraic operations: algorithmic approach. In: Kłopotek, M., Wierzchoń, S.T., Michalewicz, M. (eds.) Proc. IIS 2002, Intelligent Information Systems 2002, June 3-6, pp. 311–320. Physica Verlag, Poland (2002)
24. Kosiński, W., Prokopowicz, P., Ślęzak, D.: On algebraic operations on fuzzy reals. In: Rutkowski, L., Kacprzyk, J. (eds.) Proc. of the Sixth Int. Conference on Neural Network and Soft Computing, Zakopane, Poland, June 11-15, 2002. Advances in Soft Computing, pp. 54–61. Physica-Verlag (2003)
25. Kosiński, W., Prokopowicz, P., Ślęzak, D.: Ordered fuzzy numbers. Bulletin of the Polish Academy of Sciences, Sér. Sci. Math. 51(3), 327–338 (2003)
26. Kosiński, W., Prokopowicz, P., Ślęzak, D.: On algebraic operations on fuzzy numbers. In: Kłopotek, M., Wierzchoń, S.T., Trojanowski, K. (eds.) Proc. of Int. IIS: IIPWM 2003, Intelligent Information Processing and Web Mining, Poland, June 2-5, pp. 353–362. Physica-Verlag (2003)
27. Kosiński, W., Prokopowicz, P., Ślęzak, D.: Calculus with Fuzzy Numbers. In: Bolc, L., Michalewicz, Z., Nishida, T. (eds.) IMTCI 2004. LNCS (LNAI), vol. 3490, pp. 21–28. Springer, Heidelberg (2005)
28. Kosiński, W., Weigl, M.: General mapping approximation problems solving by neural networks and fuzzy inference systems. Systems Analysis Modelling Simulation 30 (1), 11–28 (1998)
29. Kosiński, W., Wilczyńska-Sztyma, D.: Defuzzification and implication within ordered fuzzy numbers. In: WCCI 2010 IEEE World Congress on Computational Intelligence - CCIB, Barcelona, Spain, July 18-23, pp. 1073–1079 (2010)
30. Kościeński, K.: Modul of step ordered fuzzy numbers in control of material point motion. PJWSTk, Warszawa (2010) (in Polish)

31. Łojasiewicz, S.: Introduction to the Theory of Real Functions. Biblioteka Matematyczna, vol. 46. PWN, Warszawa (1973) (in Polish)
32. Martos, B.: Nonlinear Programming - Theory and methods. PWN, Warszawa (1983) (Polish translation of the English original published by Akadémiai Kiadó, Budapest, 1975)
33. Nguyen, H.T.: A note on the extension principle for fuzzy sets. J. Math. Anal. Appl. 64, 369–380 (1978)
34. Prokopowcz, P.: Algorithmisation of operations on fuzzy numbers and its applications, Ph. D. Thesis, IPPT PAN, Warszawa (2005) (in Polish)
35. Prokopowicz, P.: Using ordered fuzzy numbers arithmetic. In: Cader, A., Rutkowski, L., Tadeusiewicz, R., Zurada, J. (eds.) Proc. of the 8th International Conference on Artificial Intelligence and Soft Computing, Zakopane, Polska, June 25-29. Fuzzy Control in Artificial Intelligence and Soft Computing, pp. 156–162. Academic Publishing House EXIT, Warsaw (2006)
36. Sanchez, E.: Solutions of fuzzy equations with extended operations. Fuzzy Sets and Systems 12, 237–248 (1984)
37. Van Leekwijck, W., Kerre, E.E.: Defuzzification: criteria and classification. Fuzzy Sets and Systems 108, 159–178 (1999)
38. Wagenknecht, M.: On the approximate treatment of fuzzy arithmetics by inclusion, linear regression and information content estimation. In: Chojcan, J., Łęski, J. (eds.) Fuzzy Sets and Applications, pp. 291–310. Wydawnictwo Politechniki Śląskiej, Gliwice (2001)
39. Wagenknecht, M., Hampel, R., Schneider, V.: Computational aspects of fuzzy arithmetic based on Archimedean *t*-norms. Fuzzy Sets and Systems 123/1, 49–62 (2001)
40. Wilczyńska, D.: On control aspects within ordered fuzzy numbers in MATLAB environment, Msc Thesis, WMFiT, Kaziemierz-Wielki University, Bydgoszcz (2007) (in Polish)
41. Zadeh, L.A.: Fuzzy sets. Information and Control 8, 338–353 (1965)
42. Zadeh, L.A.: The concept of a linguistic variable and its application to approximate reasoning, Part I. Information Sciences 8, 199–249 (1975)

Part II
Parallel and Multiobjective Evo-Based Techniques and Architectures in Large-Scale Global Optimization

Chapter 3
Parallel Approaches in MOACOs for Solving the Bi-criteria TSP: A Preliminary Study

A.M. Mora, P.A. Castillo, M.G. Arenas,
P. García-Sánchez, J.L.J. Laredo, and J.J. Merelo

Abstract. This work presents two parallelization schemes applied to three different Multi-Objective Ant Colony Optimization (MOACO) algorithms. The aim is to get a better performance, improving the quality, quantity and the spread of solutions over the Pareto Front (the ideal set of solutions), rather than just reduce the running time. Colony-level (coarse-grained) implementations have been tested for solving two instances of the Bi-criteria TSP problem, yielding better sets of solutions, in the mentioned sense, than the correspondent sequential approach.

3.1 Introduction

The main objective when the parallelization of an algorithm is addressed is normally to improve its running time, while the quality of the solutions remains. But in some metaheuristics the parallelization may imply a different searching scheme (and different solutions), as in the case of the Ant Colony Optimization (ACO) [7, 8] algorithms. These metaheuristics are based in a set of artificial agents (ants) which explore the searching space cooperating to get the solution for a problem. In addition, the main feature of a good multi-objective (MO) algorithm [2] (devoted to find solutions for more than one objective function) is to get the maximal set of non-dominated solutions, the so-called Pareto Set (PS), which includes those solutions that optimize all the functions in the problem. The ACO algorithms implemented to deal with several objectives are known as MOACOs (see [10] for a survey).

A.M. Mora · P.A. Castillo · M.G. Arenas · P. García-Sánchez · J.L.J. Laredo · J.J. Merelo
Departamento de Arquitectura y Tecnología de Computadores,
Universidad de Granada, Spain
e-mail: {amorag,pedro,maribel,pgarcia,juanlu,jmerelo}@geneura.ugr.es

J. Kołodziej et al. (Eds.): Advances in Intelligent Modelling and Simulation, SCI 422, pp. 45–66.
springerlink.com

The idea addressed in this study has been the distribution of the ants (grouped in colonies) into several computing nodes, being each one of these nodes focused into a different area of the searching space. Such a structure contributes to yield a better set of results, by promoting the explorative behaviour. Thus a bigger amount of non-dominated solutions, with higher quality (closer to the optimal ones), and with a better spread or distribution over the ideal set of solutions will be obtained.

Two different parallelization approaches have been tested, considering in the study three algorithms: two of the most known MOACOs in the literature, Iredi et al. *BIANT* [13], and Barán et al. *MOACS* [1]. In addition, we have adapted an own algorithm called CHAC [16].

This parallelization has been performed, as stated, at the colony level. In addition, the three methods have been applied to solve the same problem: a *Bi-criteria Travelling Salesman Problem* (Bi-TSP) [19], which is the transformation into a multi-criteria problem of the classical TSP [15]. In the single-objective TSP the target is to minimize distance, while in this version there is a set of different costs between each pair of connected cities, which could correspond, for instance to distance and travel time. Those distributed implementations have been compared with the correspondent sequential approach, and the profits in running time and in the quality of the solutions yielded have been analyzed.

This work presents a novel study, since as far as we know there are in the literature some distributed or parallel ant colony algorithms [14], but no one which deals with multi-objective problems.

The rest of the chapter is structured as follows: Section 3.2 introduces some concepts which are dealt in the work. Then Section 3.3 describes the parallelization approaches considered for the study. Section 3.4 presents the MOACOs to analyze. The performed experiments are shown in Section 3.5. Finally, conclusions and future lines of work are commented in Section 3.6.

3.2 Preliminary Concepts

In this section, the Ant Colony Optimization metaheuristic is introduced and the Multi-Objective optimization concepts are briefly presented.

3.2.1 Ant Colony Optimization

The Ant Colony Optimization (ACO) is a meta-heuristic inspired by the naturally observed fact that some species of ants are able to find the shortest path from nest to food sources after a short time of exploring different paths between them [4]. This behaviour has been explained through the concept of *stigmergy* [11], that is, communication between agents using the environment, so every ant, while walking, deposits on the ground a substance called *pheromone* which the others can smell and which is evaporated after some time. One ant tends to follow strong concentrations of pheromone caused by repeated passes of ants; a pheromone trail is then formed from nest to food source, so in intersections between several trails an ant moves with high probability following the highest pheromone level.

ACO algorithms, introduced by Dorigo et al. [8] in 1991, take this model of ant behaviour as inspiration to solve combinatorial optimization problems using a colony of "artificial ants", which are computational agents that communicate with each other through the environment using *pheromones*.

Every problem to be solved using ACO must be transformed into a graph with weighted edges. In every iteration, each ant builds a complete path (solution), by travelling through the graph. At the end of this construction (and in some versions, during it), each ant leaves a trail in the visited edges depending on the fitness of the solution it has found. This is a measure of desirability for that edge and it will be considered by the following ants. In order to guide its movement, each ant uses two kinds of information that will be combined: *pheromone trails*, which correspond to "learnt information" changed during the algorithm run, denoted by τ; and *heuristic knowledge*, which is a measure of the desirability of moving to the next node, based in previous knowledge about the problem (it does not change during the algorithm run), denoted by η. The ants usually choose edges with better values in both properties, but sometimes they may "explore" new zones in the graph because the algorithm has a stochastic component, that broadens the searching space to regions not previously explored. Due to all these properties, all ants cooperate in order to find the best solution for the problem (the best path in the graph), resulting in a global emergent behaviour.

ACOs initially took two different shapes: Ant System (AS) (which will be presented next) and Ant Colony System (ACS) (presented in Section 3.2.1.2). Nowadays there are lots of variants and new approaches, based on these original algorithms, but applying other terms or parameters.

3.2.1.1 The Ant System

This was the initial model proposed by Dorigo et al. [5] for solving the classical Travelling Salesman Problem (TSP) [15].

The building of solutions is strongly based in the *State Transition Rule* (STR), since every ant uses it to decide which node j is the next in the construction of a solution (path), when the ant is at the node i. This formula calculates the probability associated to every node in the neighbourhood of i, as follows:

$$P(i,j) = \begin{cases} \dfrac{\tau(i,j)^{\alpha} \cdot \eta(i,j)^{\beta}}{\sum_{u \in N_i} \tau(i,u)^{\alpha} \cdot \eta(i,u)^{\beta}} & if\ j \in N_i \\ 0 & otherwise \end{cases} \tag{3.1}$$

where α and β are weighting parameters to set the relative importance of pheromone and heuristic information respectively, and N_i is the current feasible neighbourhood for the node i.

In the *AS*, the pheromone update is performed once all the ants have built their solutions. This updating is made at a global level by every ant, which retraces its solution path. It consists of an evaporation (left term) and a contribution (right term):

$$\tau(i,j)^t = (1-\rho) \cdot \tau(i,j)^{t-1} + \Delta\tau(i,j)^{t-1} \tag{3.2}$$

t marks the new pheromone value and *t-1* the old one. ρ in the range [0,1] is the common evaporation factor, and $\Delta\tau$ is the deposited amount of pheromone.

As it can be seen in Equation 3.2, there are two steps: first, all the pheromone trails are reduced by a constant factor (evaporation), after this, every ant deposits an amount of pheromone in its path (in the edges) depending on the quality of its solution (the better solution the ant has found, the higher amount of pheromone is added).

3.2.1.2 The Ant Colony System

The *ACS* is the successor of the AS, and was presented by the same authors [6]. It has three differences with regard to the AS.

The first one is the application of a different state transition rule (called *pseudo-random proportional state transition rule*), defined as:

If $(q \leq q_0)$

$$j = \arg\max_{j \in N_i} \left\{ \sum_{u \in N_i} \tau(i,u)^{\alpha} \cdot \eta(i,u)^{\beta} \right\} \tag{3.3}$$

Else

$$P(i,j) = \begin{cases} \dfrac{\tau(i,j)^{\alpha} \cdot \eta(i,j)^{\beta}}{\sum_{u \in N_i} \tau(i,u)^{\alpha} \cdot \eta(i,u)^{\beta}} & if\ j \in N_i \\ 0 & otherwise \end{cases} \tag{3.4}$$

where q is a random number in the range [0,1] and q_0 is a parameter which set the balance between exploration and exploitation. If $q \leq q_0$, the best node is chosen as the next one (exploitation), otherwise one of the feasible neighbours is selected, considering different probabilities for each one (exploration). The rest of the parameters are the same as in Equation 3.1.

The second difference is that there is a *global pheromone updating*, which is only performed for the edges of the global best solution:

$$\tau(i,j)^t = (1-\rho)\cdot\tau(i,j)^{t-1} + \rho\cdot\Delta\tau(i,j)_{GlobalBest} \qquad \forall(i,j)\ in\ S_{GlobalBest} \tag{3.5}$$

Finally, there is also a *local pheromone updating*, which is performed by every ant every time a node j is added to the path which it is building. This formula is:

$$\tau(i,j)^t = (1-\varphi)\cdot\tau(i,j)^{t-1} + \varphi\cdot\tau_0 \tag{3.6}$$

being φ in the range [0,1] the local evaporation factor, and τ_0 the initial amount of pheromone (it corresponds to a lower trail limit). This formula results in an additional exploration technique, because it makes the edges traversed by an ant less attractive to the following ants and helps to avoid that many ants follow the same path.

Both of these approaches, but specially ACS, have been adapted to particular problems, mainly through the heuristic functions, and possibly also by changing the number of colonies or pheromone matrices. In particular, they are possible ways of approaching multi-objective optimization problems, which will be introduced below. The ACOs dealing with multiple optimization criteria are known as Multi-Objective Ant Colony Optimization algorithms (or MOACOs). See [10] for a survey.

3.2.2 Multi-Objective Optimization

Multi-Criteria or Multi-Objective Optimization Problems (MOP) [17] are those where several objectives have to be simultaneously optimized. So, there is usually not a single best solution, that is, better than any other with regard to every objective. Moreover, frequently improving the solution for one objective implies a worsening for other one.

Solving a multi-objective problem implies maximizing or minimizing a function $f(x)$ composed by k other cost functions (one per objective) and considering n parameters (decision variables):

$$\begin{aligned} f(x) &= (C_1(x), C_2(x), ..., C_k(x)) \\ x &= (x_1, x_2, ..., x_n) \in X \end{aligned} \tag{3.7}$$

In a multi-objective optimization problem (MOP), there is a set of solutions that are better than the remainder considering all the objectives, which is known as the *Pareto Set* (PS) [18]. They try to approach the ideal set of solutions which is named the *Pareto Front* (PF). It is related to an important concept, the *dominance*, defined as follows (a dominates b):

$$a \prec b \ if: \\ \forall i \in 1,2,...,k \mid C_i(a) \leq C_i(b) \ \wedge \ \exists j \in 1,2,...k \mid C_j(a) < C_j(b) \tag{3.8}$$

where $a \in X$ and $b \in X$ are two different decision vectors of n values, and every C is a cost function (one per objective). If it intends to minimize the cost and Equation 3.8 is true, then *b is dominated by a.*

Hence, the solutions in the PS are known as *non-dominated solutions*, while the remainder are known as dominated solutions. Figure 3.1 shows the representation of the Pareto Front of a multi-objective (bi-criteria) problem, where the domination concept can be clearly seen.

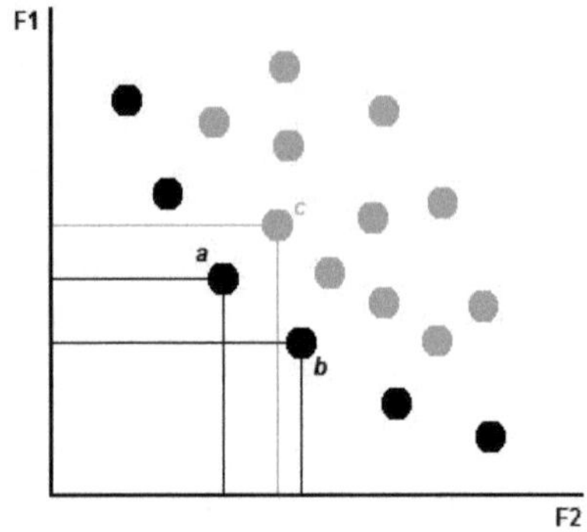

Fig. 3.1. Example of Pareto Front in a problem with two objectives to minimize (F1 and F2). Black dots correspond to solutions in the Pareto Front (*a* and *b* among them). Dominated solutions are shown in grey (*c* for instance).

Since none of the solutions in the Pareto Set is absolutely better than the other non-dominated solutions, all of them are equally acceptable as regards the satisfaction of all the objectives. The interested reader is directed to [2] for a deeper examination of these concepts.

3.3 Parallel Approaches

Since an ACO algorithm works using independent agents (ants), it can be adapted to a parallel architecture in a direct way. The ants communicate with each other through the so-called pheromone matrix (which simulates the real environment for the stigmergy effect [11]), which can be updated asynchronously, so they do not require to pass continuous synchronization nor information data through the net, as many of the parallel implementations of other methods (such as the parallel Genetic Algorithms).

There are several parallel ACO approaches [14] which mainly distribute the ants into several computing nodes following a different parallelization grain. In fine-grained implementations, every ant corresponds to one computational node, while in coarse-grained implementations every node contains a set of ants.

Typically, these implementations are centralized (following master/slave architecture), that means there is one node, called master process, which collects the solutions or the pheromone information from all the other nodes. After this, it performs the pheromone updating and computes the new pheromone matrix, which is then sent to the other nodes (called slaves processes). In a decentralized approach every node has to compute the pheromone update by itself, using information that it has received from other nodes. The main goal of these approaches is to improve the running time without changing the optimization behaviour of the algorithm.

In contrast, ACO algorithms designed for parallel architectures try to change the standard ACO algorithm, thus the parallel version works more efficiently. One direct approach consists in to divide the whole population of ants into several subsets which do information exchange between them every few number of iterations, not in every one. This can also have a positive effect on the optimization behaviour, because the subset in each node may get specialized in different regions of the searching space.

ACO algorithms composed by several colonies of ants, where each of them uses their own (and different) pheromone matrix, are called *multi colony ACO algorithms*. They are suitable for parallelization, since a processor can host just a colony of ants, and normally there will be less information exchange among the colonies as would have between groups of ants in standard ACO. They are typically decentralized. In addition, these kind of algorithms are normally used for addressing MOPs, having each colony specialized in the optimization of one specific objective. With respect to this question, it has been demonstrated in the literature [13, 9, 1] that in MOACOs, the use of specialized colonies (or ants) for each objective or even each area of the searching space, yields very good results [10], getting a big set of non-dominated solutions which 'covers' almost all the ideal Pareto Front.

The aim in this work is to get mainly one profit: *improve the quality of the solutions obtained in solving the Bi-TSP problem, rather than just improve the running time as usual in parallel approaches.* That is, obtaining a good (large) set of non-dominated solutions with a good distribution along the Pareto Front, which is the main task of any MO algorithm.

This way, the proposal in this chapter implies adapting some models to a parallel environment with the advantages that can be expected, and taking a *coarse-grain* parallelization approach, that is, a parallelization at colony level, so every computational node will contain a set of ants. We propose two different distributed approaches:

- *Space Specialized Colonies* (SSC): it consists on a group of independent colonies, each of them searching in a different area of the space of solutions. At the end of the process they merge their solutions (their Pareto sub-sets) to constitute a single Pareto Set (considering dominance criteria to build it, since the non-dominated solutions of a colony may be dominated by the solutions yielded by another colony). The split of the space is made through the use of some parameters which weights the objectives in the search for each ant in every colony.

- *Objective Specialized Colonies* (OSC): it also consists on a group of independent colonies, but this time, each one tries to optimize only one of the objectives. Every colony does not consider the other objectives in the search, but all of them are taken into account when the solutions are evaluated, so the colonies search, as in the previous model, in a multi-objective space of solutions. Again, at the end, all the PSs are merged (considering the dominance criterion) into the final (or global) PS.

3.4 MOACOs to Study

As commented, the parallelization approaches described in the previous section have been applied to two state of the art MOACO algorithms, and to the adaptation of an own one.

Each method was designed for solving a different MOP, so the three all have been adapted in this work to solve the Bi-criteria TSP.

All of them consider the same *Heuristic Functions* (one per objective), described as:

$$\eta_1(i,j) = \frac{1}{d_1(i,j)} \tag{3.9}$$

$$\eta_2(i,j) = \frac{1}{d_2(i,j)} \tag{3.10}$$

where $d_x(i,j)$ is the correspondent distance between nodes i and j. They depend on the problem definition, and this time both are Euclidean Distances.

The *Evaluation Functions* (one per objective) have been also common to all the approaches, and are defined as:

$$F_1(S_p) = \sum_{n \in S_p} [d_1(n-1,n)] \tag{3.11}$$

$$F_2(S_p) = \sum_{n \in S_p} [d_2(n-1,n)] \tag{3.12}$$

where S_p is the solution path to evaluate and n is a node in that path. Each one of the d_x is again the Euclidean Distance.

Moreover, in the three methods it has been used a key parameter in the search, inside the correspondent State Transition Rule (the main element in a ACO algorithm), $\lambda \in [0,1]$, which let us to focus in a specific area of the searching space to explore as will be shown in the next STR formulas.

The first two algorithms (BIANT and MOACS) were initially defined with a 'variable' λ policy, which consists in assign a different value for the parameter to each ant h, following the expression:

$$\lambda_h = \frac{h-1}{m-1} \qquad \forall h \in [1,m] \tag{3.13}$$

Considering that there are m ants, the parameter takes an increasing value that goes from 0 for the first ant to 1 for the last one. This way, the algorithms search in all the possible areas of the space of solutions (each ant is devoted to a zone of the PF).

As stated, in this work this parameter has been used to determine the area of the searching space that each colony has to explore, so will be constant for all the ants in a colony (and different to the rest of the colonies).

Finally, the three approaches have been improved by means of a local search (LS) application: the 2-OPT method [3], since it has been widely demonstrated in the literature that this hybridization mechanism yields a better performance in ACOs.

3.4.1 BIANT

The first considered and adapted algorithm is ***BIANT*** (Bi-Criterion Ant), which was proposed by Iredi et al. [13] as a solution for a multi-objective problem with two criteria (the Single Machine Total Tardiness Problem, SMTTP). It is an Ant System (AS) which uses just one colony, and two pheromone matrices and two heuristic functions (one per objective).

The *STR* for choosing the next node j when an ant is in the node i is as follows:

$$P(i,j) = \begin{cases} \dfrac{\tau_1(i,j)^{\alpha\cdot\lambda}\cdot\tau_2(i,j)^{\alpha\cdot(1-\lambda)}\cdot\eta_1(i,j)^{\beta\cdot\lambda}\cdot\eta_2(i,j)^{\beta\cdot(1-\lambda)}}{\sum_{u\in N_i}\tau_1(i,u)^{\alpha\cdot\lambda}\cdot\tau_2(i,u)^{\alpha\cdot(1-\lambda)}\cdot\eta_1(i,u)^{\beta\cdot\lambda}\cdot\eta_2(i,u)^{\beta\cdot(1-\lambda)}} & if\ j\in N_i \\ 0 \quad otherwise \end{cases} \tag{3.14}$$

where the terms and parameters are the named as in Equation 3.1, but there are one τ and one η per objective (those described in Equations 3.9 and 3.10). In addition, it can be seen that this rule considers the λ parameter to weight the objectives in the search.

This expression calculates the probability $P(i,j)$ associated to each of the feasible nodes; then, the algorithm uses a roulette wheel to choose the next node in the solution path that the ant is building.

Since BIANT is an AS, it is just performed a *Global Pheromone Updating*, including evaporation in all nodes and contribution just in the edges of the best paths to the moment (those included in the Pareto Set (PS)). The Equations for the evaporation are:

$$\tau_1(i,j)^t = (1-\rho)\cdot\tau_2(i,j)^{t-1} \tag{3.15}$$

$$\tau_2(i,j)^t = (1-\rho)\cdot\tau_2(i,j)^{t-1} \tag{3.16}$$

and for the contribution (in the edges of the solution paths):

$$\tau_1(i,j)^t = \tau_1(i,j)^{t-1} + 1/F_1 \tag{3.17}$$

$$\tau_2(i,j)^t = \tau_2(i,j)^{t-1} + 1/F_2 \tag{3.18}$$

as usual, ρ is the evaporation factor, t marks the new pheromone value and $t-1$ the preceding one; F_1 and F_2 are the evaluation functions for the two objectives (Equations 3.11 and 3.12).

As a detail, the *initial amounts of pheromone* are calculated following the expressions:

$$\tau_{0,1} = \frac{1}{L_1} \tag{3.19}$$

$$\tau_{0,2} = \frac{1}{L_2} \tag{3.20}$$

being L_x the cost obtained using a Greedy algorithm for solving the problem defined by each one of the objective functions. It is a usual initialization value in ACOs.

3.4.2 MOACS

The second algorithm to study is ***MOACS*** (Multi-Objective Ant Colony System), which was proposed by Barán et al. [1], to solve the Vehicle Routing Problem with Time Windows (VRPTW). It uses a single pheromone matrix for both objectives.

The *STR* is defined as follows:

If $(q \leq q_0)$

$$j = \arg\max_{j \in N_i} \left\{ \tau(i,j) \cdot \eta_1(i,j)^{\beta \cdot \lambda} \cdot \eta_2(i,j)^{\beta \cdot (1-\lambda)} \right\} \tag{3.21}$$

Else

$$P(i,j) = \begin{cases} \dfrac{\tau(i,j) \cdot \eta_1(i,j)^{\beta \cdot \lambda} \cdot \eta_2(i,j)^{\beta \cdot (1-\lambda)}}{\sum\limits_{u \in N_i} \tau(i,u) \cdot \eta_1(i,u)^{\beta \cdot \lambda} \cdot \eta_2(i,u)^{\beta \cdot (1-\lambda)}} & \text{if } j \in N_i \\ 0 & \text{otherwise} \end{cases} \tag{3.22}$$

where the terms and parameters are the same as in Equation 3.3, but this time there are two heuristic functions: η_1 and η_2 (defined in Equations 3.9 and 3.10). This rule applies again λ to balance the relative importance of the objectives in the search.

This STR works as follows: when an ant is building a solution path and is placed at one node i, a random number q in the range [0,1] is generated, if $q \leq q_0$ the best neighbour j is selected as the next node in the path (Equation 3.21). Otherwise, the algorithm decides which node is the next by using a roulette wheel considering $P(i,j)$ as probability for every feasible neighbour j (Equation 3.22).

Since *MOACS* is an ACS, there are two levels of pheromone updating, local and global. The equation for *Local Pheromone Updating* is:

$$\tau(i,j)^t = (1-\rho)\cdot\tau(i,j)^{t-1} + \rho\cdot\tau_0 \tag{3.23}$$

considering:

$$\tau_0 = \frac{1}{L_1 \cdot L_2} \tag{3.24}$$

as previously, ρ is the evaporation factor, t marks the new pheromone value and $t-1$ the preceding one. Again L_1 and L_2 are the cost obtained in each objective using a Greedy algorithm for solving the problem.

MOACS applies a *pheromone reinitialization mechanism*, so the τ_0 value is not constant during the algorithm run, as usual in ACS, but it undergoes adaptation. Every time an ant h builds a complete solution, it is compared to the Pareto set P generated until the moment, to check if the former is a non-dominated solution. At the end of each iteration, τ'_0 is calculated following the formula:

$$\tau'_0 = \frac{1}{\bar{F_1}\cdot\bar{F_2}} \tag{3.25}$$

where $\bar{F_1}$ and $\bar{F_2}$ are respectively the average costs in each objective for the solution paths currently included in the Pareto set.

Then, if $\tau'_0 > \tau_0$ (the current initial pheromone value), a better Pareto Set has been found, and the pheromone trails are reinitialized considering the new value for $\tau_0 \leftarrow \tau'_0$.

Otherwise the *Global Pheromone Updating* is performed for every solution in the PS:

$$\tau(i,j)^t = (1-\rho)\cdot\tau(i,j)^{t-1} + \frac{\rho}{F_1\cdot F_2} \tag{3.26}$$

where, as in previous equations, F_1 and F_2 are the evaluation functions for each objective, ρ is the common evaporation factor; and t and $t-1$ stands for actual and previous pheromone values.

3.4.3 CHAC

The third adapted algorithm is the so-called ***CHAC*** (Compañía de Hormigas ACorazadas), which was proposed by the authors in [16], to solve the Bicriteria Military Unit Pathfinding Problem (MUPFP-2C).

It is an ACS adapted to deal with two objectives, and it was designed considering some features of both BIANT (STR) and MOACS (pheromone updating scheme, without reinitialization mechanism).

It uses two pheromone matrices and two heuristic functions (one per objective), and considered a single colony originally. Two different approaches for CHAC were defined, using two STRs: the (*Combined State Transition Rule, CSTR*), which combines the pheromone and heuristic information for each objective weighted using α, β and λ parameters; and the (*Dominance State Transition Rule, DSTR*), which ranks neighboring cells according to how many they dominate [16].

In the present adaptation, the elements of CHAC are common to those of the previously commented algorithms, applying the same probability term as BIANT in its STR, but considering the ACS scheme (as MOACS does). So, the *STR* is defined as:

```
If (q ≤ q0)
```

$$j = \arg\max_{j \in N_i} \left\{ \tau_1(i,j)^{\alpha\cdot\lambda} \cdot \tau_2(i,j)^{\alpha\cdot(1-\lambda)} \cdot \eta_1(i,j)^{\beta\cdot\lambda} \cdot \eta_2(i,j)^{\beta\cdot(1-\lambda)} \right\} \tag{3.27}$$

```
Else
```

$$P(i,j) = \begin{cases} \dfrac{\tau_1(i,j)^{\alpha\cdot\lambda} \cdot \tau_2(i,j)^{\alpha\cdot(1-\lambda)} \cdot \eta_1(i,j)^{\beta\cdot\lambda} \cdot \eta_2(i,j)^{\beta\cdot(1-\lambda)}}{\sum_{u \in N_i} \tau_1(i,u)^{\alpha\cdot\lambda} \cdot \tau_2(i,u)^{\alpha\cdot(1-\lambda)} \cdot \eta_1(i,u)^{\beta\cdot\lambda} \cdot \eta_2(i,u)^{\beta\cdot(1-\lambda)}} & if\ j \in N_i \\ 0 \quad otherwise & \end{cases} \tag{3.28}$$

where all the terms are the same as in Equations 3.14, 3.21 and 3.22. It works as described in MOACS algorithm section.

The use of the parameter $\lambda \in (0,1)$, was defined as constant during the algorithm for all ants, so CHAC always searched in the same zone of the space of solutions (the zone related to the chosen value for λ).

The DSTR was designed to perform a global exploration of the space of solutions (it is not guided by λ), so, in principle, it is not useful in the present study.

As stated, CHAC is an ACS and considers the same pheromone updating scheme as MOACS, so there is a *Local Pheromone Updating* step, defined in Equation 3.23, and a *Global Pheromone Updating* process performed for every solution in the PS and formulated in Equation 3.26.

3.5 Experiments and Results

Several experiments have been performed to validate the proposed methods. Firstly, a sequential implementation of each of them have been tested (in a single processor). Then, experiments were conducted for the two parallel approaches in a different number of processors going from 2 to 16.

The parallelization has been implemented using MPI [12] and tested in a 16 processors cluster, with shared memory.

All the experiments have been performed for solving the *Krolak/Felts/Nelson problem* [19], in two sizes: 100 and 200 cities, and with two different instances (A and B) corresponding to the two objectives.

The experimental setup can be found in Table 3.1, having obtained the set of configuration parameters starting from the standard values and performing systematic experimentation for tuning some of them. These parameters have been shared by all the algorithms.

Table 3.1. Parameters of every colony used in the three algorithms.

Number of ants in the colony	35
Number of iterations	500
Number of iterations in Local Search (2-OPT)	15
α	1
β	2
q_0 *(used in MOACS and CHAC)*	0.8

In addition, the same random seed have been used in all the experiments in order to make a fairer comparison, avoiding a stochastic level.

The following list describes the different sets of experiments performed using every one of the algorithms (BIANT, MOACS and CHAC):

- the sequential approach is run in one processor, considering the previously commented variable λ policy (originally proposed by the authors for BIANT and MOACS), that is, a different value for that parameter per ant in a single colony.
- the two processor experiments consider the OSC approach, since there are two objectives, so two colonies are used.
- the experiments in 4, 8 and 16 processors apply the SSC algorithm, so they consider a different number of colonies (one per processor), with a different value for λ in each one (the same for all the ants in the colony).

It would be important to clarify that the aim (classical in multi-objective optimization) is to get the biggest amount of non-dominated solutions (PS), which should be as better distributed as possible along the ideal Pareto Front, 'covering' it in the best way.

Once the experiments have been performed, the resulting PSs for the BIANT algorithm distribution approaches are shown in Figure 3.2.

As it is shown, each colony explores over a different area of the searching space yielding a different set of solutions from the rest. In addition, it can be seen that the distribution of solutions along the PF is better for a higher

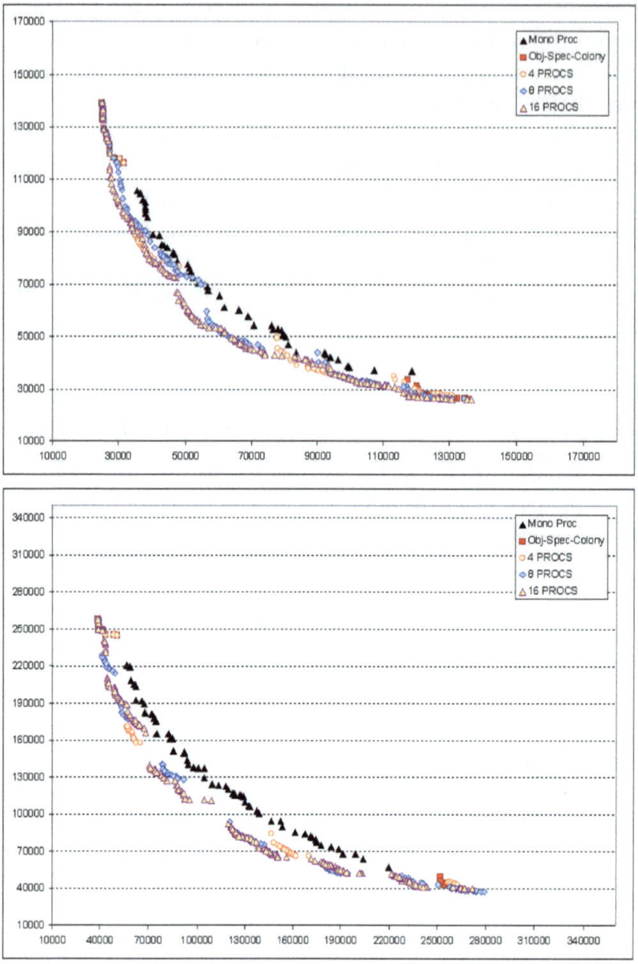

Fig. 3.2. Results for BIANT algorithm solving the Bi-TSP: KROA+B 100 cities (top), KROA+B 200 cities (bottom). They are distributed in objective specialized colonies, and in space specialized colonies, from 2 to 16 processors. They are compared with those obtained by the mono-processor (or sequential) approach.

number of colonies (and processors), yielding better (smaller) values in both objectives, and covering a wider area. Sometimes an approach with a smaller number of colonies gets some better solutions in a concrete zone, since they explore in a more restricted area of the space of solutions (due to the λ value), so a higher exploitation factor is applied in that area.

The OSC approach (in two processors) does not yield very good solutions, since these colonies explore just in the edges of the PF, obtaining good solutions there, but not in a wide central area.

Considering the different problems (KROA+B 100 and 200 cities), it can be seen that in the second case (Figure 3.2-bottom) the spread of solutions is worse, even in the 16 processors case. The reason could be the small number of ants which remains the same for both problems (for fair comparison reasons), but being the second much harder. An increasing in the number of ants would mean a higher exploration factor, and thus a wider PS.

The MOACS experiments are represented in Figure 3.3. It shows that yielded results by each of the approaches are distributed in a similar shape

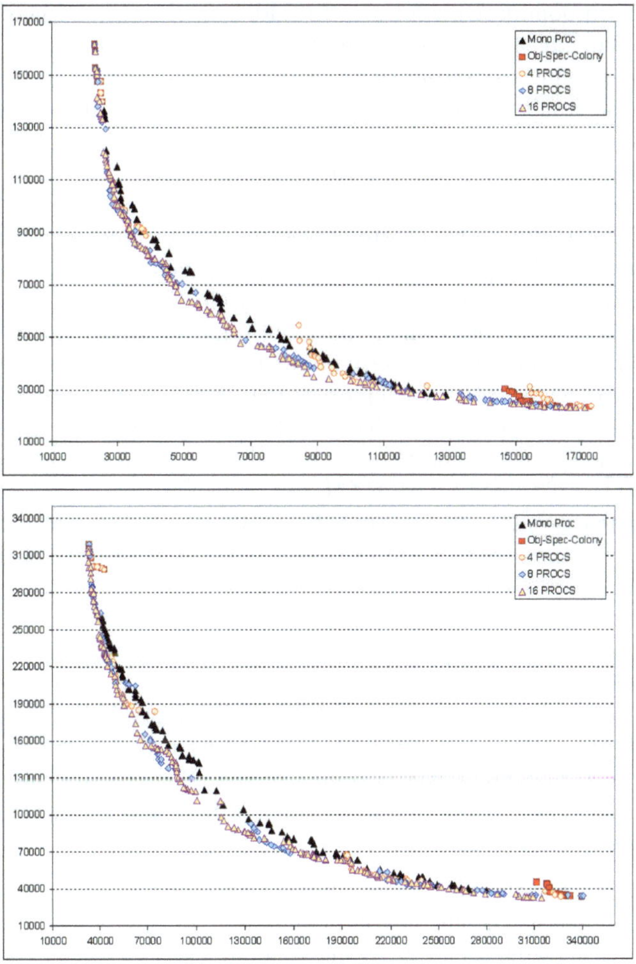

Fig. 3.3. Results for MOACS algorithm solving the Bi-TSP: KROA+B 100 cities (top), KROA+B 200 cities (bottom). They are distributed in objective specialized colonies, and in space specialized colonies, from 2 to 16 processors. They are compared with those obtained by a mono-processor (sequential) approach.

to the BIANT experiments. This time the PSs are closer between them than in the previous case, but again, the 16 processors approach yields the best global set: better solutions and better distribution along the PF. The mono-processor run is quite competitive applying MOACS algorithm, but its solutions are dominated by most of the 8 and 16 processors approaches. As can be seen in Figure 3.3, this algorithm (in all the approaches) shows flaw in the spread of solutions, having a bad coverage in the central areas. This is more evident looking at the 200 cities problem experiment (graph at the bottom of the figure), where the small number of ants makes it harder to yield a high amount of non-dominated solutions. In addition, the q_0 parameter, used in ACSs for balancing the exploration and exploitation factors in the search has been set for tending to exploitation.

Finally, the results for CHAC algorithm are presented in Figure 3.4. The graphics show a very good amount of solutions, well distributed in the 16 processor approach above all, but there are also some gap zones when the number of colonies is slower, which is more evident in the 2 processors case. The flaws in distribution which are present in BIANT and MOACS PSs, are softer in these results, showing less gaps in the central areas.

As a preliminary comparison, the results yielded by MOACS and CHAC can be considered as better than those obtained by BIANT, since the PSs are closer to the minimum in several points (most of them in the sides, not in the center), are wider (they reach more values in the edges), and also shows a better distribution in every case (they fit better on a bigger area of the PF). Looking at the CHAC results (Figure 3.4), they could be considered as quite similar to those obtained by MOACS, but showing a better PF covering in almost all the cases. The flaw is that its solutions are a bit worse than those of BIANT from the quality point of view (above all in the central part of the PF).

The final idea when one performs this kind of parallelization is to get a unique PS, so once all the colonies have yielded its own set of non-dominated solutions, all these sets are merged (just considering the whole set of non-dominated solutions), getting a global PS. Following this idea and in order to make a better comparison, we have built the best global PS of each one of the algorithms, that are, those obtained in the 16 processors approaches in both problems. They are plotted in Figure 3.5. It clearly shows the strengths and weaknesses of every method:

- BIANT reaches the best set of central solutions (closer to the minimum in both objectives), but on the other hand the solutions are quite concentrated, showing a bad distribution along the PF.
- MOACS performs better in the sense of distribution, but the quality of the solutions is worse, above all in the central area of the PF.
- CHAC presents a very good agreement, yielding quality solutions (the best in several areas), while they are well distributed along the PF.

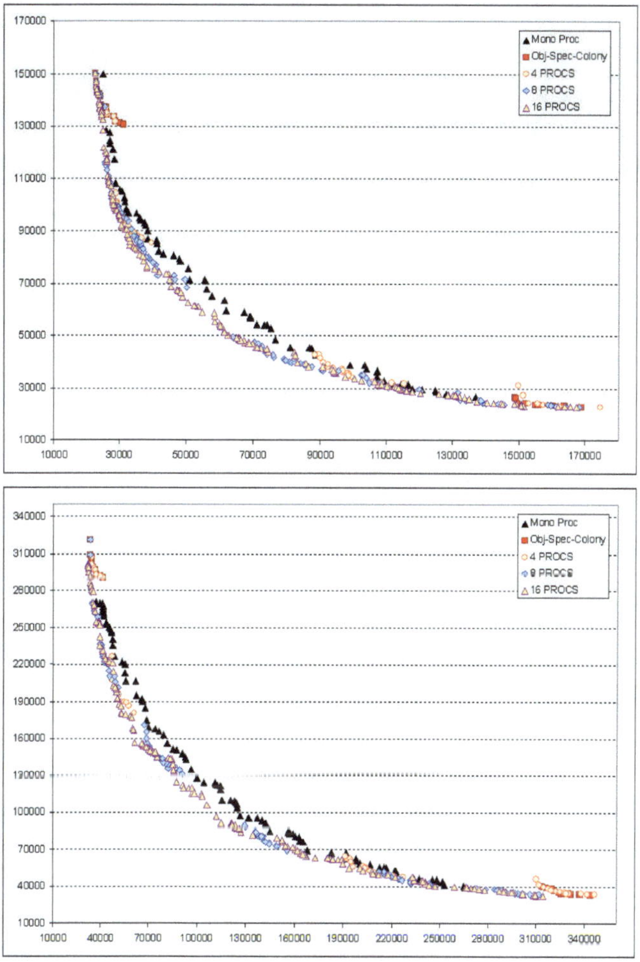

Fig. 3.4. Results for CHAC algorithm solving the Bi-TSP: KROA+B 100 cities (top), KROA+B 200 cities (bottom). They are distributed in objective specialized colonies, and in space specialized colonies, from 2 to 16 processors. They are compared with those obtained by a mono-processor (sequential) approach.

Then, in Table 3.2 the number of non-dominated solutions in each of the global PSs per experiment is shown.

Looking at those results, it can be noticed that BIANT yields more non-dominated solutions than the other methods in every case. The reason is that it is more explorative (is an AS) than MOACS and CHAC (both are ACSs using q_0 and tending to exploitation in this case). But looking at the previous figure (3.5), the two latter show a better distribution as stated,

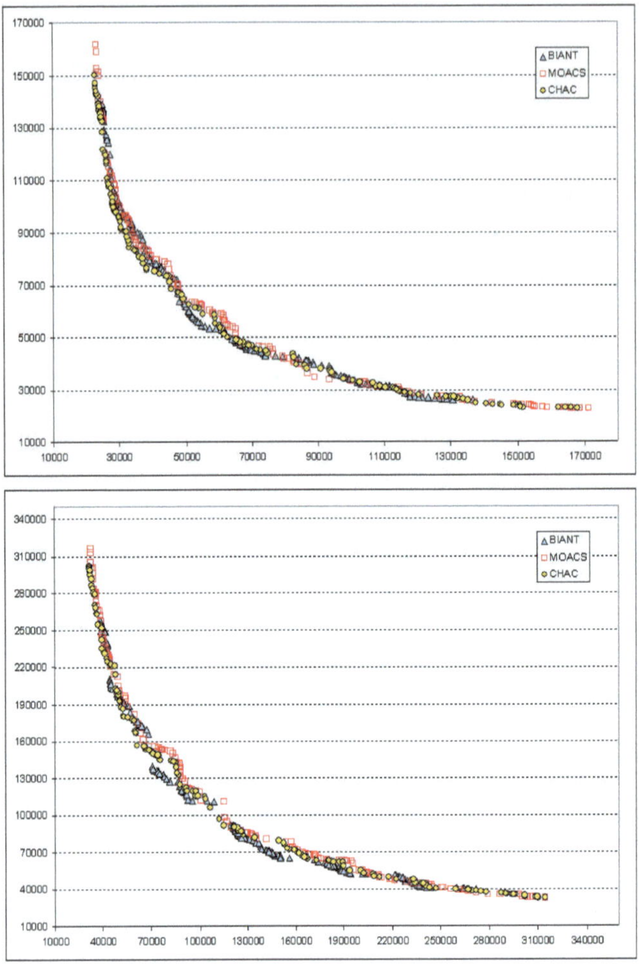

Fig. 3.5. Pareto Sets comparison of the three algorithms for the 16 processors approach, solving the Bi-TSP: KROA+B 100 cities (top), KROA+B 200 cities (bottom).

while BIANT concentrates a big amount of solutions in some specific areas, so it can be considered that MOACS and CHAC performs globally better. In the comparison between them, CHAC and MOACs yield similar numbers of non-dominated solutions.

Finally, an analysis on the time performance has been carried out. Despite our proposed parallelization approaches focus on improving the quality of solutions, it has also to be considered that parallelization uses to report speedups in time. Figure 3.6 presents the performance in time of the different approaches when they scale from 1 to 16 processors. Results show that a

Table 3.2. Number of solutions in the whole Pareto Set in each of the experiments and problems.

	Mono	2 Procs	4 Procs	8 Procs	16 Procs
	KROA+B 100 cities				
BIANT	46	50	89	158	202
MOACS	68	28	63	97	142
CHAC	69	33	74	118	168
	KROA+B 200 cities				
	Mono	2 Procs	4 Procs	8 Procs	16 Procs
BIANT	59	25	69	126	205
MOACS	92	24	50	99	153
CHAC	98	26	61	90	158

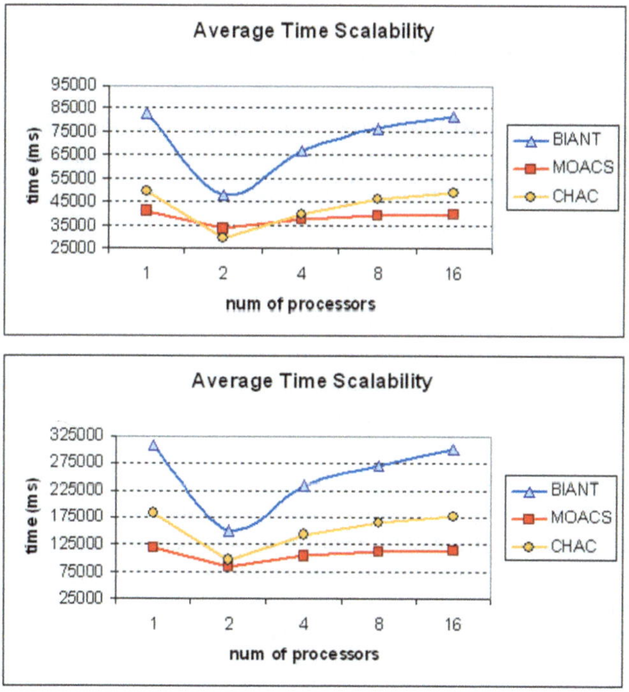

Fig. 3.6. Time scalability in average for MOACS and BIANT in a different number of processors, solving the Bi-TSP: KROA+B 100 cities (top), KROA+B 200 cities (bottom).

certain speedup can be expected with a peak in the 2 processor setting and a smooth degradation for increasing numbers of processors. Therefore, up to 16 nodes the approaches are taking less time than the sequential one in addition to reporting a higher quality on the solution sets.

As it can be seen, all the algorithms follow the same progression, taking much less time (in average) to get the results for a number of processors smaller than 16. In the last case, the average time is closer (but smaller) to the time taken by the sequential approach, but the quality of the set of solutions makes it worthwhile this parallelization. Again MOACS and CHAC show a better performance than BIANT, being the first one the fastest in almost all the cases. The reason is the use of a single pheromone matrix, rather than the two that consider CHAC and BIANT, which means much less calculation time when the pheromone updating steps have to be performed.

3.6 Conclusions and Future Work

In this work, three Multi-Objective Ant Colony Optimization algorithms (MOACOs) have been implemented in a distributed shape. They are two well-known state of the art approaches: BIANT [13] and MOACS [1], and an own algorithm named CHAC, previously presented in [16].

Two different parallelization approaches have been tested in each case; one considering a different ant colony specialized in a concrete area of the searching space (SSC), and another where there is a colony specialized in each one of the objectives of the problem (OSC).

All of the algorithms use a parameter called λ which sets the relative importance of the objectives in the search, guiding all the ants in a colony to explore the same zone of the searching space.

Experiments for solving two instances of the Bi-criteria TSP have been performed. The colonies have been distributed into a different number of processors (from 1 to 16), showing that the best set of solutions is obtained for the 16 processors approach, which yields a well distributed along the Pareto Front and crowded set (the aim of any multi-objective algorithm).

In the comparison between them, MOACS and CHAC yield better results from the distribution point of view than BIANT, but the latter obtains better quality solutions in the central zone of the ideal Pareto Front. CHAC presents the best option, since it yields a Pareto Set which is wide, fits quite well the ideal Pareto Front and shows good quality solutions.

Finally, looking at the running time profit, it is as good as expected, being improved when the process is distributed along a bigger number of processors until 16, when it is closer (but lower) than a mono-processor run time. Considering the quality of the yielded non-dominated solutions, it is clearly worthwhile.

The results yielded in this work are very promising, thus several future lines of work arise. Firstly we would like to test these approaches to solve some other multi-objective problems (such as the Vehicle Routing Problem with Time Window). Other line guides us to implement other approaches, such as heterogeneous colonies (a different algorithm running in each processor), to compensate the flaws of one with the strengths of another.

After this, the next objective could be to implement a fine-grained parallelization approach (at ant level), in order to improve the performance in time. The aim is to deal with very large instances of multi-objective problems.

Acknowledgements. This work has been supported in part by HPC-Europa 2 project (with the support of the European Commission - Capacities Area - Research Infrastructures), the CEI BioTIC GENIL (CEB09-0010) of the MICINN CEI Program PYR-2010-13 and PYR-2010-29 projects, the TIN2011-28627-C04-02 and UGR PR-PP2011-5 projects, and the P08-TIC-03903 project awarded by the Andalusian Regional Government.

References

1. Barán, B., Schaerer, M.: A multiobjective ant colony system for vehicle routing problem with time windows. In: IASTED International Multi-Conference on Applied Informatics. IASTED IMCAI, vol. (21), pp. 97–102 (2003)
2. Coello, C.A.C., Veldhuizen, D.A.V., Lamont, G.B.: Evolutionary Algorithms for Solving Multi-Objective Problems. Kluwer Academic Publishers (2002)
3. Croes, G.: A method for solving traveling salesman problems. Operations Res. 6, 791–812 (1958)
4. Deneubourg, J.L., Pasteels, J.M., Verhaeghe, J.C.: Probabilistic behaviour in ants: a strategy of errors? J. Theor. Biol. 105, 259–271 (1983)
5. Dorigo, M., Maniezzo, V., Colorni, A.: The ant system: Optimization by a colony of cooperating agents. IEEE Transactions on Systems, Man, and Cybernetics Part B: Cybernetics 26(1), 29–41 (1996)
6. Dorigo, M., Gambardella, L.M.: Ant colony system: A cooperative learning approach to the traveling salesman problem. IEEE Transactions on Evolutionary Computation 1(1), 53–66 (1997)
7. Dorigo, M., Caro, G.D.: The ant colony optimization meta-heuristic. In: Corne, D., Dorigo, M., Glover, F. (eds.) New Ideas in Optimization, pp. 11–32. McGraw-Hill (1999)
8. Dorigo, M., Stützle, T.: The ant colony optimization metaheuristic: Algorithms, applications, and advances. In: Glover, F., Kochenberber, G.A. (eds.) Handbook of Metaheuristics, pp. 251–285. Kluwer (2002)
9. Gambardella, L., Taillard, E., Agazzi, G.: Macs-vrptw: A multiple ant colony system for vehicle routing problems with time windows. In: Corne, D., Dorigo, M., Glover, F. (eds.) New Ideas in Optimization, pp. 73–76. McGraw-Hill (1999)
10. García-Martínez, C., Cordón, Ó., Herrera, F.: An Empirical Analysis of Multiple Objective Ant Colony Optimization Algorithms for the Bi-criteria TSP. In: Dorigo, M., Birattari, M., Blum, C., Gambardella, L.M., Mondada, F., Stützle, T. (eds.) ANTS 2004. LNCS, vol. 3172, pp. 61–72. Springer, Heidelberg (2004)

11. Grassé, P.-P.: La reconstruction du nid et les coordinations inter-individuelles chez bellicositermes natalensis et cubitermes sp. la theorie de la stigmerie. Insects Soc. 6, 41–80 (1959)
12. Gropp, W., Lusk, E., Doss, N., Skjellum, A.: A high-performance, portable implementation of the MPI message passing interface standard. Parallel Computing 22(6), 789–828 (1996)
13. Iredi, S., Merkle, D., Middendorf, M.: Bi-Criterion Optimization with Multi Colony Ant Algorithms. In: Zitzler, E., Deb, K., Thiele, L., Coello Coello, C.A., Corne, D.W. (eds.) EMO 2001. LNCS, vol. 1993, pp. 359–372. Springer, Heidelberg (2001)
14. Janson, S., Merkle, D., Middendorf, M.: Parallel Metaheuristics. In: Parallel Ant Algorithms. Wiley, London (2005)
15. Lawler, E.L., Lenstra, J.K., Kan, A.H.G.R., Shmoys, D.B. (eds.): The Traveling Salesman Problem. John Wiley & Sons Ltd. (1985)
16. Mora, A.M., Merelo, J.J., Laredo, J.L.J., Millán, C., Torrecillas, J.: CHAC, a MOACO algorithm for computation of bi-criteria military unit path in the battlefield: Presentation and first results. International Journal of Intelligent Systems 24(7), 818–843 (2009)
17. Osyczka, A.: Multicriteria optimization for engineering design. In: Gero, J.S. (ed.) Design Optimization, pp. 193–227. Academic Press (2005)
18. Pareto, V.: Cours d'economie politique, volume I and II. F. Rouge, Lausanne (1896)
19. Reinelt, G.: TSPLIB (2004), `http://www.iwr.uni-heidelberg.de/groups/comopt/software/TSPLIB95/`

Chapter 4
Island Injection Genetic Algorithm with Relaxed Coordination for the Multiple Sequence Alignment Problem

Lidia Araujo Miranda, Marcos Fagundes Caetano, Luiza Jaques,
Jan Mendonca Correa, Alba Cristina Magalhaes Alves de Melo,
and Jacir Luiz Bordim

Abstract. Multiple sequence alignment (MSA) is an important problem in Bioinformatics since it is often used to identify evolutionary relationships and predict secondary/tertiary structure, among others. MSAs are usually scored with the Sum-of-Pairs (SP) function and the exact SP MSA is known to be NP-Hard. Therefore, heuristic methods are used to tackle this problem. In this chapter, we propose and evaluate a parallel island injection genetic algorithm to solve the MSA problem. Unlike the other strategies, our parallel solution uses two types of interconnected archipelagoes, each with distinct types of individuals. Also, we added a relaxed coordination mechanism among the archipelagoes that contributes to reduce the execution time of our strategy. The results obtained with real protein data sets show that our strategy is able to obtain better results, when compared to the traditional island model. Also, we were able to reduce considerably the execution time, when compared to the sequential version.

4.1 Introduction

In the last decade, genome projects have produced an enormous amount of biological data. In order to better understand newly sequenced organisms, biologists compare their sequences against other organisms contained in genomic databases. Nowadays, this comparison is done millions of times a day, all over the world.

Lidia Araujo Miranda · Marcos Fagundes Caetano · Luiza Jaques ·
Jan Mendonca Correa · Alba Cristina Magalhaes Alves de Melo · Jacir Luiz Bordim
Department of Computer Science, University of Brasilia (UnB), Brazil
e-mail: lidiamir@gmail.com, caetano@cic.unb.br, luizajaques@gmail.com,
jan@cic.unb.br, albamm@cic.unb.br, bordim@unb.br

J. Kołodziej et al. (Eds.): Advances in Intelligent Modelling and Simulation, SCI 422, pp. 67–87.
springerlink.com

Sequence comparison is in fact a problem of finding an approximate pattern matching between the sequences [8]. It can involve only two sequences (Pairwise Alignment) or more than two sequences (Multiple Sequence Alignment). In a Multiple Sequence Alignment (MSA), similar residues among a set of n sequences are aligned together.

Multiple Sequence Alignments are often used as a building block to solve important and complex problems in Molecular Biology, such as the identification of conserved motifs in a family of proteins, definition of phylogenetic relationships and 3D homology modeling, among others. In all these cases, the quality of the solutions relies heavily on the quality of the underlying multiple alignment.

MSAs are often scored with the Sum-of-Pairs (SP) objective function and the exact SP MSA problem is known to be NP-hard [15]. Therefore, heuristic methods are often used to solve this problem, even when the number of sequences is small.

In the literature, a great number of algorithms and methods were proposed to tackle the Multiple Sequence Alignment problem. In a general way, they fall into two categories: progressive and iterative. A progressive MSA method initially generates all pairwise alignments and ranks them. The closest sequences are aligned first and then an MSA is built by adding the other sequences. Iterative methods create an initial MSA of groups of sequences and then modify it, until a reasonable result is attained.

Each of the MSA algorithms/tools are more adequate for some classes of problems. For instance, ClustalW [12] is a progressive method that is able to find adequate alignments of a large number of closely related sequences. T-COFFEE [6] uses ClustalW to obtain global and local pairwise alignments and then creates a weighting system that will be used to guide the MSA. In general, T-COFFEE seems to produce better alignments than ClustalW for closely related proteins, but it is much slower. SAGA [11] is the first and most representative MSA algorithm that uses a genetic algorithm to create appropriate sequence alignments. SAGA seems to perform better with small/medium sequences of low identity. Nevertheless, methods based on genetic algorithms are known to be computationally intensive and, thus, are usually slower than the other methods.

In order to accelerate MSA algorithms based on genetic algorithms, some parallel approaches were proposed in the literature [1, 10, 14]. GenAlignRefine [14] uses a parallel genetic algorithm to refine MSAs generated by alignment tools such as ClustalW. Inside the MSA, fuzzy (poorly aligned) regions composed by up to 100 residues are identified. Each fuzzy region is sent to a processor, that will execute a genetic algorithm to evolve it. At the end of the computation, the evolved fuzzy regions are collected and an MSA is constructed. An island parallel genetic algorithm is proposed in [1] to solve the MSA problem. In this case, the a set of initial MSAs is distributed to processors (islands) that evolve them separately, exchanging the best solutions periodically. PHGA [10][10] is a parallel island genetic algorithm that runs on multiprocessors. Unlike the previously described parallel approaches, it converts an MSA into a weighted directed acyclic nseq-dimensional graph and tries to find the shortest path in this graph. RAGA [17] is an extension of SAGA [11] that aligns 2 RNA sequences, taking into account the secondary structure, and PRAGA [17] is its parallel extension. The parallel hierarchical island model is used, where several identical Genetic Algorithms (GAs) run independently, exchanging the best individuals periodically.

After analyzing several tools and methods for MSAs, we noticed that the average best results are obtained with very elaborate methods that make many assumptions about the characteristics of the sequences and the structure of the alignments. Nevertheless, it is still an open question whether a generic optimization method is able to solve the MSA problem, making little or no assumptions about the characteristics of the problem.

In this book chapter, we propose and evaluate a Parallel Island Injection Genetic Algorithm with relaxed coordination to solve the MSA problem. It extends the work in [18] by adding asynchronous message primitives for intra and inter-archipelagoes communication. In our strategy, there are two different types of archipelagoes: one root archipelago and p leaf archipelagoes. The root archipelago evolves one population where each individual is a complete alignment, and each leaf archipelago divides the population of alignments in parts and evolves these parts separately. Periodically, the archipelagoes migrate their best individuals among themselves.

Inside the root archipelago and among the archipelagoes, the coordination is relaxed in the sense that every island migrates its best individuals at fixed migration intervals but does not wait to receive the best individuals from other islands to proceed, augmenting the parallelism.

The relaxed coordination algorithm proposed in this chapter was implemented in C and MPI (Message Passing Interface) and executed in a 12-core cluster. The protein sequences used to obtain the results were retrieved from real test sets from BAliBASE [13] and PFAM [5]. The results show that our strategy is able to obtain better scores than the traditional island model in all tests. Also, for the same test set, we were able to reduce the execution time from more than 24 hours (one core) to 36 minutes (12 cores).

The remainder of this chapter is organized as follows. Section 4.2 discusses serial and parallel genetic algorithms. The Multiple Sequence Alignment Problem is presented in Section 4.3. Section 4.4 discusses some proposals of parallel genetic algorithms for MSA. Section 4.5 presents our parallel island injection algorithm for MSA. Experimental results are discussed in Section 4.6.

4.2 Serial and Parallel Genetic Algorithms

A Genetic Algorithm (GA) is a stochastic search method that is based upon the principles of evolution [7]. In a basic GA, a set of possible solutions is coded as individuals. Genetic operations such as mutation and crossing-over are applied to the population of solutions thus generating new individuals. An objective function (OF) is used to characterize an individual's performance to the problem. This function is applied to the existing individual's and generally only the fittest ones survive to compose a new population, in which genetic operations will be applied again. This process continues until a certain number of generations or a reasonable level of fitness is reached.

Algorithm 1 illustrates a basic GA. In line 1, the initial population is created. The crossover/mutation genetic operators are applied to some individuals that compose the population, in order to generate new individuals (line 3). All individuals are evaluated using the objective function (line 4). A new population is created by the selection function (line 5). In this case, either the best individuals, according to the objective function, will compose the next population (elitism) or a fraction of randomly chosen individuals will belong to the next population, in addition to the best individuals. In either case, the population size usually does not vary. This process continues until a termination condition is reached (line 2).

Algorithm 1. Basic Genetic Algorithm (BGA)

```
Input: -
Output: pop;
1: pop=create_initial_population();
2: while termination_condition_not_reached(pop) do
3:    children=operators_crossover_mutation(pop);
4:    evaluate (pop, children);
5:    pop=select(pop, children);
6: end while
```

Genetic Algorithms have been used to solve many types of problems in scientific/industrial domains. Nevertheless, they are known to be computationally intensive and, for this reason, parallel strategies are often used to accelerate them.

4.2.1 Parallel Genetic Algorithms

Parallel GAs can be designed according to many models. The simplest one is called Master-Slave model [3], which evolves a single global population.

In this case, the master processor executes most operations of the algorithm depicted in Algorithm 1. The slave processors are only used in the evaluation phase (line 4 in Algorithm 1). The master distributes a subset of individuals that compose the population to each slave, that will evaluate the individuals assigned to it. At the end of this phase, the results are sent to the master, that proceeds with the selection. This solution is rarely used since it involves a high communication overhead. Figure 4.1(b) illustrates the Master-Slave Model.

A more elaborate model consists of $p+1$ complete GAs (lines 2-6 in Algorithm 1) that are executed separately by all processors. They often start with distinct individuals from the same initial population and evolve them independently. This model is known as the Island Model (iGA) [3]. Usually, each island (processor) executes the same Genetic Algorithm on its own population, each trying to optimize the same objective function, using the same representation for the individuals. The island model can be a) connected, where the best individual are migrated to other processors periodically; or b) partitioned, where no interaction occurs among the islands. Figure 4.1(c) illustrates the connected island model with 5 processors.

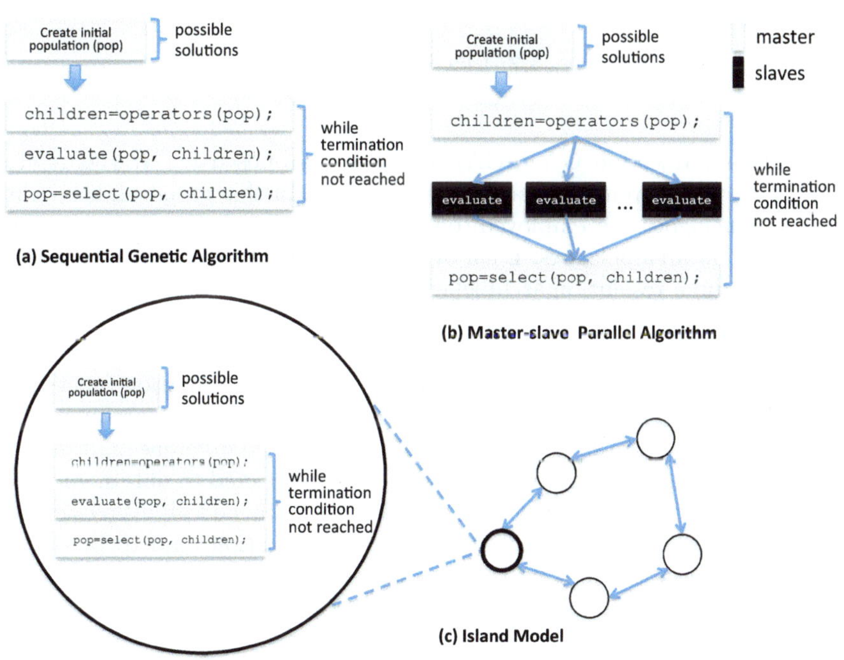

Fig. 4.1. Genetic Algorithm Models: Sequential (a), Master-Slave (b) and Island (c)

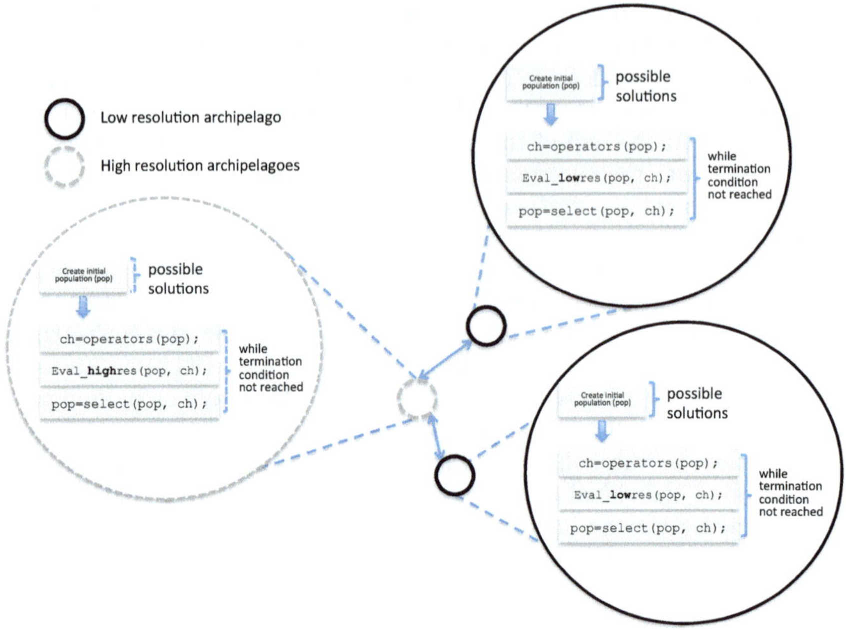

Fig. 4.2. Island Injection Model

The Island Injection GA (iiGA) model was proposed by [2] and is also used to search different populations (Figure 4.2). In this proposal, three distinct archipelagoes are employed: one high resolution archipelago and two low resolution archipelagoes. Each archipelago executes a Connected Island model. The high resolution archipelago solves the original problem using the original individual representation, applying a complex and time-consuming evaluation function. The low resolution archipelagoes solve the same problem using also the individual representation, but a simpler evaluation function is used. Therefore, both the high resolution and the low resolution archipelagoes execute the algorithm illustrated in Algorithm 1. The difference is that the high resolution archipelago evolves slower but is expected to obtain good individuals in few generations whereas the low resolution archipelagoes evolve quickly but the individuals produced in each generation tend to be not as good as the ones produced by the high resolution archipelago. At migration intervals, the best individuals are exchanged among the archipelagoes.

4.3 Multiple Sequence Alignment

A global Multiple Sequence Alignment (MSA) of $k > 2$ sequences $S = S_1, S_2, \ldots, S_k$ is obtained in such a way that spaces (gaps) are inserted into each of the k sequences so that the resulting sequences have the same length l. Then, the sequences are arranged in k rows of l columns each, so that each character or space of each sequence is in a unique column [8]. Figure 4.3 shows an example of an MSA with 3 DNA sequences.

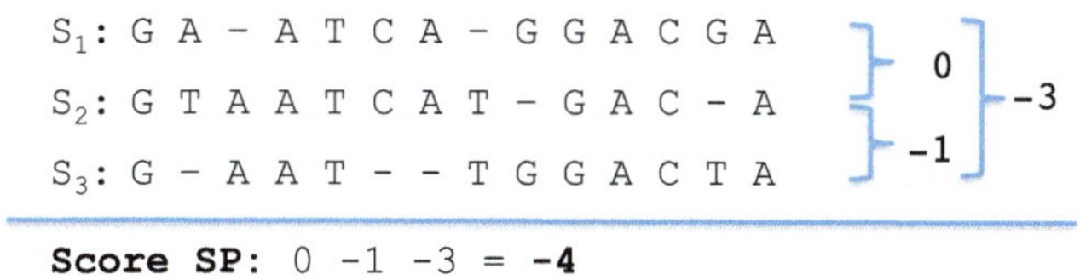

Fig. 4.3. The Sum-of-Pairs scoring function

Usually, MSAs are scored with the Sum-of-Pairs (SP) function and the exact SP MSA problem is known to be NP-hard [15]. In SP, every pair of bases is scored with the pairwise scoring function and the resultant score is the addition of all these values [8]. For instance, considering that the punctuation for matches (similar characters), mismatches (different characters) and gaps are +1, −1 and −2, respectively, the score generated by pairwise comparison of sequences S_1 and S_2 is $+1-1-2+1+1+1+1-2-2+1+1+1-2+1=0$. When comparing proteins, a substitution matrix is used to score matches/mismatches. The most common substitution matrices are PAM and BLOSUM [8].

In a general way, an MSA problem can be solved by progressive or iterative methods. Progressive methods are executed in three steps. First, an exact algorithm often based on dynamic programming is used to perform pairwise alignments with all sequences. After that, a phylogenetic tree is constructed with the information obtained in phase 1 and, finally, the tree is used to guide the alignment of the sequences sequentially, from the most closely related to the less related ones. ClustalW [12] is an example of MSA method that use progressive alignment. Iterative methods can also use dynamic programming but, unlike the progressive methods, iterative methods periodically evaluate the quality of the scores produced and realign subgroups of already aligned sequences. SAGA [17] and DIALIGN [9] are examples of iterative MSA methods that use GAs and dynamic programming, respectively.

SAGA [17] is considered the most successful example of serial genetic algorithm for the MSA problem. Its initial population is composed by randomly-generated individuals, where each individual is a possible MSA of the input sequences. The population is evaluated with a Weighted Sum-of-Pairs (WSP) function [8] and the fittest 50% of the individuals will compose the next generation. The remaining individuals are chosen for crossover/mutation using the roulette wheel method [7], respecting the crossover/mutation rates. SAGA offers a set of crossover/mutation operators and the selection of which operator will be used in each case is also made by the roulette wheel method. Both the parents and the child individuals are evaluated with the WSP function and the best ones are chosen to compose the second half of the population. This process is iterated until the termination condition is attained.

Even though SAGA is able to obtain good results, it is known as a slow method. For instance, to align 12 protein sequences of average length 281, SAGA took 5 hours and 42 minutes [17].

4.4 Parallel Genetic Algorithms for MSA

Parallel processing is a good alternative to accelerate the production of results for MSA when using strategies based on genetic algorithms. For this reason, different parallel strategies were proposed in the literature.

iPGA (island Parallel Genetic Algorithm) [1] is a parallel genetic algorithm that is based on SAGA [17] that employs the Island Model (section 4.2). Each processor evolves a subpopulation separately and the best individuals are exchanged between neighbors in predefined migration intervals. The scoring function is WSP and the initial population is randomly chosen. Two crossover operators and one mutation operator are provided. These operators are a subset of the ones provided by SAGA. iPGA was implemented in C and PVM (Parallel Virtual Machine) and executed in a 120-processor cluster. The results obtained with 17 protein sequences of average size 169 amino acids show that the parallel version is able to produce better alignments than the sequential one.

PHGA (Parallel Hybrid Genetic Algorithm) [10] is also a parallel GA that uses the Island Model (section 4.2), migrating the best individuals in each sub-population at predefined migration intervals. The particularity of this approach is that the individuals are represented as k-dimension weighted Directed Acyclic Graphs (DAGs), where k is the number of sequences and the algorithm tries to find the shortest path in this graph. The crossover and mutation operators are totally different from the other GAs for MSA since they deal with this converted representation. The initial population is generated by a progressive heuristic. PHGA was evaluated in a 4-processor shared memory machine and it took 1hour:41min:20sec to align 24 protein sequences with a maximum length of 473 amino acids.

The goal of GenAlignRefine [14] is to refine alignments produced by other MSA tools. Having an alignment, GenAlignRefine identifies regions containing gaps surrounded by gap-free regions of at least 20 nucleotides. These regions are treated by a Parallel GA that uses the Partitioned Island Model (section 4.2). Each region (which is part of an alignment) generates a population that evolves separately in each processor. At the end, the best solution for each region is sent to the master processor, that reconstructs the alignment. The results obtained in a 32-node cluster show that GenAlignRefine was able to augment the number of matches in at least 7%, when compared to DIALIGN [9].

Parallel AlineaGA [16] uses a connected island model with a star master-slave topology. In this case, each island evolves a different population and the best individuals are sent from the master to the slaves and from the slaves to the master at predefined migration intervals. Variations of the SAGA operators are provided for crossover and mutation. The mutation operators are only applied when the resultant individual presents a better fitness than the original one. Results obtained with 4 datasets from BAliBASE and tested in 4 and 8 island versions show that Parallel AlineaGA is able to produce better alignments than ClustalW and T-COFFEE for some datasets.

All approaches in Table 4.1 use the Island Parallel Model. Nevertheless, while GenAlignRefine evolves each population separately, without communication (Partitioned Island), iPGA, PHGA and Parallel AlineaGA migrate the best individuals in predetermined periods of time (Connected Island). Parallel AlineaGA uses a hybrid version of the island model where the best individuals are only migrated to/from the master node.

Table 4.1. Parallel Genetic Algorithms for the Multiple Sequence Alignment Problem

Name	**Parallel Model**	**Representation**	**Initial Population**	**Scoring Function**
iPGA	connected island	alignment	random	Weighted Sum-of-Pairs
PHGA	connected island	DAG	progressive heuristic	Sum-of-Pairs
GenAlign Refine	partition. island	parts of alignment	other MSA tools	COFFEE
Parallel AlineaGA	connected island (star topology)	alignment	random	Sum-of-Pairs

The most intuitive representation for an individual is the Multiple Sequence Alignment itself and that is the representation chosen by iPGA and Parallel AlineaGA. PHGA represents each individual as a Directed Acyclic Graph (DAG) in order to save space and to use specialized genetic operators. Since GenAlignRefine tries to ameliorate parts of the alignments, this is, thus, the representation used.

In two approaches (PHGA and GenAlignRefine), the initial population is composed by alignments produced by other tools/algorithms. In iPGA and Parallel AlineaGA, a random initialization is used. As scoring function, iPGA uses Weighted Sum-of-Pairs (section 4.3), PHGA and Parallel AlineaGA use Sum-of-Pairs (section 4.3) and GenAlignRefine uses the COFFEE function [6].

4.5 MSA Island Injection Algorithm with Relaxed Coordination

Using Genetic Algorithms to solve MSA Problems is a good alternative to progressive methods since GA-based solutions are able to re-evaluate previously aligned sequences and potentially escape from local optima. Nevertheless, GAs are known to be time consuming and, therefore, parallel GAs are often used to accelerate the production of results.

Unlike the previous parallel GA proposals discussed in section 4.3, we propose the adaptation of the Island Injection GA described in [2] (section 4.2) to the MSA problem. Our goals here are twofold: to accelerate the production of results and to potentially generate better alignments.

4.5.1 General Overview

As in [2], our solution contains one high resolution archipelago and two low resolution archipelagoes. However, unlike [2], all archipelagoes execute the same objective function. The difference between these two kinds of archipelagoes in our approach is the representation of an individual and the parallel island model used in each case.

The High Resolution Archipelago (HRA) solves the original problem with the original individual representation, that corresponds to a complete MSA. The Low Resolution Archipelagoes (LRA) solve part of the original problem, where the individuals are only part of MSAs. Therefore, both the HRA and the LRAs execute the algorithm illustrated in Algorithm 1. The difference is that, while the HRA evolves the complete individuals, the LRAs evolve only parts of them. Periodically, the best individuals found by the LRAs are migrated to the HRA and incorporated to its population. An overview of our solution is shown in Figure 4.4.

Our parallel algorithm receives a set of biological sequences to be aligned as input. These sequences are sent to all archipelagoes (HRA and LRAs).

In the High Resolution Archipelago (HRA), each MSA is an individual. The HRM (Figure 4.4) receives the input sequences and generates a random initial population of complete MSAs. The initial population is split into b_{HRA} subsets, where b_{HRA} is the number of Basic Genetic Algorithms (BGA) that compose the HRA. Each BGA evolves the population received and migrates the fittest individuals to the right neighbor at predefined internal intervals. Therefore, the HRA executes a traditional Connected Island model with one

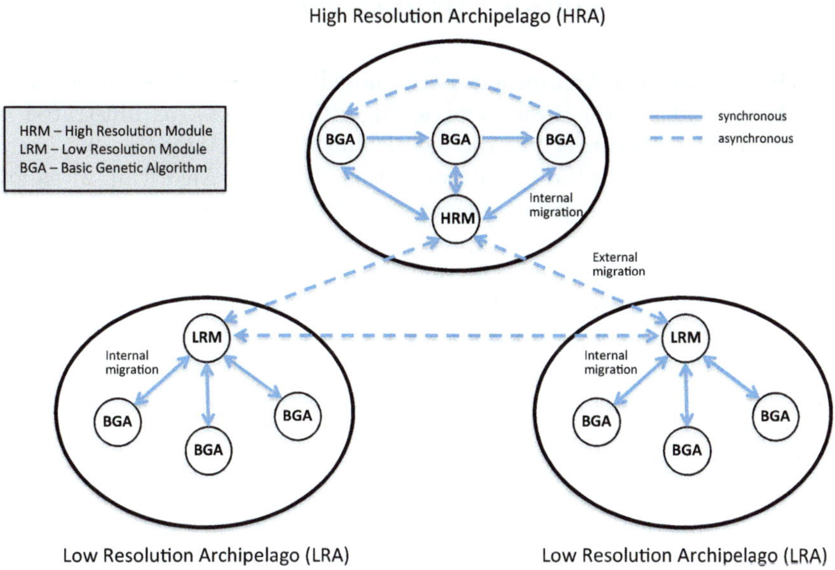

Fig. 4.4. Overview of the Parallel island injection algorithm.

difference. At fixed time intervals, called *external migration intervals*, the HRA receives the fittest individuals from the Low Resolution Archipelagoes, combines them and incorporates the resultant individuals into its own population. After doing that, some individuals are sent back to the LRM.

In the Low Resolution Archipelagoes (LRA), each individual is part of an MSA. The LRM (Figure 4.4) also generates a random initial population of MSAs but, instead of dividing the population into subsets, as the HRM does, the LRM cuts each MSA into x $parts(pa_1, pa_2, ..., pa_x)$, where pa1 is the set containing the first part of the multiple sequence alignment and pa_x is the set containing the last part of the MSA [18].

The LRM then sends the set containing sub-alinment pa_1 to the first BGA, pa_2 to the second BGA and pa_3 to the third BGA. Each sub-alignment will evolve separately. The LRA employs, thus, the Partitioned Island model. At internal migration intervals, the entire population is sent to the LRM, which will connect all sub-alignments to generate complete MSAs. These alignments are evaluated and the fittest will compose the next generation, which will be used to generate sub-alignments again. Each time the LRM generates sub-alignments, the cutting points are different. At external migration intervals, the LRMs send their fittest individuals to the HRM. This process continues until the maximum number of generations is attained.

4.5.2 High Resolution Archipelago

The High Resolution Archipelago is composed by a High Resolution Module (HRM) and several Basic Genetic Algorithm (BGA), as illustrated in Figure 4.4. The HRM executes the algorithm shown in Algorithm 2.

The HRM generates the initial population (line 2 in Algorithm 2) as follows. First, an initial alignment is created from the input sequences by adding terminal gaps. After that, the Gap Generator operator is used to insert random gaps into the sequences. The number of gaps inserted depends on the size of the longest sequence. The alignments generated this way are evaluated with the SP function (section 4.3) and alignments with the same score are eliminated. This process continues until the size of the initial population is attained. The population is then partitioned and the same number of individuals is distributed to the BGAs.

After this point, the HRM coordinates the external migration among the archipelagoes and the internal migration from its BGAs (Algorithm 2). At given periods of time, it asynchronously receives the best *i* individuals from its BGAs (line 5) and migrates the best *i* individuals to the Low Resolution Modules (LRMs) (line 6). It then asynchronously receives the best *i* individuals from the LRMs (line 7) and distributes them to its BGAs (line 8). When the termination criterion is attained, it receives the best *i* individuals from its BGAs for the last time (line 10) and outputs the best one among them (line 11).

Algorithm 2. Algorithm executed in the High Resolution Module

Input: seqs, BGAs, individuals, i, LRMs;
Output: output;
1: receive_input_sequences(seqs);
2: pop=generate_initial_population(seqs);
3: distribute_initial_population(pop, BGAs);
4: **while** termination_condition_not_reached **do**
5:    async_receive(individuals, BGAs, i);
6:    migrate_individuals (i, LRMs);
7:    async_receive(individuals, LRMs
8:    distribute_individuals (i, BGAs);
9: **end while**
10: receive(individuals, BGAs, i);
11: output = obtain_best_individual();

4.5.3 Low Resolution Archipelago

The Low Resolution Archipelago is composed of a Low Resolution Module (LRM) and several Basic Genetic Algorithm (BGA), as shown in Figure 4.4. The LRM executes the algorithm shown in Algorithm 3.

The LRM generates the initial population (line 2 in Algorithm 3) as described in section 4.5.2. At predefined periods of time (migration interval), the best individuals are migrated to the other LRM and to the HRM (lines 5 and 6).

After that, the best individuals are received from these two archipelagoes (lines 7 and 8). All individuals are evaluated with the SP function and the best ones are selected to compose the next generation (lines 9 and 10). Having the current population, each individual is cut into parts (line 12).

Note that, in order to avoid local optima, we cut the individuals in different points each time. The parts are migrated to the BGAs (line 13) that will evolve each part separately. The LRM then receives the best individuals from all BGAs and merges the parts, generating individuals that are complete multiple sequence alignments (lines 14, 15 and 16). These individuals are evaluated and the best ones will compose the next generation (lines 17 and 18).

Algorithm 3. Algorithm executed in the Low Resolution Module

```
Input: seqs, individuals, LRMs, HRM, BGAs;
Output: pop;
 1: receive_input_sequences(seqs);
 2: pop=generate_initial_population(seqs);
 3: while termination_condition_not_attained do
 4:    if external_migration_interval then
 5:       migrate(individuals, LRMs);
 6:       migrate(individuals, HRM);
 7:       async_receive(individuals, LRMs);
 8:       async_receive(individuals, HRM);
 9:       select_best_individuals();
10:       pop=generate_new_population();
11:    end if
12:    cut_individuals_into_parts();
13:    migrate_parts(BGAs);
14:    receive_best_parts(BGAs);
15:    merge_parts();
16:    apply_gap_operators();
17:    evaluate_individuals();
18:    generate_new_population();
19: end while
```

4.5.4 Basic Genetic Algorithm

The Basic GA (BGA) that is executed in the archipelagoes uses the generic structure shown in Figure 1. The main difference is that, at internal migration intervals, the best individuals are sent: a) to the right neighbour and to the HRM, in the HRA, or b) to the LRM, in the LRA. Moreover, the initial population is not generated by the BGA since it is received from the HRM or LRM.

The crossover, mutation and population breeding methods used in the BGAs are inspired in SAGA (section 4.3). The best 50% percent individuals are automatically put in the next generation. Crossover and mutation operators are applied to part of the remaining individuals. The choice of a particular crossover/mutation operator is done by the roulette wheel method, as in SAGA.

4.6 Experimental Results

The algorithm proposed in section 4.5 was implemented in C and MPI. The tests were conducted in one 12-core cluster with 6 nodes, each one containing a Intel Core 2 Duo 2.93GHz, 4GB RAM and 160GB disk. A non-dedicated Gigabit Ethernet network was used to interconnect the nodes.

In our tests, the following strategies of parallel and sequential GAs were used: a) Island Injection two-way ($i2$); b) Island Injection one-way ($i1$); c) Nine Islands ($p9$); d) Three islands ($p3$); and e) Sequential (seq). The strategy $i2$ (Figure 4.5.a) is the implementation of the algorithm described in section 4.5, with 3 archipelagoes composed of 3 islands each. The strategy $i1$ (Figure 4.5.b) uses the same number of archipelagoes and islands. The only difference is that the results are not redistributed to the LRAs. The next two strategies ($p9$ and $p3$) implement the Connected Island Model with 9 and 3 islands, respectively, (Figures 4.5.c and 4.5.d). The coordinator process is only used to start/terminate the computation. Finally, seq is the implementation of a sequential GA (Figure 4.5.e). All these strategies execute the same BGA (section 4.5.4).

For the parallel strategies ($i2$, $i1$, $p9$ and $p3$), the internal migration interval was set to 10 generations and for the island injection strategies ($i2$ and $i1$), the external migration interval was set to 50 generations. These values were chosen empirically. The PAM 250 matrix was used in all the tests in order to score matches/mismatches and SP with affine gap was used as the objective function.

In our tests, we used three sets of real sequences that were retrieved from from BaliBASE [13] and PFAM [5]. The characteristics of these sets are illustrated in Table 4.2.

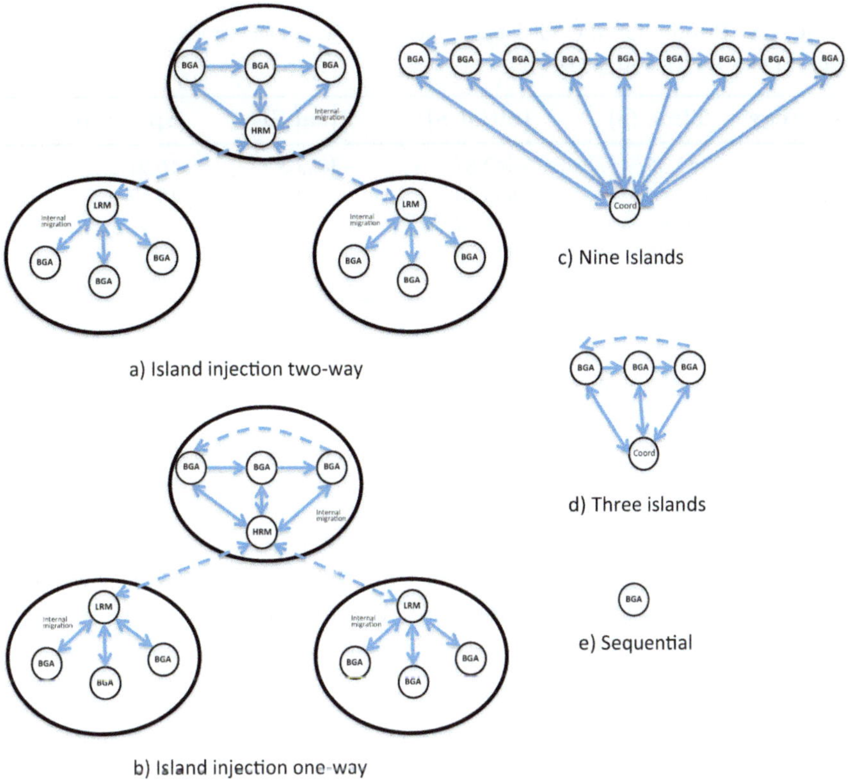

Fig. 4.5. Strategies compared in the tests.

Table 4.2. Characteristics of the real datasets used in the tests

Name	Base	# Seq	Avg Size	Characteristic
1ac5	BAliBASE	4	445	difficult set, local optima
ttkrsyedq	PFAM	3	513	easy set, similar sequences
virul fac	PFAM	15	820	more sequences, medium-sized

4.6.1 Results for 1ac5

The results obtained for the 1*ac*5 base are shown in Table 4.3. In all the configurations, the total number of generations was 5,000 and the population size was 180 MSAs.

Table 4.3 presents the results obtained with 1*ac*5 by the five strategies. In this table, the baseline for the speedup computation is the sequential execution time and the best results are highlighted in gray. For the 1*ac*5

Table 4.3. Results for the 1*ac*5 base, including wallclock times, initial and final sum-of-pairs scores and speedups.

Strategy	Time (s)	Initial SP	Final SP	Speedup
*i*2	253.35	–442.87	–308.65	16.91
*i*1	273.44	–435.24	–340.91	15.67
*p*9	504.60	–438.17	–331.98	8.49
*p*3	831.61	–449.73	–330.54	5.15
seq	4,285.52	–482.75	–326.88	1.00

base, the strategy *i*2 obtained the best execution times and the alignments with the highest score.

Considering the quality of the alignments, *i*2 was able to obtain a multiple sequence alignment with score −308.65. This is 5.58% better than the second best strategy (*seq*) and 9.44% better than the strategy that obtained the worst result (*i*1).

The best speedups were also obtained by the *i*2 strategy, which was able to reduce the execution time from 4,285.52 seconds (1 hour and 11 minutes) to 253.35 seconds (4 minutes and 13 seconds).

4.6.2 Results for ttkrsyedq

The results obtained for the *ttkrsyedq* base are shown in Table 4.4. As in the previous case, the total number of generations was 5,000 and the population size was 180 MSAs, for all configurations.

Table 4.4. Results for the *ttkrsyedq* base, including wallclock times, initial and final sum-of-pairs scores and speedups.

Strategy	Time (s)	Initial SP	Final SP	Speedup
*i*2	364.48	–187.72	168.27	21.01
*i*1	387.11	–164.81	127.05	19.82
*p*9	696.69	–147.45	142.89	11.01
*p*3	1,176.76	–102.31	166.49	6.52
seq	7,675.55	–278.80	192.26	1.00

For the *ttkrsyedq* base, the strategy *i*2 obtained the best execution times but the alignment with the highest score was obtained by the sequential strategy.

Considering the quality of the alignments, seq was able to obtain a multiple sequence alignment with score 192.26. This is 12.50% better than the second best strategy (*i*2). Nevertheless, the sequential strategy obtained this result in 7,7675.55 seconds (two hours and 8 minutes) and the *i*2 strategy obtained its result in 364.48 seconds (six minutes and 3 seconds). Since the scores obtained by seq and *i*2 are close, we believe that the *i*2 strategy would easily outperform *seq*, if a greater number of generations is considered.

4.6.3 Results for virul fac

The results obtained for the *virulfac* base are shown in Table 4.5. In all the configurations, the total number of generations was 5,000 and the population size was 180 MSAs. For the virul fac base, the strategy *i*2 obtained the best execution times and a multiple sequence alignment with a score that was very close to the best score retrieved.

Considering the quality of the alignments, *i*2 was able to obtain a multiple sequence alignment with score –705.84. This score is very close to the alignment obtained in this test by the seq policy (score = –704.73).

Table 4.5. Results for the *virulfac* base, including wallclock times, initial and final sum-of-pairs scores and speedups.

Strategy	Time (s)	Initial SP	Final SP	Speedup
*i*2	2,208.64	–825.96	–705.84	39.76
*i*1	2,546.64	–861.98	–734.38	34.48
*p*9	5,243.87	–814.39	–715.24	16.74
*p*3	8,194.05	–826.04	–732.93	10.71
seq	87,811.11	–815.97	–704.73	1.00

As in the previous cases, the best speedups were obtained by the *i*2 strategy, which was able to reduce the execution time from 87,811.11 seconds (24 hours and 23 minutes) to 2,208.64 seconds (36 minutes and 48 seconds).

4.6.4 Overall Evaluation

In the three tests presented above, all parallel strategies were able to obtain superlinear speedups, when compared to the sequential execution. This happened because the sequential execution evolved 180 individuals, using one core, and the other strategies (*p*3, *p*9, *i*1 and *i*2) evolved the same number of individuals, that were distributed to many cores. Therefore, each core treated less individuals, resulting in a smaller number of cache misses. In the virul fac case, the sequences were longer and, for this reason, the sequential version also experienced an augmented paging activity, which had a considerable influence at the execution times.

The best execution times were obtained for the *i*2 strategy in all the test cases. The times obtained by the *i*1 strategy were very close to the ones achieved by *i*2. This can be easily explained since both strategies used 12 cores (9 for genetic algorithms and 3 for the low and high resolution modules). Even though the *p*9 strategy executed with 8 cores, its running times were close to the double of the execution times presented by *i*2 and *i*1. The reason for that is that the island model is connected as a ring (Figure 4.5.c) and that generates an overhead at migration time.

The *i*2 strategy was able to obtain very good alignments for the 1*ac*5 and *virulfac* data sets. These sets are challenging since 1*ac*5 is a data set that can be considered difficult, with several local optima, and *virulfac* contains a considerable number of sequences with medium size. For the *ttkrsyedq* set, which is considered easy, the alignment retrieved by *i*2 was not very good. As already observed in SAGA [17], multiple sequence alignment methods that are based on genetic algorithms are more effective for data sets with high/medium complexity.

4.6.5 Comparison with the Strong Coordination Strategy

In this section, we compare our relaxed coordination iiGA with the strong coordination iiGA described in [18]. The difference between both strategies is that the first one uses asynchronous communication primitives among/inside the archipelagoes whereas the second one uses synchronous primitives. This relaxed coordination mechanism was added to improve performance but it created one side effect: at a given moment, two islands may be evolving different generations. For instance, at moment x, island 1 may be evolving generation 100 while island 2 may be evolving generation 101.

Table 4.6 presents the performance gain obtained by the relaxed approach, when compared to the strong coordination one. Also, we present the best MSA retrieved in each case.

The best gains in execution time for the relaxed approach were achieved for the 1*ac*5 test set. For the 4-core execution, the execution time was reduced from 1,267.93s (strong) to 487.73s (relaxed). Also, we can notice that the relaxed iiGA was able to retrieve the alignment with the highest score for the same base. This shows that the effect of having different generations being evolved in different islands did not compromise the quality of the results.

Table 4.6. Performance gains obtained by the relaxed approach and the best MSA retrieved.

		1ac5	
Processors	**Gain (execution time)**	**Final SP (relaxed)**	**Final SP (strong)**
12	36.1%	−308.65	−329.73
8	56.2%	−331.67	−324.13
4	61.5%	−326.98	−341.06
2	55.2%	−335.50	−333.54
		ttkrsyedq	
Processors	**Gain (execution time)**	**Final SP (relaxed)**	**Final SP (strong)**
12	4.5%	163.16	168.27
8	0.5%	166.91	170.23
4	33.2%	170.40	194.48
2	33.3%	119.59	111.02
		virulfac	
Processors	**Gain (execution time)**	**Final SP (relaxed)**	**Final SP (strong)**
12	11.9%	−705.84	−663.96
8	9.8%	−670.53	−629.80
4	7.2%	−634.10	−669.24
2	4.8%	−666.96	−654.15

4.7 Conclusion

In this chapter, we proposed and evaluated a parallel injection island genetic algorithm with relaxed coordination for the Multiple Sequence Alignment problem. Our solution is composed of three archipelagoes and periodically the best solutions found by the low resolution archipelagoes are injected into the high resolution one. The coordination among the archipelagoes is relaxed in such a way that the individuals that are exchanged can be integrated in posterior generations. This allows our strategy to take profit of a higher degree of parallelism.

The results obtained with real protein sequences in two separate clusters show that our proposal is able to find better results, when compared to the traditional parallel island model. Also, our parallel algorithm was able to reduce significantly the execution time, when compared to the sequential genetic algorithm.

As future work, we intend to compare our algorithm to other Multiple Sequence Alignment tools that are not based on genetic algorithms. In order to do so, we will implement and use the Weighted Sum-of-Pairs (WSP) function.

Also, we intend to study in detail the impact of the migration intervals and population size on the behavior of our algorithm. We also intend to investigate new merge operators. Finally, we intend to implement our strategy in GPUs (Graphics Processing Units).

References

1. Ambarasu, L.A., Narayanasamy, P., Sundararajan, V.: Multiple Molecular Sequence Alignment by Island Parallel Genetic Algorithm. Current Science 78(7), 858–863 (2000)
2. Babbar, M., Minsker, B.S., Goldberg, D.: A Multiscale Island Injection Genetic Algorithm for Optimal Groundwater Remediation Design. Journal of Water Res. Plan & Man 132(5), 341–350 (2006)
3. Cantu-Paz, E.: Implementing Fast and Flexible Parallel Genetic Algorithms. Practical Handbook of Genetic Algorithms 3, 65–84 (1998)
4. Eddy, S.: HMMER User's Guide v. 2.3.2. Washington University School of Medicine (2003)
5. Finn, R., et al.: The Pfam Protein Families Database. Nucleic Acids Research 36, D281–D288 (2008)
6. Heringa, J., Notredame, C., Higgins, D.G.: T-COFFEE: a Novel Method for Fast and Accurate Multiple Sequence Alignment. Journal of Mol. Biol. 302(1), 205–217 (2000)
7. Michalewicz, Z.: Genetic Algorithms + Data Structures = Evolution Programs, 387p. Springer (1999)
8. Mount, D.: Bioinformatics: Sequence and Genome Analysis, 692p. C. S. Harbor Lab Press (2004)
9. Morgenstern, B., et al.: Multiple DNA and Protein Sequence Alignment Based on Segment-to-Segment Comparison. Proc. of Natl. Acad. Sci., USA, 12098–12103 (1996)
10. Nguyen, H.D., Yoshihara, I., Yamamori, K., Yasunaga, M.: Aligning Multiple Protein Sequences by Parallel Hybrid Genetic Algorithm. Genome Informatics 13, 123–132 (2002)
11. Notredame, C., Higgins, D.G.: SAGA: sequence alignment by genetic algorithm. Nucleic Acids Research 24(8), 407–422 (1996)
12. Thompson, J.D., Higgins, D.G., Gibson, T.J.: Clustal W: Improving the Sensitivity of Progressive Multiple Sequence Alignment Through Sequences Weighting, Position-Specific Gap Penalties and Weight Matrix Choice. Nucleic Acids Research 22(22), 4673–4680 (1994)
13. Thompson, J.D., Koehl, P., Ripp, R., Poch, O.: BAliBASE 3.0: Latest Developments of Multiple Sequence Alignment Benchmark. Proteins: Structure, Function and Bioinformatics 61(1), 127–136 (2005)
14. Wang, C., Lefkowitz, E.J.: Genomic Multiple Sequence Alignments: "Refinements using a Genetic Algorithm". BMC Bioinformatics 6 (2005)

15. Wang, T., Jiang, T.: On the Complexity of Multiple Sequence Alignment. J. Comp. Biol. 1(4), 337–348 (1994)
16. Silva, F.J.M., Perez, J.M.S., Pulido, J.A.G., Rodriguez, M.A.V.: Parallel AlineaGA: an Island Parallel Evolutionary Algorithm for Multiple Sequence Alignment. In: Proc. of the Int. Conf. on Soft Computing and Pattern Recognition (SoCPar), pp. 279–284 (2010)
17. Notredame, C., Obrien, E.A., Higgins, D.G.: RAGA: RNA Sequence Alignment by Genetic Algorithm. Nucleid Acids Research 25(22), 4570–4580 (1997)
18. Miranda, L.A., Caetano, M.F., Melo, A.C.M.A., Correa, J.M., Bordim, J.L.: Multiple Biological Sequence Alignment with a Parallel Island Injection Genetic Algorithm. In: Proc. of the IEEE Int. Conf. on High Performance Computing and Communications (HPCC), pp. 314–321 (2010)

Chapter 5
Distributed Evolutionary Computation Using SOAP and REST Web Services

P.A. Castillo, P. García-Sánchez, M.G. Arenas,
J.L. Bernier, and J.J. Merelo

Abstract. In this chapter, a high-level comparison of both SOAP (Simple Object Access Protocol) and REST (Representational State Transfer) is made. These are the two main approaches for interfacing to the web with web services. Both approaches are different and present some advantages and disadvantages for interfacing to web services: SOAP is conceptually more difficult (has a steeper learning curve) and more "heavy-weight" than REST, although it lacks of standards support for security. In order to test their efficiency (in time), three experiments have been performed using both technologies: first a basic client-server model implementation to test communications has been implemented; then, a master-slave based genetic algorithm (GA) to solve an optimization problem has been used; and finally, as third experiment, an approach to evolutionary distributed optimization of multilayer perceptrons (MLP) using REST and language Perl has been done. In these experiments, a master-slave based evolutionary algorithm (EA) has been implemented, where slave processes evaluate the costly fitness function (training a MLP to solve a classification problem). As expected, the parallel version of the developed programs obtains similar or better results using much less time than the sequential version, obtaining a good speedup. The results obtained have shown that both SOAP and REST can be used as communication protocol for distributed evolutionary computation, obtaining a good speedup. Results obtained are comparable, and only for large amounts of data (big messages), REST communications take longer than SOAP communications.

P.A. Castillo · P. García-Sánchez · M.G. Arenas · J.L. Bernier · J.J. Merelo
GeNeura, Department of Architecture and Computer Technology,
CITIC (University of Granada)
e-mail: {pedro,pgarcia,mgarenas,jbernier,jmerelo}@geneura.ugr.es

J. Kołodziej et al. (Eds.): Advances in Intelligent Modelling and Simulation, SCI 422, pp. 89–111.
springerlink.com

5.1 Introduction

Service Oriented Architecture (SOA) [46] is a paradigm for organizing and utilizing distributed computational resources, called services. Using this paradigm, the service providers publish the descriptions (or interfaces) of the services they offer in a service registry, so the service requesters can discover them and bind to the correspondent service provider. The Web Services are the key point of integration for different applications belonging to different platforms, languages, systems since they are based in a set of standards that make them independent of the underlying technologies used for providing them.

Although there are several technologies for developing web services (SOAP, REST or XMLRPC among others [44, 45]), nowadays the main approaches are SOAP (Simple Object Access Protocol) [6, 4] and REST (Representational State Transfer) [57]. Both implementations are suitable for designing Web Services, however, it is important to understand the pros and cons of each one.

SOAP is the traditional, standards-based approach, but the majority of the web services with public API offer REST interfaces, while some of them offer both REST and SOAP and very few offer just SOAP. All of the major Web Services providers use REST [1]: Twitter, Yahoo's, Flickr, del.icio.us, pubsub, bloglines, technorati, and several others. Both eBay and Amazon have Web Services for both REST and SOAP.

On the other hand, SOAP Web Services are used in lots of enterprise software as well; for example, Google implements their Web Services using SOAP, with the exception of Blogger, which uses XML-RPC, an early and simpler pre-standard of SOAP.

The philosophies of SOAP and RESTful Web Services are very different. Strictly, SOAP is a protocol for distributed computing, whereas REST adheres much more closely to a web-based design. SOAP requires a greater implementation and understanding effort of the client side to difference of REST based APIs, that focus these efforts on the server side. Table 5.1 shows the main strengths and weaknesses for both SOAP and REST.

It is important to note that one of the advantages of SOAP is the use of the "generic" transport. While REST today uses HTTP/HTTPS, SOAP can use almost any transport to send the request. However, one perceived disadvantage is the use of XML because of its verbosity, and the time necessary to parse it.

In this chapter we propose using these two interfacing approaches for distributed computation, demonstrating how it could be used for evolutionary computation.

In order to determine the efficiency of these two interfacing approaches, we have performed three experiments in which both a SOAP and REST

[1] `http://www.infoq.com/news/2011/06/Is-REST-Successful`

Table 5.1. Strengths and weaknesses for both SOAP (above) and REST (below). Adapted from `http://ajaxonomy.com/2008/xml/web-services-part-1-soap-vs-rest`

SOAP	
Strengths (pros)	Weaknesses (cons)
+ Handle distributed computing environments	- More verbose
+ Built-in error handling	- Harder to develop, requires tools
+ Extensibility	- Conceptually more difficult, more "heavy-weight" than REST
+ Language, platform, and transport agnostic	
+ Prevailing standard for web services	
+ Support from other standards (WSDL, WS-*)	
REST	
Strengths (pros)	Weaknesses (cons)
+ Language and platform agnostic	- Assumes a point-to-point communication model
+ Much simpler to develop than SOAP	- Not usable for distributed computing environment
+ Small learning curve, less reliance on tools	- Lack of standards support for security, etc.
+ Concise, no need for additional messaging layer	- Tied to the HTTP/HTTPS transport model
+ Closer in design and philosophy to the Web	

implementations are evaluated. As first experiment a client-server model is implemented, in which the server process runs on a machine and the client processes send and receive text strings. Next, a master-slave based GA is implemented, running on the master process the GA and the fitness evaluation on the slave processes. Finally, as third experiment, we implement a distributed evolutionary algorithm (EA) using Perl and REST, to solve a costly problem: tuning learning parameters and to set the initial weights and hidden layer size of a multilayer perceptron (MLP), based on an EA and Quick Propagation [17] (QP) to solve classification problems.

This work continues with our previous research in service oriented algorithms, as previously stated in [26], where a service-oriented platform was presented, or evolutionary optimization of MLP (*G-Prop* method) presented in [10, 11]. *G-Prop* leverages the capabilities of two classes of algorithms: the ability of EA to find a solution close to the global optimum, and the ability of the back-propagation algorithm (BP) to tune a solution and reach the nearest local minimum by means of local search from the solution found by the EA.

Instead of using a pre-established topology, the population is initialized with different hidden layer sizes, with some specific operators designed to change them (mutation, multi-point crossover, addition and elimination of hidden units, and QP training applied as operator). The EA searches and optimizes the architecture (number of hidden units), the initial weight setting for that architecture and the learning rate for that net.

The main idea of this chapter, which is basically a proof of concept, is to see what are the possibilities of this setup as a meta-computer by implementing an EA using it, and then measuring the speedup when several computers are used at the same time. The problem we will attempt to solve is a costly classification problem, so that it takes time enough to get some improvement from parallelization. We will only try to measure how running time scales when new (heterogeneous) nodes are added to the system, being the main objective to test if this kind of system is suitable for scientific computation.

This chapter is structured as follows: In sections 5.2 and 5.3 a comprehensive description of SOAP and REST technologies are provided, respectively. Section 5.6 describes the experiments. In concrete, the client-server and master-slave models implemented for testing are described, so the experimental configuration, the methodology considered in the study ; finally, the results obtained are shown. Last section (Section 5.7), throws some conclusions and presents the proposed future work.

5.2 SOAP: Simple Object Access Protocol

SOAP is a standard protocol proposed by the W3C ([6], [4]) to interface Web Services, and that extends the remote procedure call (XML-RPC). Thus, SOAP can be considered as an evolution of XML-RPC protocol, much more complete and mature, that allows to perform remote procedure calls to distributed routines (services) based on an XML interface as interfacing language. Thus, SOAP clients can access to objects and methods that are residing in remote servers, using an standard mechanism that makes transparent the details of implementation, such us the programming language of the routines, the operating system or the platform used by the provider of the service. At the moment, there exist complete implementations of SOAP for Perl, Java, Python, C++ and other languages [51]. In opposite to other remote procedure call methods, such as RMI (*remote method invocation*, used by the Java language) or XML-RPC, SOAP has two main advantages: it can be used with any programming language, and it can use any type of transport (HTTP, SHTTP, TCP, SMTP, POP and other protocols).

SOAP sends and receives messages using XML [49, 30, 5], wrapped HTTP-in headings. The interfaces of the methods that can be accessed using SOAP services are specified by a Web Services Description Language (WSDL) [50, 55]. The WSDL of an Web Service consists in an XML description of its interface, i.e., it is a file that describes the name of the methods, the parameters and type of data, the type of response that the Web Service may return, etc. Using a WSDL file, that it is based on a neutral language such as XML, the service can be specified for different languages, so that a Java client can access a Perl server.

In this way, SOAP constitutes a high level protocol, making easy the task of distributing objects among different servers, and avoiding the difficulties derived of defining the message formats, nor the explicit call to remote servers.

5.3 REST: Representational State Transfer

After some years, Internet architects have found an alternative method for building web services in the form of Representational State Transfer (REST) [57] .

REST is a style of software architecture for distributed hypermedia systems such as the World Wide Web. The term Representational State Transfer was introduced and defined in 2000 by Roy Fielding in his doctoral dissertation [18, 19]. Fielding is one of the principal authors of the Hypertext Transfer Protocol (HTTP) specification versions 1.0 and 1.1 [32, 33].

REST-style architectures consist of clients and servers. Clients initiate requests to servers; servers process requests and return appropriate responses. Requests and responses are built around the transfer of representations of resources. A resource can be essentially any coherent and meaningful concept that may be addressed.

Although REST was initially described in the context of HTTP, is not limited to that protocol. RESTful architectures can be based on other Application Layer protocols if they already provide a rich and uniform vocabulary for applications based on the transfer of meaningful representational state. RESTful applications maximize the use of the pre-existing, well-defined interface and other built-in capabilities provided by the chosen network protocol, and minimize the addition of new application-specific features on top of it.

In a REST environment, clients are not concerned with data storage, which remains internal to each server, so that the portability of client code is improved. Servers are not concerned with the user interface or user state, so that servers can be simpler and more scalable. Servers and clients may also be replaced and developed independently, as long as the interface is not altered. Finally, servers are able to temporarily extend or customize the functionality of a client by transferring logic to it that it can execute.

The client-server communication is further constrained by no client context being stored on the server between requests. Each request from a client contains all of the information necessary to serve the request, and any session state is held in the client. The server can be stateful; this constraint merely requires that server-side state be addressable by URL as a resource. This not only makes servers more visible for monitoring, but also makes them more reliable in the face of partial or network failures as well as further enhancing their scalability.

Main REST web services features are:

- Simple and lightweight (not a lot of extra XML markup)
- Human readable format
- Easy to build (no toolkits required)
- High performance

5.4 Parallel and Distributed EA Using Web Services

There are many implementations of distributed genetic algorithms [27], usually using PVM or MPI [9], thus this chapter does not intend to innovate in that sense, but in the implementation.

There are many ways to implement a distributed genetic algorithm, one of which is the global parallelization (*farming*), in which, as Fogarty and Huang propose [22], Abramson and Abela [1], or Hauser and Männer [31], individual evaluation and/or genetic operator application are parallelized. A master processor can supervise the population and select individuals to mate; then slave processors receive the individuals to evaluate them and to apply genetic operators.

Another way to implement parallelization is the migration: the population is divided into small subpopulations of the same size assigned to different processors. From time to time each processor selects the best individuals in its subpopulation and it sends them to his nearer processors, receiving as well copies of the best individuals of his neighbors (migration of individuals). All processors replace the worst individuals of their populations. This kind of algorithms is also known as distributed evolutionary algorithms (Tanese [54], Pettey et al. [47], Cantú-Paz and Goldberg [7]).

An ideal client-server implementation of a distributed evolutionary algorithm could be a server process with several threads. Each thread would include a population, and would communicate with other threads through the shared code among them. Each thread would use an own tail of individuals to send to other threads. Each thread would evaluate its individuals in different remote computers, carrying out the communication using a server.

5.5 Comparing SOAP and REST Programming Models

REST-based programming model leverages the universality of the HTTP protocol and uses a URI addressing scheme.

Another differentiating factor is that SOAP Services tend to be operation-based, while REST services are said to be resource-based. That means clients are effectively going to call methods on a SOAP service, while client of a REST service are going to send HTTP requests to a URI and expect to get some resource in return.

From an architectural perspective what this means is that service operations will be inherently more constrained than resources that are freely accessible.

To make data available to clients (especially clients that may not understand or care about SOAP) focusing on the speed, a REST-based service is more efficient.

The SOAP model was implemented using the SOAP::Lite[2] [36] module, while the REST implementation was carried out using the Perl Dancer[3] module [53, 52], for their stability. In addition, servers developed using these modules are easy to implement and deployed using the computer infrastructure available to us in our department.

Figures 5.1 and 5.2 show the Perl source code of the client-server SOAP and REST implementations developed for the first experiment (subsection 5.6.1).

5.6 Experimental Setup and Results

In this chapter we carry out three experiments to compare two parallel models implemented using SOAP and REST technologies in Perl language (due to the familiarity of the authors with this language [38, 40, 43]).

The Experiment 1 (subsection 5.6.1) consisted in the implementation of a client-server model. In this case, the server process runs on a machine that attends client requests, involving different lengths of text strings. The experiment 2 (subsection 5.6.2) implements a master-slave based GA. In this case, a master process runs the GA, while different slave processes evaluate the fitness function. The experiment 3 (subsection 5.6.3) implements a distributed EA to optimize MLPs: *G-Prop* method [10, 11] is adapted as a distributed EA using SOAP and REST using a master-slave model.

Up to 4 computers have been used to run the algorithm and to obtain results both in sequential and parallel versions of the program. Experiments were conducted running the programs on Ubuntu/Linux machines. Computer speeds range from 1.5 Ghz to 2 Ghz and are connected using the Ethernet network of the university (with a high communication latency, i.e. an average *ping* of 7 ms). No experiments using homogeneous computer network have been done, because our aim is to demonstrate potential of distributed EA using web services.

---

[2] `http://www.soaplite.com`

[3] `http://perldancer.org`

```
use SOAP::Transport::HTTP;
my $daemon = SOAP::Transport::HTTP::Daemon
    -> new (LocalPort => 80)
    -> dispatch_to('Demo');
$daemon->handle;

package Demo;
our $src="";
sub push {
    my ($class, $cad) = @_;
    $self->src = $cad;
    return "ok";
};
sub pop {
    my $tmp = $self->src;
    $self->src = " ";
    return $tmp;
};
```

```
use Time::HiRes qw( gettimeofday tv_interval);
use SOAP::Lite;
my $i=0;
my $tmp_it = [gettimeofday()];
for ($i=0; $i<100 ; $i++) {
    my $cad="01234567890 ... 01234567890";
    print SOAP::Lite
        -> uri('http://www.soaplite.com/Demo')
        -> proxy('http://vaio/')
        -> push($cad) -> result;
    print SOAP::Lite
        -> uri('http://www.soaplite.com/Demo')
        -> proxy('http://vaio/')
        -> pop() -> result;
};
print "TIME: ", tv_interval( $tmp_it );
```

Fig. 5.1. SOAP programming example: server (left) and client (right). The string $cad value varies from 100 to 10000 chars in order to configure different loads.

5.6.1 Proof of Concept: Client-Server Efficiency Comparison

A classic client-server model is implemented in which clients can send and receive a text string. Different string lengths (100, 1000, 5000 and 10000 chars) have been configured in order to probe with different loads. In this way, we have tried to determine how the string length (the amount of data) affects the running time (due to communications).

The experiment consisted in sending 100 times an string of chars to the server (SOAP/REST). The experiment was repeated for 50 times for each case, measuring the time spent using "gettimeofday" function (in order to achieve a good precision).

As shown in Table 5.2, the SOAP version takes more time to complete the communications than the REST implementation.

```
use Dancer;
my $src = "";
get '/pop/' => sub {
    my $tmp = $src;
    $src = " ";
    return $tmp;
};
get '/push/:cad' => sub {
    my ($class, $cad) = @_;
    $src = $cad;
    return "ok";
};
Dancer->dance;
```

```
use Time::HiRes qw( gettimeofday tv_interval);
use LWP;
my $nav = new LWP::UserAgent;
$nav->agent("RESTzilla");
my $i=0;
my $tmp_it = [gettimeofday()];
for ($i=0; $i<100 ; $i++) {
    my $cad="01234567890 ... 01234567890";
    my $rpush = new HTTP::Request GET
        => 'http://127.0.0.1:3000/push/'."$cad";
    my $upush = $nav->request($rpush);
    my $rpop = new HTTP::Request GET
        => 'http://127.0.0.1:3000/pop/';
    my $upop = $nav->request($rpop);
};
print "TIME: ", tv_interval( $tmp_it );
```

Fig. 5.2. REST programming example: server (left) and client (right). The string $cad value varies from 100 to 10000 chars in order to configure different loads.

Table 5.2. Results obtained on the first experiment (client-server implementations). REST is faster when sending small amount of data (100 and 1000 chars), while SOAP version takes similar time whenever the amount of data (differences between sending a 100 chars string and a 10000 chars string are smaller).

	sending 100 chars	sending 1000 chars	sending 5000 chars	sending 10000 chars
SOAP	5.64 ± 0.17	5.83 ± 0.17	5.61 ± 0.12	5.44 ± 0.08
REST	2.56 ± 0.10	3.45 ± 0.10	5.82 ± 0.55	7.68 ± 0.59

REST is faster when sending small amount of data (100 and 1000 chars), while SOAP version takes similar time whenever the amount of data (differences between sending a 100 chars string and a 10000 chars string are smaller). The limit is about 5000 chars, when both implementations need similar time. Time taken to parse the XML messages in SOAP does not increment with the amount of data sent (as seen in Figure 5.3).

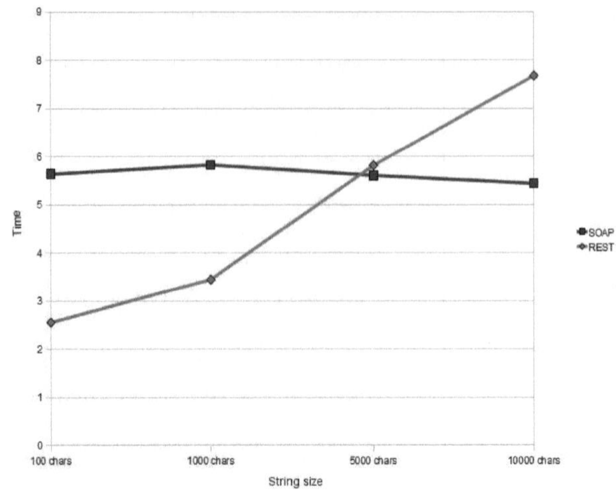

Fig. 5.3. Time taken to complete the first experiment as the string size (amount of data) increases.

5.6.2 Master-Slave Based GA Implementation

In the Experiment 2, we have parallelized a GA following a master-slave model. We do not intend to innovate in terms of the parallel model, but in the implementation (because implementation matters [43]).

As stated before, an ideal client-server implementation of a distributed evolutionary algorithm could be a server process with several threads, in which each thread would include a population. However, as we cannot use a threaded version of the Perl modules, our implementation will focus on the fitness function evaluation.

Thus, the simplest way of task distribution along this model is to evaluate the individual fitness function on the clients and to do the other steps on a master process (as shown in Figure 5.4); this scheme is usually called *farming.*

The evolutionary algorithm has been implemented using the Algorithm::Evolutionary (A::E) library [38, 39]. Version 0.76.2 is used in this work, available under GPL license at `http://opeal.sourceforge.net`

In this experiment, the fitness function is devoted to optimize the function given by equation 5.1, which is plotted in Figure 5.5. Our aim is to find the optimum ($f(0,0) = 1$) with an accuracy of 10^{-6}.

$$f(x,y) = 1 + \frac{sin(\sqrt{x^2+y^2})}{\sqrt{x^2+y^2}} \quad (5.1)$$

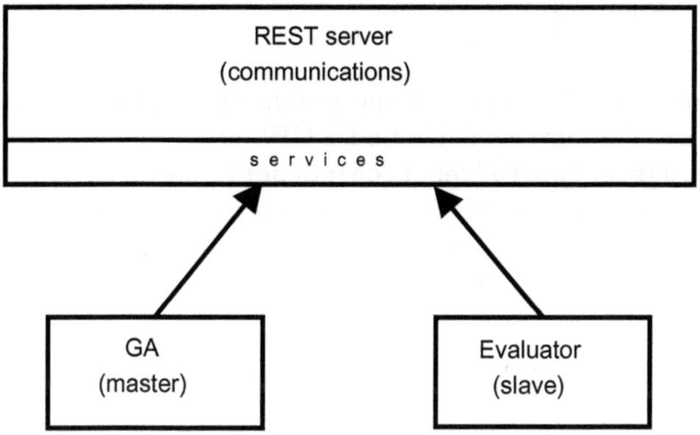

Fig. 5.4. Schema of the master-slave based GA implemented in the second experiment. The master process runs the GA and the slave processes evaluate the fitness function

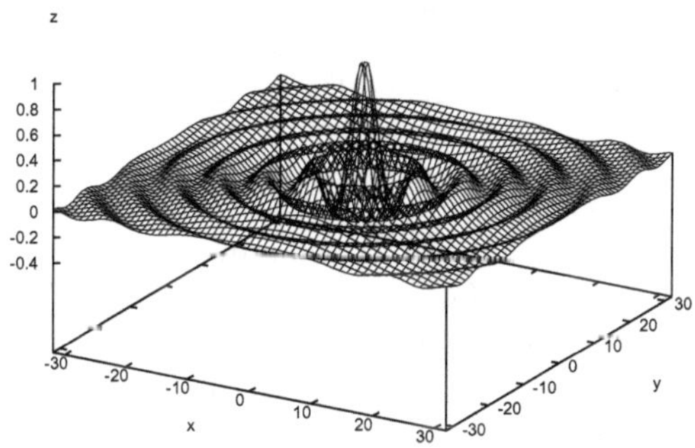

Fig. 5.5. Fitness function representation (given by equation 5.1). The optimum of this function is $f(0,0)=1$

GA individuals are represented using bitstrings (data type A::E::Individual::bitstring). As genetic operators, a bitflip mutation (A::E::Op::Mutation) and a two points crossover (A::E::Op::Crossover) are used.

Remainder GA parameter values are set as follows (default values are used, since we do not intend to find the optimal ones, but to prove feasibility of the implementation, and carry out a comparison):

- Population size = 50
- Generations = 20
- Individual length = 64 bits
- Mutation rate = 20%

- Crossover rate = 80%
- Selection rate = 40%

The full source code (servers, GA and evaluators), links to the A::E library and experiment data are available under GPL at:
`http://atc.ugr.es/pedro/research/webservices/`

As seen in Table 5.3, the REST implementation is faster, due to the SOAP verbosity and time taken to decode the XML messages.

Table 5.3. Results obtained on the second experiment (master-slave implementations). Both implementations obtain good results using even a small number of generations and population size. As far as the running time is concerned, REST implementation is faster in all configurations (10 gen./10 indiv. ; 20 gen./50 indiv. ; 50 gen./50 indiv. ; 100 gen./100 indiv.) as the message size is small.

	10 generat. 10 individ.	20 generat. 50 individ.	50 generat. 50 individ.	100 generat. 100 individ.
SOAP				
accuracy	0.997942 ± 0.000762	0.999867 ± 0.000101	1	1
time (sec.)	3.79 ± 0.42	31.03 ± 1.89	133.08 ± 0.91	264.87 ± 0.39
REST				
accuracy	0.996092 ± 0.004081	0.999976 ± 0.000003	1	1
time (sec.)	2.06 ± 0.08	15.05 ± 1.17	40.75 ± 2.81	100.84 ± 0.55

Both implementations obtain good results in terms of accuracy (both find the optimum with an accuracy of 10^{-6}) using even a small number of generations and population size. As far as the running time is concerned, REST implementation is faster in all configurations (see Table 5.3 and Figure 5.6). It might be due to the XML verboseness of SOAP communications (that increases the time taken to parse the messages). This result was expected taking into account the results obtained in the first experiment (subsection 5.6.1), as the message size in this experiment is small (64 chars).

5.6.3 Master-Slave Based EA Implementation Using Web-Services

In this experiment, as in the previous one, our implementation will focus on the most time consuming operation in G-Prop: the fitness function evaluation. The whole evolutionary algorithm is run on a master process and only the objective function is sent to the slaves for evaluation, following a *farming* model (as shown in Figure 5.4).

The whole system can be sketched as follows:

1. The EA process sends the fitness function code to the REST server and creates the EA population.

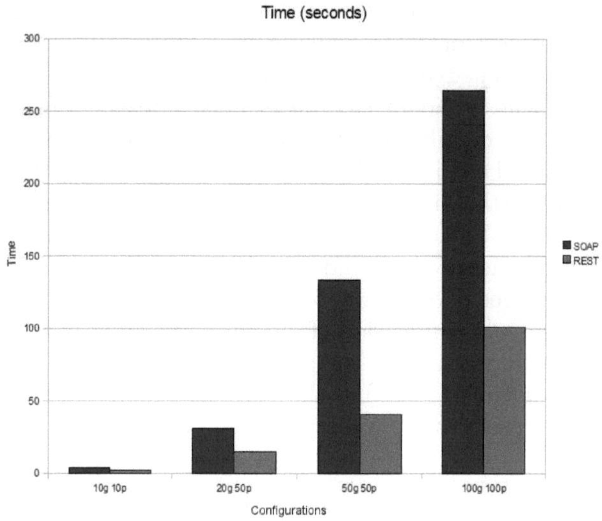

Fig. 5.6. Comparing time (seconds) taken to complete the second experiment for each configuration (10 gen./10 indiv. ; 20 gen./50 indiv. ; 50 gen./50 indiv. ; 100 gen./100 indiv.). As the message size is small, REST implementation yields better results as far as running time is concerned

2. Some clients connect the REST server and load the fitness function sent as Perl code from the server (Figure 5.2 shows an example of server and client processes implementations to upload and download the fitness function source code).
3. The EA process sends non-evaluated individuals to the server.
4. The clients ask for individuals to the server in order to evaluate them.
5. The clients evaluate individuals and send the result back to the server.
6. The EA process obtains evaluated individuals from the server and continues the evolutionary loop.
7. The EA terminates after a fixed number of generations (it sends a termination message throughout the REST server to the clients that remain ready to attend new workloads).

The server in these experiments is mainly used for scheduling and balancing the tasks among the different clients; the network itself is used for communication, but all the interchange of information among clients must be cleared by the central server. However, one of the objectives of the work presented in this chapter has been to create an infrastructure that would get rid of the bottleneck represented by the central server in these experiments.

As in the previous experiments, the EA has been implemented using the Algorithm::Evolutionary (A::E) library.

Again, the full source code (servers, EA and evaluators), links to the A::E library and experiment data are available under GPL at:
`http://atc.ugr.es/pedro/research/webservices/`

In this experiment, we adapt *G-Prop* as a distributed EA using SOAP and REST following the detailed structure. *G-Prop* method has been fully described and analyzed out in previous chapters (see [10, 11]), thus we refer to these chapters for further details. In most cases, evolved MLP should be coded into chromosomes to be handled by the genetic operators, however, G-Prop uses no binary codification, instead, the initial parameters of the network are evolved using specific variation operators such as mutation, multi-point crossover, addition and elimination of hidden units, and QP training applied as operator to the individuals of the population. The EA optimizes the classification ability of the MLP, and at the same time it searches for the number of hidden units (architecture), the initial weight setting and the learning rate for that net.

5.6.3.1 The Experiment

The tests used to assess the accuracy (obtained error) of a method must be carefully selected, since some synthetic problems (also called "toy problems") are not suitable for certain capacities of the BP algorithm, such as generalization [16]. We agree with the view put forward by Prechelt [48], stating that real problems should be used in order to test an algorithm. In real life problems, the division between classes is not as clear as it is in synthetic problems. The dispersion of samples within a single class is also greater, due to noise [41, 42]. In any case, the best way to test the algorithm ability as well as its limitations is to use it to resolve real world problems.

In this chapter, the Glass pattern classification problem is used. This problem was put forward by Prechelt in his chapter *"PROBEN1 – A Set of Benchmarks and Benchmarking Rules for Neural Network Training Algorithms"* [48].

This dataset is based on the glass problem dataset from the UCI library of machine learning databases. This task was prompted by the needs of forensic scientists involved in criminal investigation. The results of a chemical analysis of glass splinters (content of 8 different elements in percentage terms) together with a refractive index, are used in the classification of the sample as either float-processed or non-float processed building windows, vehicle windows, containers, tableware or head lamps. It contains 214 entries. Each sample has 9 attributes plus the class attribute (type of glass): refractive index, sodium, magnesium, aluminum, silicon, potassium, calcium, barium, and iron. This dataset is very difficult to classify due to two important features. First, the number of available patterns is low (214) for six different classes.

Second, the number of patterns in each class is very unbalanced, ranging from 76 (building windows non-float processed) to 9 (tableware).

Dataset was divided into three disjoint parts: one for training, one for validating, and one for testing, as proposed in [48]. In order to determine the fitness of an individual, the MLP was trained by the training set and its fitness was established from the classification error with the validating set. After the EA has finished, i.e. when it has reached the limit of generations, we obtain the generalization ability by using the testing set (previously unseen patterns). This generalization value is shown in tables.

On the other hand, it is very important to know which parameter values involved in the design of an EA have the greatest influence on its behavior and performance. Instead of setting those values by hand, or using the most common values found in the bibliography, or by trial and error [35, 28], authors carried out a statistical study [12] in order to determine the most important parameters (regarding their influence on the results), and to establish the most suitable values for such parameters (thus obtaining an optimal operation). In that study, the ANOVA (ANalysis Of the VAriance) [20, 21] statistical method was used. As a result, running parameter values shown in Table 5.4 were obtained.

Table 5.4. Parameters set using statistical methods (see [12] for details). The number of generations and the population size needed for greater diversity should, of course, be higher or lower depending on the difficulty of the problem.

Parameter	Value
number of generations	500
population size	500
selection rate	20%
initial weights range	[-0.05,0.05]
mutation operator priority	2.0
crossover operator priority	0.5
addition operator priority	1.0
elimination operator priority	0.5
training operator priority	0.5
mutation probability	0.4
weight mutation range	[-0.001,0.001]
learning constant mutation range	[-0.010,0.010]

Time was measured using the "gettimeofday" function in order to achieve a good precision. Time taken to run the EA is reported in Table 5.5. Sequential version of the program was run in the faster machine; and in parallel runs, the EA (master process) was run on the faster machine while the evaluators were run on slower machines.

5.6.3.2 Obtained Results

In this experiment we are not interested on comparing results against other authors, but in using a costly problem that justifies using a farming model.

Results obtained can be shown in Table 5.5.

Table 5.5. Results (error % and time) obtained using both the sequential and the parallel versions (up to 4 evaluators-slaves are used in the farming model). Comparable classification ability is obtained, while time is improved as the number of evaluators is increased.

Model	Error (%)	Time (seconds)
Sequential	33 ± 2	1215 ± 104
Master-slave (using SOAP)		
1 eval.	32.8 ± 1.3	1343 ± 116
2 eval.	32.2 ± 1.5	717 ± 81
3 eval.	31.8 ± 1.3	508 ± 91
4 eval.	31.0 ± 1.6	404 ± 89
Master-slave (using REST)		
1 eval.	32.2 ± 1.6	1517 ± 109
2 eval.	32.4 ± 1.5	804 ± 89
3 eval.	31.4 ± 2.1	566 ± 97
4 eval.	31.6 ± 1.8	447 ± 82

Classification errors show a comparable algorithmic result (see Table 5.5 and Figure 5.7). However, better results in time are obtained parallelizing the problem between several computers.

As can be seen in Figure 5.8, SOAP version is slightly faster REST version, as the amount of data sent in each communication is very high (more than 6000 chars on average). This result is in agreement with the first experiment, where for large amounts of data (big messages), REST communications take longer than SOAP communications.

Figure 5.9 shows that speedup does not equals the number of computers used; however, running time is improved using several computers. Thus, as adding new evaluators (heterogeneous computers running a Perl process) is an easy and costless task, we could take advantage of this system structure to solve costly optimization problems.

In the case of SOAP implementation, speedup is slightly better than in the case of REST, due to the fact that SOAP communications are faster for big messages.

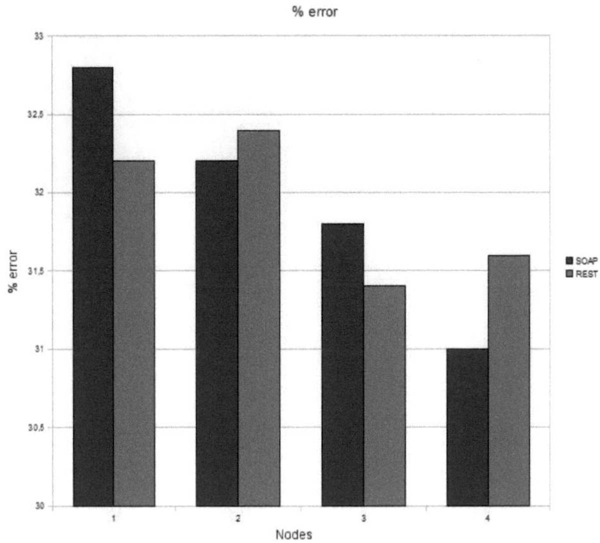

Fig. 5.7. Plot of the classification error (%). As can be seen, the results in error are very similar

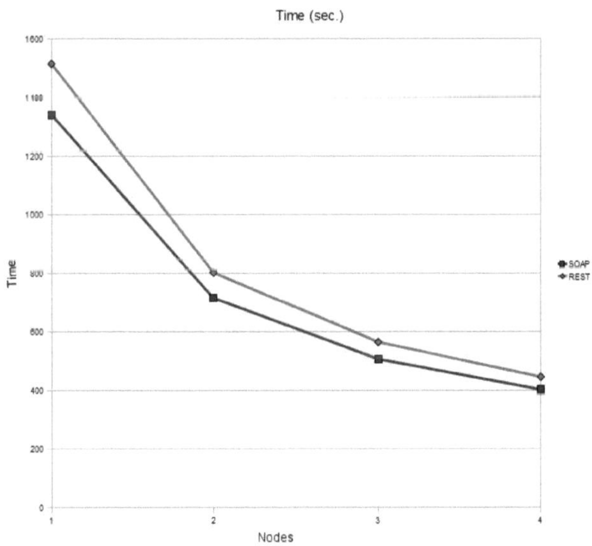

Fig. 5.8. Plot of the running time (seconds) for both the SOAP and REST implementations. As can be seen, the REST version takes a slightly higher time as each communication takes longer than in the SOAP version (the message size is high in this experiment)

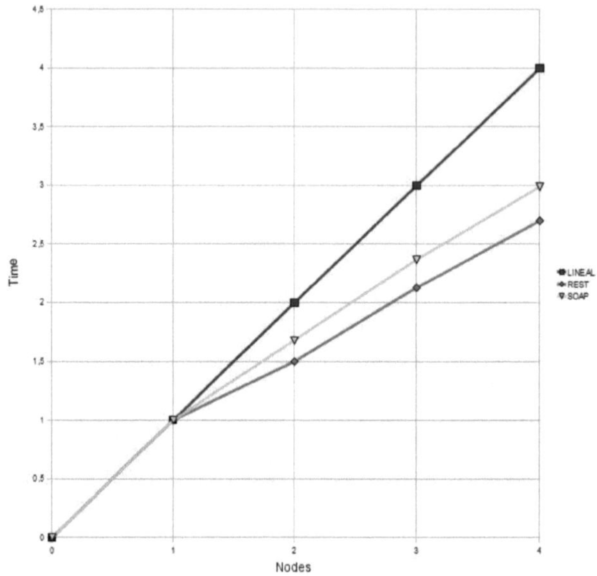

Fig. 5.9. Plot of the speedup obtained using SOAP and REST implementations, and $f(x) = x$ function (solid line). Although speedup is not lineal, it can be seen that running time is improved using several computers as clients dedicated to evaluate the individual fitness functions increase. REST-speedup is a bit lower compared to the SOAP-speedup, due to the fact that REST version takes a slightly higher time

Table 5.6. Classification ability obtained using other methods for the Glass problem.

Scyth [13]	ADAMENN [13]	K-NN [14]	C4.5 [14]	Coop.Ens. [24]	Dzeroski [23]
27.1	24.8	28.0	31.8	22.9±4.8	25.2
Prechelt [48]	**Grönroos [29]**	**ARGEN+ AREPO[37]**	**Cascade Ens.[25]**	**MOBNET [23]**	**Cantú-Paz [8]**
32.08	32±0.5	32.33	27±3	29.6±3.1	32.9

As in this chapter no improvement is made in terms of the algorithm, but in terms of the implementation and the technologies, we are not interested on comparing results against other authors. However, results obtained are comparable to those found in the bibliography [48, 29, 23, 8, 15, 37, 25, 24] (see Table 5.6). In some cases, authors report comparable errors on these problems [13, 23, 14], although no information on the experimental setup is given. Moreover, standard deviation is not reported, and therefore, no comparison can be made in order to apply statistical tests.

5.7 Conclusions

As reported in the experiments provided, both techniques are suitable for developing parallel systems. However, depending on the amount of data sent, one might be heavier than the other.

On another hand, REST technology could not be used to implement a distributed GA following the island model as it does not support asynchronous processing and invocation, while SOAP does support it.

We can conclude that each technology approach has their uses. Moreover, they both have pros and cons. However we can devise some applications/situations where one of them might work better than the other:

- REST is more suitable if...
  - bandwidth and resources are limited
  - stateless *CRUD* (Create, Read, Update, and Delete) operations are needed (operation does not need to be continued)
  - the information can be cached because of the totally stateless operation of the REST approach
- SOAP is a good solution if...
  - the application needs asynchronous processing and invocation
  - the application needs a guaranteed level of reliability and security
  - both sides (provider and consumer) have to agree on the exchange format (rigid specifications)
  - the application needs contextual information and conversational state management

This chapter presents a new parallel-distributed computation implementation using SOAP and REST web services that shows the useful these technologies can be in the field of evolutionary computation.

To implement and use communications using SOAP and REST it is not necessary running virtual machines (as in Java programming), nor daemons, just only to install several libraries available for almost any programming language. Moreover, an arbitrary number of computers (clients-evaluators) can be added to the system, making it more efficient.

In these experiments, we have demonstrated that both can be used as communication protocol for distributed evolutionary computation, obtaining a good speedup. Results obtained are comparable, and only for big messages, REST communications take longer than SOAP communications. Moreover, they provide a common interface that can be called from almost any programming language. Thus, programs can be written in any language and can share data without the need of worrying about the message formats or communication protocols.

At the same time, it does not overload too much the network. Using other distributed systems, such as Jini [56, 3], the network traffic is so high that when a high number of computers are used, communication becomes difficult.

From these implementations, several paths for improvement are devised: changing the models so that more computation is moved to the clients, leaving the server as just a hub for information interchange among clients; that information interchange will have to be reduced to the minimum. That will make this model closer to the island model, with just the migration policies regulated by the server. That way, the server bottleneck is almost eliminated.

As future research, it could be of interest adding support for SOAP and REST to existing distributed evolutionary algorithm libraries, such as JEO [2], EO [34], and libraries in other languages, in order to allow the implementation of multi-language evolutionary algorithms. Another possibility is to test P2P architectures, where each computer communicates only with one or two computers in the network. It would be very interesting to parallelize the proposed method using random topologies, in such a way that a "servent" (server/client) can enter or leave the network at any moment.

Acknowledgements. This work has been supported in part by the CEI BioTIC GENIL (CEB09-0010) MICINN CEI Program (PYR-2010-13 and PYR-2010-29) project, UGR PR-PP2011-5, the FPU Grant 2009-2942, the Junta de Andalucía TIC-3903 and P08-TIC-03928 projects, and the Jaén University UJA-08-16-30 project.

References

1. Abramson, J., Abela, A.: Parallel genetic algorithm for solving the school timetabling problem. In: Proceedings of the Fifteenth Australian Computer Science Conference (ACSC-15), vol. 14, pp. 1–11 (1992)
2. Arenas, M., Foucart, L., Merelo-Guervós, J.J., Castillo, P.A.: JEO: a framework for Evolving Objects in Java. Actas Jornadas de Paralelismo, UPV, Universidad Politécnica de Valencia, 185–191 (2001)
3. Atienza, J., García, M., González, J., Merelo-Guervós, J.J.: Jenetic: a distributed, fine-grained, asynchronous evolutionary algorithm using Jini. In: Wang, P.P. (ed.) Proc. JCIS 2000 Joint Conference on Information Sciences, vol. I, pp. 1087–1089 (2000) ISBN: 0-9643456-9-2
4. Ávila, P.: SOAP: revolución en la red. Linux Actual 19, 55–59 (2001)
5. Box, D.: Inside SOAP (2011), `http://www.xml.com/pub/a/2000/02/09/feature/index.html`
6. Box, D., Ehnebuske, D., Kakivaya, G., Layman, A., Mendelsohn, N., Nielsen, H., Thatte, S., Winer, D.: Simple Object Access Protocol (SOAP) 1.1, W3C Note (May 08, 2000), `http://www.w3.org/TR/SOAP`

7. Cantú-Paz, E., Goldberg, D.E.: Modeling idealized bounding cases of parallel genetic algorithms. In: Koza, J., Deb, K., Dorigo, M., Fogel, D., Garzón, M., Iba, H., Riolo, R. (eds.) Proceedings of the Second Annual Conference on Genetic Programming 1997. Morgan Kaufmann, San Francisco (1997)
8. Cantú-Paz, E., Kamath, C.: Inducing oblique decision trees with evolutionary algorithms. IEEE Trans. Evolutionary Computation 7(1), 54–68 (2003)
9. Castellano, J.G., García-Arenas, M., Castillo, P.A., Carpio, J., Cillero, M., Merelo-Guervós, J.J., Prieto, A., Rivas, V., Romero, G.: Objetos evolutivos paralelos. In: de Granada, D.A.U. (ed.) XI Jornadas de Paralelismo, pp. 247–252 (2000)
10. Castillo, P.A., Carpio, J., Merelo, J.J., Rivas, V., Romero, G., Prieto, A.: Evolving multilayer perceptrons. Neural Processing Letters 12, 115–127 (2000)
11. Castillo, P.A., Merelo-Guervós, J.J., Prieto, A., Rivas, V., Romero, G.: G-Prop: Global optimization of multilayer perceptrons using GAs. Neurocomputing 35, 149–163 (2000)
12. Castillo, P., Merelo, J., Romero, G., Prieto, A., Rojas, I.: Statistical Analysis of the Parameters of a Neuro-Genetic Algorithm. IEEE Transactions on Neural Networks 13(6), 1374–1394 (2002)
13. Domeniconi, C., Peng, J., Gunopulos, D.: Adaptive metric nearest neighbor classification. In: Proc. of IEEE Conf. on CVPR, Hilton Head Island, South Carolina, pp. 517–522 (2000)
14. Duch, W.: Datasets used for classification: comparison of results (2004), `http://www.phys.uni.torun.pl/kmk/projects/datasets.html`
15. Dzeroski, S., Zenko, B.: Is combining classifiers with stacking better than selecting the best one? Machine Learning 54(3), 255–273 (2004)
16. Fahlman, S.: An empirical study of learning speed in back-propagation networks, Tech. rep., Carnegie Mellon University (1988)
17. Fahlman, S.: Faster-Learning Variations on Back-Propagation: An Empirical Study. In: Proceedings of the 1988 Connectionist Models Summer School. Morgan Kaufmann (1988)
18. Fielding, R.: Architectural Styles and the Design of Network-based Software Architectures, Doctoral dissertation, University of California, Irvine (2000)
19. Fielding, R., Taylor, R.: Principled Design of the Modern Web Architecture. ACM Transactions on Internet Technology (TOIT) 2(2), 115–150 (2002)
20. Fisher, R.: Theory of Statistical Estimation. Proceedings of the Cambridge Philosophical Society 22, 700–725 (1925)
21. Fisher, R.: The Comparison of Samples with Possibly Unequal Variances. Annals of Eugenics 9, 174–180 (1936)
22. Fogarty, T., Huang, R.: Implementing the genetic algorithm on transputer based parallel processing systems. Parallel Problem Solving From Nature, 145–149 (1991)
23. García-Pedrajas, N., Hervás-Martínez, C., Munoz-Pérez, J.: Multiobjective cooperative coevolution of artificial neural networks. Neural Networks 15(10), 1255–1274 (2002)
24. García-Pedrajas, N., Hervás-Martínez, C., Ortiz, D.: Cooperative coevolution of artificial neural network ensembles for pattern classification. IEEE Transactions on Evolutionary Computation 9(3), 271–302 (2005)

25. García-Pedrajas, N., Ortíz-Boyer, D., del Castillo-Gomariz, R., Hervás-Martínez, C.: Cascade Ensembles. In: Cabestany, J., Prieto, A.G., Sandoval, F. (eds.) IWANN 2005. LNCS, vol. 3512, pp. 598–603. Springer, Heidelberg (2005)
26. García-Sánchez, P., González, J., Castillo, P.A., Merelo, J.J., Mora, A.M., Laredo, J.L.J., Arenas, M.G.: A Distributed Service Oriented Framework for Metaheuristics Using a Public Standard. In: González, J.R., Pelta, D.A., Cruz, C., Terrazas, G., Krasnogor, N. (eds.) NICSO 2010. SCI, vol. 284, pp. 211–222. Springer, Heidelberg (2010)
27. Goldberg, D.E.: Genetic Algorithms in search, optimization and machine learning. Addison Wesley (1989)
28. Grefenstette, J.: Optimization of control parameters for genetic algorithms. IEEE Trans. Systems, Man, and Cybernetics 16(1), 122–128 (1986)
29. Grönroos, M.: Evolutionary Design of Neural Networks, Master of Science Thesis in Computer Science. Dept. of Mathematical Sciences. University of Turku (1998)
30. Harold, E.R.: XML Bible. IDG Books worldwide (1991)
31. Hauser, R., Männer, R.: Implementation of Standard Genetic Algorithm on Mimd Machines. In: Davidor, Y., Männer, R., Schwefel, H.-P. (eds.) PPSN 1994. LNCS, vol. 866, pp. 504–513. Springer, Heidelberg (1994)
32. IETF: RFC 1945 (2011), `http://tools.ietf.org/html/rfc1945`
33. IETF: RFC 2616 (2011), `http://tools.ietf.org/html/rfc2616`
34. Keijzer, M., Merelo, J.J., Romero, G., Schoenauer, M.: Evolving Objects: A General Purpose Evolutionary Computation Library. In: Collet, P., Fonlupt, C., Hao, J.-K., Lutton, E., Schoenauer, M. (eds.) EA 2001. LNCS, vol. 2310, pp. 231–244. Springer, Heidelberg (2002)
35. Kim, D., Kim, C.: Forecasting time series with genetic fuzzy predictor ensemble. IEEE Transactions on Fuzzy Systems 5(4), 523–535 (1997)
36. Kuchenko, P.: SOAP: Lite (2011), `http://www.soaplite.com`
37. León-Barranco, A., Reyes-García, C.A.: ARGEN + AREPO: Improving the Search Process with Artificial Genetic Engineering. In: Cabestany, J., Prieto, A.G., Sandoval, F. (eds.) IWANN 2005. LNCS, vol. 3512, pp. 637–645. Springer, Heidelberg (2005)
38. Merelo, J.J.: A perl primer for evolutionary algorithm practitioners. Sigevolution 4(4), 12–19 (2010), Doi `http://doi.acm.org/10.1145/1810136.1810138`
39. Merelo-Guervós, J.J., Castillo, P.A., Alba, E.: Algorithm:Evolutionary, a flexible Perl module for evolutionary computation. In: Soft Computing. A Fusion of Foundations, Methodologies and Applications, vol. 14(10), pp. 1091–1109. Springer, Heidelberg (2010)
40. Merelo, J.J., Mora, A., Castillo, P., Laredo, J.L.J., Fernandes, C.: Optimizing evolutionary algorithms at program level. In: Proceedings META 2010, International Conference on Metaheuristics and Nature Inspired Computing (2010)
41. Merelo-Guervós, J.J., Prieto, A.: G-LVQ, a combination of genetic algorithms and LVQ. In: Pearson, N.D.W., Albrecht, R.F. (eds.) Artificial Neural Nets and Genetic Algorithms, pp. 92–95. Springer (1995)
42. Merelo-Guervós, J.J., Prieto, A., Morán, F.: Optimization of classifiers using genetic algorithms, ch. 4, pp. 91–108. MIT press (2001) ISBN: 0262162016

43. Merelo, J.J., Romero, G., Arenas, M.G., Castillo, P.A., Mora, A.M., Laredo, J.L.J.: Implementation Matters: Programming Best Practices for Evolutionary Algorithms. In: Cabestany, J., Rojas, I., Joya, G. (eds.) IWANN 2011, Part II. LNCS, vol. 6692, pp. 333–340. Springer, Heidelberg (2011)
44. Michaeldehaan: XMLRPC vs REST vs SOAP vs all your RPC options (2011), http://bit.ly/MPRXA
45. Oluyede: If XMLRPC is really better than REST its not for there reasons (2011), http://bit.ly/bThgss
46. Papazoglou, M., van den Heuvel, W.J.: Service oriented architectures: approaches, technologies and research issues. The VLDB Journal 16, 389–415 (2007), doi:10.1007/s00778-007-0044-3
47. Pettey, C.B., Leuze, M.R., Grefenstette, J.J.: A parallel genetic algorithm. In: Grefenstette, J.J. (ed.) Proceedings of the Second International Conference on Genetic Algorithms, pp. 155–162. Lawrence Erlbaum Associates (1987)
48. Prechelt, L.: PROBEN1 — A set of benchmarks and benchmarking rules for neural network training algorithms, Tech. Rep. 21/94, Fakultät für Informatik, Universität Karlsruhe, D-76128 Karlsruhe, Germany (1994)
49. Ray, E.T.: Learning XML: creating self-describing data. O´Reilly (2001)
50. Ryman, A.: Understanding web services (2011), http://bit.ly/tPHPOQ
51. soaprpc.com: SOAP software (2011), http://www.soaprpc.com/software
52. Sukrieh, A.: PerlDancer. Documentation (2011), http://perldancer.org/documentation
53. Sukrieh, A.: PerlDancer. The easiest way to write web applications with Perl (2011), http://perldancer.org
54. Tanese, R.: Parallel genetic algorithms for a hypercube. In: Grefenstette, J.J. (ed.) Proceedings of the Second International Conference on Genetic Algorithms, pp. 177–184. Lawrence Erlbaum Associates (1987)
55. Vasudevan, V.: A web services primer (2011), http://oreil.ly/u4tJM6
56. Venners, B.: Jini FAQ (frequently asked questions), http://www.artima.com/jini/faq.html
57. Wikipedia: Representational State Transfer (2011), http://bit.ly/t9Xy1c

Chapter 6
GPU Parallel Computation in Bioinspired Algorithms: A Review

M.G. Arenas, G. Romero, A.M. Mora, P.A. Castillo, and J.J. Merelo

Abstract. As bioinspired methods usually need a high amount of computational resources, parallelization is an interesting alternative in order to decrease the execution time and to provide accurate results. In this sense, recently there has been a growing interest in developing parallel algorithms using graphic processing units (GPU) also referred as GPU computation. Advances in the video gaming industry have led to the production of low-cost, high-performance graphics processing units that possess more memory bandwidth and computational capability than central processing units (CPUs). As GPUs are available in personal computers, and they are easy to use and manage through several GPU programming languages, graphics engines are being adopted widely in scientific computing applications, particularly in the fields of computational biology and bioinformatics. This chapter reviews the use of GPUs to solve scientific problems, giving an overview of current software systems.

6.1 Introduction

General-purpose computing on graphics processing units (GPGPU) is the technique of using a GPU, which typically handles computation only for computer graphics, to perform computation in applications traditionally handled by the CPU. Recently there has been a growing interest in Graphics Processing Unit (GPU) computation. The fact that this kind of processors has the ability to perform restricted parallel processing has elicited considerable interest among researchers with applications requiring intensive parallel computation.

M.G. Arenas · G. Romero · A.M. Mora · P.A. Castillo · J.J. Merelo
Department of Architecture and Computer Technology,
CITIC (University of Granada)
e-mail: {mgarenas,gustavo,amorag,pedro,jmerelo}@geneura.ugr.es

J. Kołodziej et al. (Eds.): Advances in Intelligent Modelling and Simulation, SCI 422, pp. 113–134.
springerlink.com

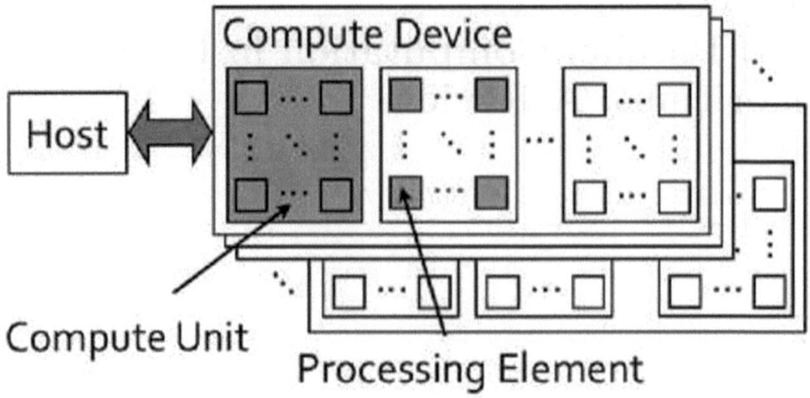

Fig. 6.1. GPUs can be seen as SIMD multi-core processors. Internally the GPU contains a number of small processors that used to perform calculations. Depending on the GPU, the number of threads that can be executed in parallel is in the order of hundreds

GPUs are specialized stream processors, initially useful for rendering graphics applications. Typically, a GPU is able to perform graphics manipulations at a much higher speed than a general purpose CPU, since the graphics processor is specifically designed to handle certain primitive operations which occur frequently in graphics applications. Internally, the GPU contains a number of small processors that are used to perform calculations. Depending on the power of a GPU, the number of threads that can be executed in parallel on such devices is currently in the order of hundreds and it is expected to multiply in a few months. Nowadays, developers can write (easily) their own high-level programs on GPU. Due to the wide availability, programmability, and high-performance of these consumer-level GPUs, they are cost-effective for, not just game playing, but also scientific computing. Now, GPUs are exposed to the programmer as a set of general-purpose shared-memory SIMD (Single Instruction Multiple Data) multi-core processors (see Figure 6.1). This makes these architectures well suited to run large computational problems, such as those from bioinformatics area.

Then, the goal of this article is to review the use of GPUs to solve bioinformatics problems, explaining the general approach to using a GPU and given an overview of currently usable software systems.

To this end, the rest of this chapter is structured as follows: Section 6.2 presents GPUs as highly parallel devices architectures. Section 6.3 gives a background on the different higher level programming languages used to profit GPUs. Finally, Section 6.4 reviews the related works in Physicaly applications on GPUs, followed by a brief conclusion (Section 6.5).

6.2 Throughput, Parallelism and GPUs

Moore's Law describes a long-term trend in the history of computing hardware: the number of transistors that can be placed inexpensively on an integrated circuit has doubled approximately every two years. The trend has continued for more than half a century and is not expected to stop (theoretically until not too many years above 2015). On 2005 Gordon Moore stated in an interview that his law cannot be sustained indefinitely because transistors would eventually reach the limits of miniaturization at atomic levels. Maybe it is time for Koomey's Law [19] to replace Moore's Law. Koomey says that energy efficiency is doubled every 18 months. For fixed computing load, the amount of battery you need will fall by a factor of two every year and a half.

Parallel computation has recently become necessary to take full advantage of the gains allowed by Moore's law. For years, processor makers consistently delivered increases in clock rates and instruction-level parallelism, so that single-threaded code executed faster on newer processors with no modification. Now, to manage CPU power dissipation, processor makers favor multi-core chip designs, and software has to be written in a multi-threaded or multi-process manner to take full advantage of the hardware.

Graphics processors have rapidly matured over the last years, leaving behind their roots as fixed function accelerators and growing into general purpose computational devices for highly parallel workloads. Some of the earliest academic work about GPUs as computational devices date back to University of Washington in 2002[40] and Stanford in 2004[5].

GPUs are similar to multi-core CPUs but with two main differences (see Figure 6.2). CPUs a made for speedup and GPUs for throughput. CPUs try to improve the execution of a single instruction stream while GPUs take the opposite route obtaining benefits from massively threaded streams of instructions and/or data (SIMD). The second difference is how threads are scheduled. The operating system schedule threads over different cores of a CPU in a pre-emptive fashion. GPUs have dedicated hardware for the cooperative scheduling of threads.

Physically GPUs are huge in comparison with CPUs, see Table 6.1. Latest microprocessors from the two main vendors, AMD and Intel, have about 1000 million transistors. Latest GPUs from AMD and NVIDIA are about 3000 million transistors. CPUs draw 130W at most, a limit established by the cost of commodity heat sink and fan. GPUs have increase power consumption and currently are in the neighborhood of 300W. This can be possible with the use of exotic cooling solutions. CPUs are built with the finest technology, read best lithography, while GPUs are made with budget in mind in more common and older processes.

The world of graphics hardware is extremely opaque and does not have standard terminology. Every company have a very different set of words to refer to the same underlying objects and principles. The marketing dictates big numbers and buzzwords instead of clear denominations. Many authors

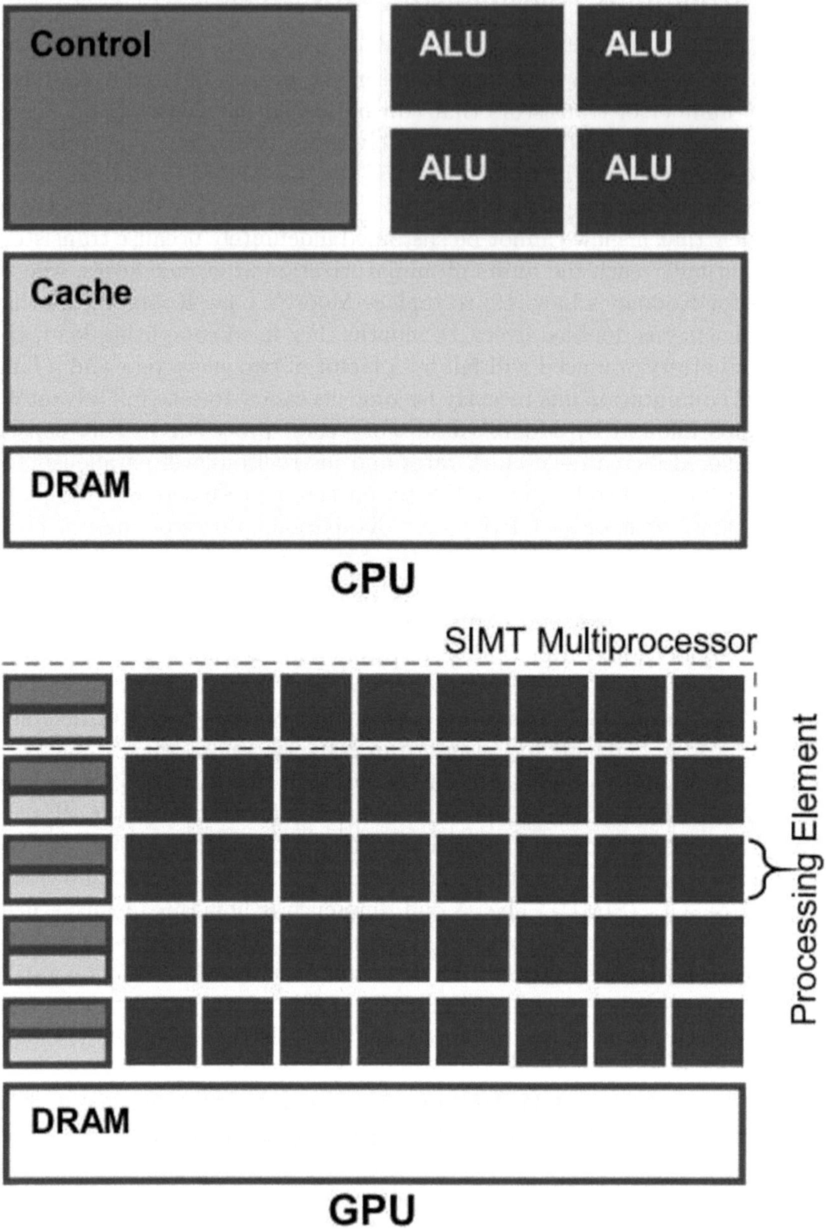

Fig. 6.2. CPU-GPU blocks arrangement: The GPU architecture devotes more transistors to data processing

try to alleviate this lack of standard terms calling 'shader core' to an Intel Execution Unit (EU), an AMD Single Instruction Multiple Data (SIMD) or an NVIDIA Streaming Multiprocessors (SM). Any of these refer to a single processor core inside of the GPU that can fetch, decode, issue and execute several instructions. The shader core is composed of several 'execution units' or EU that can execute an individual vector operation equivalent to an AVX or SSE instruction. AMD call this kind of EU streaming processor (SP) and NVIDIA CUDA core. With this in mind the next Table 6.1 can be understood more easily.

Table 6.1. CPU, GPU and APU comparison of best commodity desktop hardware available nowadays. When a slash is used it refers to CPU/GPU parts respectively.

manufacturer & model	transistor count (million)	die size (mm^2)	shader cores (ALUs)	clock rate (GHZ)	memory bandwidth (GB/s)	GFLOPS (single precision)	TDP (W)	price (€)
AMD Phenom II X6 1100T	758	258	6	2.6-3.7	15.6	57.39	125	175
Intel Core i7 990X	1170	240	6	3.46-3.73	24.5	107.58	130	950
AMD A8-3850	758	258	4/400	2.9/0.6	29.8	355	100	135
Intel Core i7 2600K	995	216	4/48	3.4/0.85	24.5	129.6	95	317
AMD Radeon HD 6970	2640	389	1536	0.88	176	2703	250	350
NVIDIA GeForce GTX 580	3000	520	512	1.544	192.4	1581.1	244	450

Nowadays NVIDIA has the biggest GPU with superior performance in some workloads. AMD choose a very different compromise in the design of the GPU. AMD has more execution units but its memory hierarchy is weaker. This way software bounded by memory bandwidth or with strong ordering interdependencies prefers NVIDIA hardware. On the other side, loads capped by pure ALU execution power use are faster on the AMD hardware.

To make things a little more "heterogeneous" now we can buy a CPU with an integrated GPU. AMD and Intel have just started selling this kind of combined processor and graphics card. The new term APU (Accelerated Processing Unit) has been coined for this kind of chips. The architectural names are Llano for AMD and Sandy Bridge for Intel. There are many reviews online these days. Most of them points that AMD's CPU cores are slower than Intel's ones but their GPU is faster. Which combination is better is not an easy question to answer. It must be backed by specific benchmarks, or better, the real application that we want it to run.

Over time the use of GPUs has pass from odd to common in our systems. Actually several time consuming processes has been parallelized inside our operating systems such as web page rendering. The only problem is that the speedup that graphics hardware can bring to us is not free. Every application that we want to accelerate must be rewritten. Furthermore, parallel software is not famous for been the easy easiest one to write.

As an image is worth a thousand word, Figure 6.3a will show the internals of NVIDIA GF100 architecture. Every green square is an NVIDIA SM. The six blue squares on the sides are memory interfaces. As every one is 64-bit, the bus is 384 bits wide and can transfer 192.4GB/s connected to GDDR5 memory chips. On the right, Figure 6.3b, we can see a disclosed SM with its 32 CUDA cores. In total there are 512 cores. The maximum theoretical throughput in simple precision is 1581.1GFLOPS.

Last AMD GPU offer has a different architecture called Cayman as can be seen in Figure 6.4a. It has 24 SIMDs processors (orange squares). Every SIMD is composed by 16 SPs (red rectangles). Finally every SP is a 4-wide Very Long Instruction Word processor or VLIW4. This way 1536 instructions can be finalized every clock. Cayman has for 64-bit dual channel memory controllers (gray rectangles at the bottom of Figure 6.4b) connected to 2 GDDR5 memory channels for a total bandwidth of 176GB/s. The maximum theoretical throughput in simple precision for this AMD design is 2703GFLOPS.

As CPU makers did some years ago, passing from single core to symmetric multiprocessing system (SMP), and more recently to multicores, GPU makers follow the same trend. We can connect more than one graphic card to our computer to improve its GPU capacities or buy a card with 2 graphic chips inside. GPUs are so much powerful than CPUs that even a small cluster of a few GPUs can be faster than classic, and much more expensive, big cluster of processors. First cluster of this kind appear in the scientific literature from 2004 [10] with big success. Even some people have small GPU clusters at home with a couple of mighty graphics cards just to game. Connecting 2, 3 or 4 graphics card is called CrossFire by AMD and Scalable Link Interface (SLI) by NVIDIA.

6.3 GPUs Programming

6.3.1 Programming Model

The way GPUs can be exploited is deeply rooted on its hardware. There exists several APIs. Every company has a proprietary one tied to their respective products. This way AMD started with Close to Metal and NVIDIA with CUDA. Over time an standard appear, OpenCL.

With respect to the programming tools which available for developers, most the Application Program Interfaces (APIs) are based on C-like languages, but having some restrictions to improve the parallel execution, such as no recursion or limited pointers. Some of them use the open source compiler LLVM [17] from University of Illinois.

From 2003 the two main GPU developers, ATI an NVIDIA, started selling hardware solutions that need to be programmed with proprietary APIs.

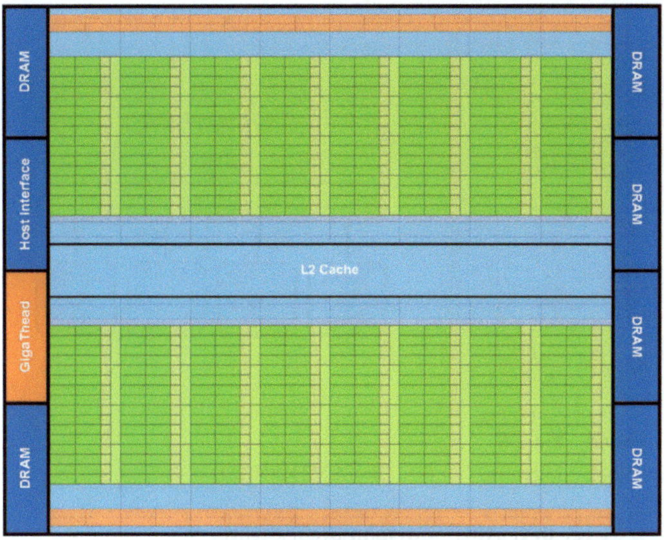

(a) NVIDIA GF100 block diagram

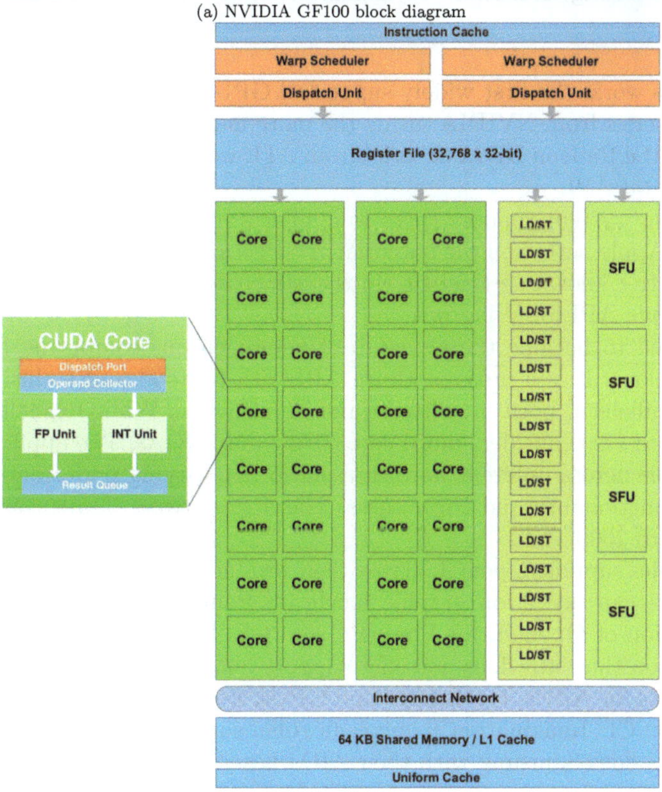

(b) NVIDIA SM

Fig. 6.3. NVIDIA GF100 architecture.

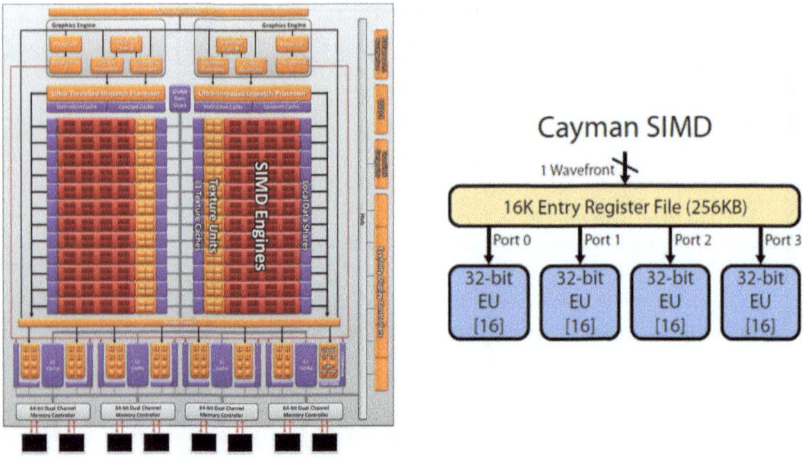

(a) AMD Cayman block diagram. (b) AMD SIMD.

Fig. 6.4. AMD Cayman architecture.

Despite previous work, the first widely supported GPUs were DX10 generation GeForce 8 series from NVIDIA, using the more mature *CUDA* API. On the other hand, the Radeon HD2xxx series from ATI, were programmed with the *Close To Metal* API.

From operating system vendors there were efforts in the same direction. Some people at Apple betted on the potential of GPUs and started developing an open API, latter known as *OpenCL*. In the same time, Microsoft created the *DirectCompute* API for Windows.

OpenCL aimed to became the OpenGL of heterogeneous computing for parallel applications. It is a cross-platform API with a broad and inclusive approach to parallelism, both in software and in hardware. While explicitly targeting GPUs, it also considers multi-core CPUs and FPGAs. The applications are portable across different hardware platforms, varying performance while keeping functionality and correctness. The first software implementations date back to 2009.

Most companies support OpenCL across their products. Apart from AMD and NVIDIA we can use it on graphic hardware from S3 and VIA. Also IBM has a version of OpenCL for PowerPC and CELL processors. Intel is the only exception that still does not offer support but they will do in their next architectures for APUs, Ivy Bridge, and GPUs. Embedded world is also interested in OpenCL. Imagination Technologies offer support for the SGX545 graphics core. As does Samsung with their ARM based microprocessors.

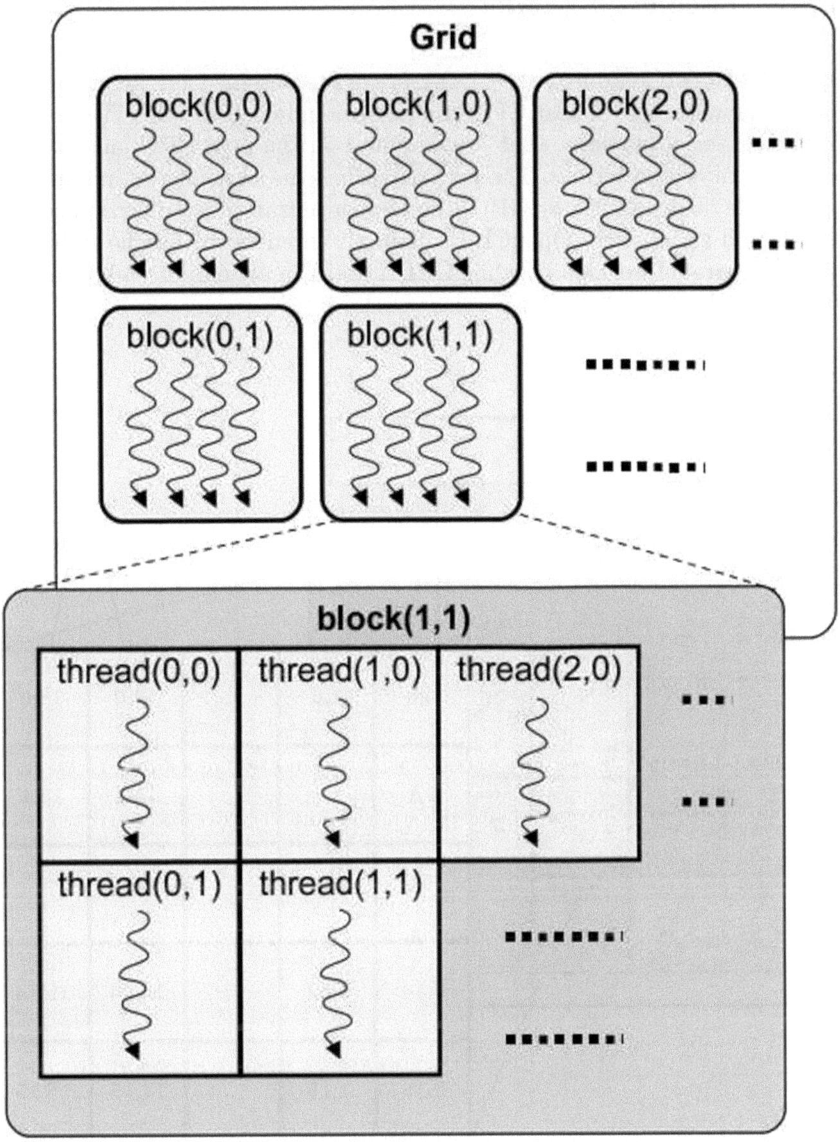

Fig. 6.5. Hierarchy of computing structure in a GPU

6.3.2 Execution Model

OpenCL, DirectCompute and CUDA are APIs designed for heterogeneous computing with both a host CPU and an optional GPU device. The applications have serial portions, that are executed on the host CPU, and parallel portions, known as *kernels*. The parallel kernels may execute on an OpenCL compatible device (CPU or GPU), but synchronization is enforced between kernels and serial code. OpenCL is distinctly intended to handle both task and data parallel workloads, while CUDA and DirectCompute are primarily focused on data parallelism.

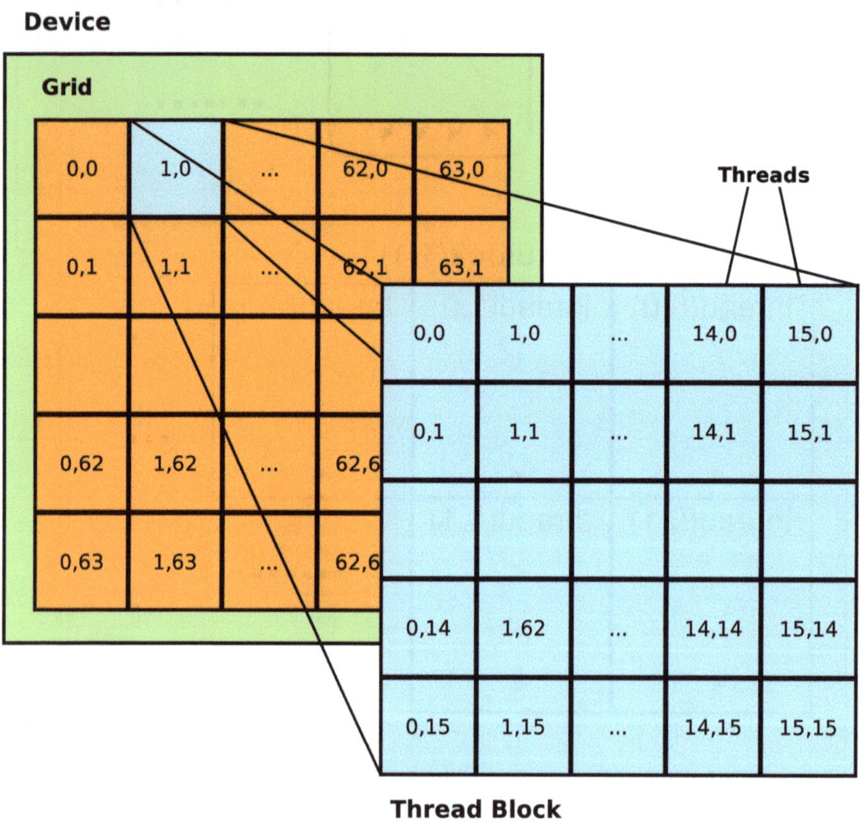

Fig. 6.6. Execution model: Each piece of data is a work-item (thread); a kernel has thousands of work-items and is organized into many work-groups (thread blocks); each work-group process many work-items

A kernel applies a single stream of instructions to vast quantities of data that are organized as a 1-3 dimensional array (see Figures 6.5 and 6.6). Each piece of data is known as a work-item in OpenCL terminology, and kernels may have hundreds or thousands of work-items. The kernel itself is organized into many work-groups that are relatively limited in size; for example a kernel could have 32K work-items, but 64 work-groups of 512 items each.

Unlike traditional computation, arbitrary communication within a kernel is strongly limited. However, communication and synchronization is generally allowed locally within a work-group. So work-groups serve two purposes: first, they break up a kernel into manageable chunks, and second, they define a limited scope for communication.

6.3.3 Memory Model

The memory model defines how data is stored and communicated within a device and between the device and the CPU. DirectCompute, CUDA and OpenCL share the same four memory types (with different terminology):

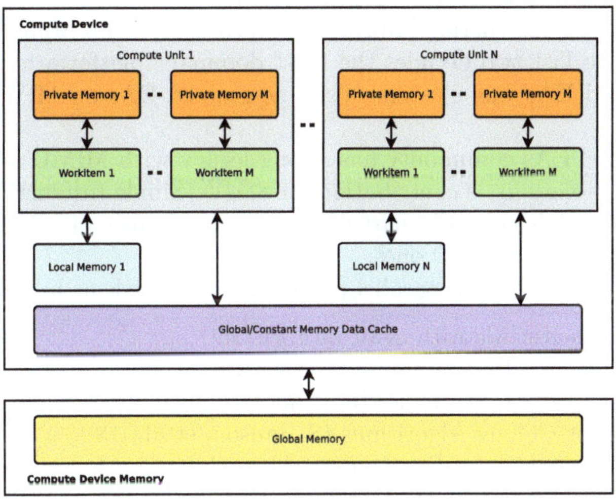

Fig. 6.7. Memory model defines how the data is stored and communicated between CPU and GPU. Global memory is RW for both CPU and work-items; constant memory is RW for CPU and RO for work-items; private memory is RW for a single work-item; local memory is RW for a work-group

- Global memory: it is available for both read and write access to any work-item and the host CPU.
- Constant memory: is a read-only region for work-items on the GPU device, but the host CPU has full read and write access. Since the region is read-only, it is freely accessible to any work-item.

- Private memory: is accessible to a single work-item for reads and writes and inaccessible for the CPU host. The vast majority of computation is done using private memory, thus in many ways it is the most critical term of performance.
- Local memory: is accessible to a single work-group for reads and writes and is inaccessible for the CPU host. It is intended for shared variables and communication between work-items and is shared between a limited number of work-items.

6.4 Bioinpired Methods on GPUs

This section reviews different bioinspired approaches using GPUs found in bibliography, focusing mainly on Evolutionary Computation (EC) and Artificial Neural Networks (ANN).

Alba et al. [1] reviewed and surveyed parallel metaheuristics on Evolutionary Computation. They identified the majority of paradigms to be hosting parallel/distributed EAs, according to Flynn's taxonomy, to fall under the MIMD (Multiple Instruction Multiple Data) category. This argument is fairly valid as in the last two decades the most dominant platform hosting parallel/distributed EAs was clusters (also fine-grained EAs on MPP are wildly used).

The parallel EAs community has a long legacy with MIMD architectures compared to a very little contribution for SIMD (Single Instruction Multiple Data) system. This comes in part due to the dominance of MIMD architectures as compared to SIMD ones.

Alba classifies the main parallel metaheuristic models as follow:

- Parallel Genetic Algorithms (Cantú Paz [6])
- Parallel Genetic Programming (F. FernÃ¡ndez, et al.[11]).
- Parallel Evolution Strategies (G. Rudolph [38]).
- Parallel Ant Colony Algorithms (S. Janson, et al. [18]).
- Parallel Estimation of Distribution Algorithms (J. Madera, et al. [25]).
- Parallel Scatter Search (F. Garcia, et al. [12]).
- Parallel Variable Neighborhood Search (F. García-lópez, et al.[13]).
- Parallel Simulated Annealing ([14]) .
- Parallel Tabu Search (T. Crainic, et al.[8]).
- Parallel Greedy Randomized Adaptive Search Procedures (M. Resende and C. Ribeiro [37]).
- Parallel Hybrid Metaheuristics (C. Cotta, et al. [7]).
- Parallel MultiObjective Optimization (A. Nebro, et al. [31]).
- Parallel Heterogeneous Metaheuristics (F. Luna, et al. [2]).

Nevertheless, when the research community use GPGPU (General-Purpose Computing on Graphics Processing Units) the authors have developed algorithms using three parallel approaches, master-slave model, fine-grained model [20], coarse-grained model [26] [35] or hybridations that use two or more of the previous parallel approaches in a hierarchical way. All the EC approaches on GPUs are parallel, thus a classification depending on the parallel model used is presented in this section. We will focus on master-slave [48], fine-grained (cellular EAs), coarse-grained (Island Model or Deme Model) approaches; and a hierarchical model [48].

As far as the ANN approaches, although the computation for ANN is inherently parallel, many algorithms require some steps that are difficult to parallelize on the GPU.

Most of these methods can be used with almost zero effort by using existing frameworks. The most complete and comprehensive review is from Parejo [33]. It is a comparative study of metaheuristic optimization frameworks. As criteria for comparison a set of 271 features grouped in 30 characteristics and 6 areas has been selected. These features include the different metaheuristic techniques covered, mechanisms for solution encoding, constraint handling, neighborhood specification, hybridization, parallel and distributed computation, software engineering best practices, documentation and user interface.

6.4.1 Master-Slave Approaches

Master-slave Evolutionary Algorithms are usually used for problems involving expensive to compute fitness function, where the master node runs the entire algorithm while the slaves execute the fitness evaluations. Hence, master-slave implementations are more efficient as the evaluations become more expensive and contribute a bigger portion in total runtime of algorithm.

Wong et al. [46] proposed an EP algorithm for solving five simple test functions, called Fast Evolutionary Programming (FEP). In this master-slave approach, some actions are executed in the CPU (main loop of the algorithm and crossover operator), while evaluation and mutation are run in the GPU (no need of external information). The competition and selection of the individuals are performed on the CPU, while mutation, reproduction and evaluation are performed on the GPU. In this case, the reproduction step implies interaction among, at least, two individuals. A maximum speedup of x5.02 is obtained when the population size increases. This is the most common organization in GPU implementations, since no interaction among individuals is required during the evaluation, so this step can be fully parallelizable.

A GP method proposed by Harding and Banzhaf [16] uses the GPU only for performing the evaluation, while the rest of the steps of the algorithm are run on the CPU. The authors tested real-coded expressions until 10000 nodes, boolean expressions until 1500 nodes, and some real world problem where they evaluate expressions until 10000 nodes. In some cases, the results yielded speedup of thousand times.

Zhang et al. [48] adapt different EAs to a GPU using CUDA. The authors implemented an hierarchical parallel genetic algorithm using a deme model at the high level, and a master-slave schema at the low level. In this implementation, the CPU initializes the populations and distributes them to thread blocks in shared memory. Then, GPU threads within each block run a GA independently (selection, crossover, mutation and evaluation), and migrates individuals to other thread blocks in its neighborhood. In this case, no speedup results were reported.

Recently, there are other papers related with this approach, like Tsutsui et al. paper [42]. This chapter uses a master-slave approach with an ACO algorithm [9] with Tabu Search [15]. Tsutsui uses an Intel Core i7 965 (3.2 GHz) processor and a single NVIDIA GeForce GTX480 GPU. They compare CPU and GPU implementations with and the results showed that GPU computation with MATA (an efficient method for thread assignment cost which they call Move-Cost Adjusted Thread Assignment) showed a promising speedup compared to computation with CPU.

6.4.2 Fine-Grained Approaches

Traditionally, fine-grained Evolutionary Algorithms or Cellular Evolutionary Algorithms (cEAs) have not received as much attention as other types of EAs. This is mainly due to the necessity of special hardware (i.e. a relatively large supply of processors in the underlying architecture). On the contrary, the legacy of parallel/distributed architectures has shown dominance of loosely coupled systems which are not adequate for fine-grained EAs. The reason behind that is the high cost of building massively parallel architectures which normally attracted fewer researchers to work on one grained EAs. A review of the trends in parallel architecture prophesizes a radical change in position of fine-grained EAs among other parallel EAs. This is due to three reasons:

- Growing trend of massive number of processors on chip or card.
- The very high inter-processors speed which is a major factor affecting the efficiency of fine-grained EAs.
- The huge drop of cost of these architectures which will attract a wide base of researchers and developers.

For this kind of algorithms, each individual is the parent at the beginning of the algorithm wile the second parent is selected by applying some selection function for its neighboring. As a result cEAs provide automatic niching effect, avoiding an early convergence.

Yu [47] recommends the use of GA with 2D structured population also called cellular Genetic Algorithm (cGA) for implementation of fine-grained GAs over a GPU (SIMT architecture). The 2D structure of cGA maps well to the GPU architecture.

In this scheme, Wong et al. [44, 45] proposed a parallel hybrid GA (HGA) where the whole evolutionary process is run on the GPU, and only the random number generation is done in CPU. Each GA individual is set to each GPU, and each one selects probabilistically an individual in its neighborhood to mate with. Just one offspring individual is generated, and replaces the old one in that GPU. The authors compare HGA with a standard GA run in a CPU and the FEP [46] algorithm. Using a new pseudo-deterministic selection method, the amount of random numbers transferred from the CPU is reduced. HGA reaches speedup of 5.30 when compared against the sequential version.

Yu et al. [47] implemented the first real cellular EA using GPU, for solving the Colville minimization problem.

They place the population in a toroidal 2D grid and use the classical Von Newmann neighborhood structure with five cells.

They store chromosomes and their fitness values in texture memory on the graphic card, and both, fitness evaluation and genetic operations, are implemented entirely with fragment programs executed in parallel on GPU. Real-coded individuals of a population are represented as a set of 2D texture maps. $BLX-\alpha$ crossover and non-uniform mutation are run as tiny programs on every pixel at each step in a SIMD-like fashion, solving some function optimization problems and reaching a speedup of x15 with a population of 512x512 individuals. They store a set of random numbers at the beginning of the evolution process to solve the random number generation problem when using GPU processors.

Luo et al. [22] implemented a cellular algorithm on GPU to solve three different SAT problems

using a greedy local search (GSAT) [39] and a cellular GA (cGA). They saved local minimums using a random walk strategy, jumping to other search space location.

The cellular GA adopts a 2D toroidal grid, using the Moore neighborhood, stored on texture GPU memory. This implementation generates the random numbers in the GPU (using a generated seed on the CPU at the beginning of the process).

The GPU version reduces in about 6 times the running time.

Li et al. [20] proposed a cellular algorithm on GPU for solving some com mon approximation functions. The authors reported experiments using big populations (up to 10000 individuals) reaching speedups of x73.6 for some implementations.

In [21] the authors propose a fine-grained parallel immune algorithm (FGIA) based on GPU acceleration, which maps parallel IA algorithm to GPU using CUDA. The results show that the proposed method (even increasing the population size) reduces running time.

Alba et al. [43] use CUDA and store individuals and their fitness values in the GPU global memory. Both, fitness evaluation and genetic operators, are run on GPU (no CPU is used). They use a pseudo random number generator provided by the SDK of CUDA named Merseinne Twister. Their experiments include some general discrete and continuous optimization problems, and they compare physical efficiency and numerical efficacy with respect to CPU implementation.

6.4.3 Coarse-Grained Approaches (Island Model)

Coarse grained algorithms are the most common among parallel EAs. Generally, coarse-grained algorithms require less tightly coupled parallel architectures, as compared to fine-grained. Coarse grained EAs divide the main population into sub-populations (also known as Islands) where the sub-populations evolve concurrently. This basic feature of coarse-grained EAs hits a physical limit of GPUs.

In order to run a coarse-grained EA using a GPU, several kernels simultaneously are run where each kernel handles a sub-population is not possible. This limitation of GPU would mean the conventional mechanics of coarse-grained EAs would need to be changed if GPU would be used.

With regard to the last topology (fine-grained EAs), one of the first island models on GPU approaches was published on the GPU competition of GECCO 2009 [34]. It presents some technical details of an island model entirely hard-coded on GPU, with a ring-like topology. Nevertheless, the evolutionary operators implemented on GPU are only specific to the GECCO competition, and the validity of the experiments just works on a small number of problems.

Tsutsui et al. [41] propose run a coarse-grained GA on GPU to solve the quadratic assignment problem (QAP) using CUDA. This is one of the hardest optimization problems in permutation domains. Their model generates the initial population on CPU and copied it to the GPU VRAM; then, each subpopulation in a GPU (NVIDIA GeForce GTX285) is evolved. At some generations, individuals in subpopulations are shuffled via the GPU VRAM. Results showed a speedup from x3 to x12 (using eight QAP instances), compared to the Intel i7 965 processor.

The model by Luong et al. [24] is based on a re-design of the island model. Three different schemes are proposed: The first one implements a coarse-grained EA using a master-slave model to run the evaluation step on GPU. The second one distributes the EA population on GPUs, while the third proposal extends the second one using fast on-chip memory. Second and third approaches reduce the CPU/GPU memory latency, although their parameters (number of islands, migration topology, frequency and number of migrants) must be adapted to the GPU features. Sequential and parallel implementations are compared, obtaining a speedup of x1757 using the third approach.

Pospíchal et al. [35, 36] propose a parallel GA with island model running on GPU. The authors map threads to individuals, thus, threads-individuals can be synchronized easily in order to maintain data consistency, and on-chip hardware scheduler can swiftly swap existing islands between multiprocessors to hide memory latency. Fast, shared memory within the multiprocessor is used to maintain populations. Since the population size is limited to 16KB per island on most GPUs, if the population is larger, slower main memory has to be used. The migration process is based on an asynchronous unidirectional ring that requires an inter-island communication (slower main memory has to be used). The authors report speedups up to 7000 times higher on GPU compared to CPU sequential version of the algorithm.

6.4.4 Hybrid Approaches

The hybrid model simply utilizes two or more of the master-slave, coarse-grained and fine-grained in a hierarchical method. At the higher level, an island model algorithm runs while at the lower level the demes (another name for sub-populations) themselves are running in parallel (fine-grained, master-slave or even another island model with high migration rates). Hybrid models are not the most common in legacy EAs due to:

- The need for additional new parameters to account for a more complex topology structure.
- The need for hierarchal parallel architectures to host hybrid algorithms, while such hierarchal parallel architectures are not so common.
- The high complexity of programming such models.

Nevertheless, for the GPUs, hybrid EAs are a perfect candidate to exploit the hierarchal memory and flexible block sizing in GPUs by well structured populations.

The design and implementation of a hybrid EA plus local search to solve MAX-SAT over GPUs was thoroughly discussed in [30]. Manuwar et al. uses a hierarchical algorithm of 2D structured sub-populations arranged as islands in a 2D grid. Therefore, each individual has 4 neighboring individuals (north, south, east and west) and each sub-population has 4 neighboring sub-populations (north, south, east and west). Instead of using a conventional algorithm for migration between the sub-populations, they introduced a new technique that they call diffusion. Diffusion is more suitable for implementation of cGAs based pGA over a GPU. In the proposed implementation, the host processor (CPU) acts as a controller while an nVidia Tesla C1060 GPU provides the required computational resources. All the configurations, memory allocations, initializations are performed over the host processor. After the initialization stage, data is transferred to the device and the code enters a loop. The loop keeps on repeating until the maximum number of generation criteria is satisfied. Results were collected over a system with nVidia Tesla C1060 GPU mounted on a motherboard with Intel $Core^{TM}i7$ 920@2.67GHz as the host CPU. C1060 have 4GB of device memory, 30 streaming MultiProcessors (MPs), and the total number of processing cores is 240. The maximum amount of shared memory per block is 16KB and clock rate is 1.30GHz. They compare the results of the algorithms over nVidia with several optimizations for local search, mutation, recombination, selection and diffusion (migration) with different implementations using serial implementation, OpenMP implementation over Intel and over Ultra Spark architectures. The found that the maximum speedup is for larger problems, and it is up to 25x if compared the serial implementation over Intel Core 2 Duo 3.3GHz (Sduo) with the NVidia implementation.

6.4.5 Artificial Neural Networks Implementations on GPUs

Artificial neural networks attempt to capture the adaptability of biological neurons in a mathematical model for information processing. ANNs are very powerful tools that are highly parallelizable but also computationally expensive and match well with the GPU computing architecture. As a workhorse of the computational intelligence field, there exists a high demand for this acceleration. As a highly analytic structure, neural networks can be reduced to a series of matrix operations, and thus are easily parallelized, as the GPU is highly optimized to perform these kinds of operations [29].

Several authors provide tips for ensuring efficient implementation of algorithms on GPU's [32, 23, 27, 28].

Zhongwen uses the GPU to first extract a set of characteristics from image data, then applies a pre-trained MLP to these characteristics for classification [23].

Bernhard [3] proposes a different approach, implementing spiking neural networks for image segmentation.

In general, significant performance gains can be elicited from implementing neural network algorithms on graphics processing units. However, these implementations are difficult to obtain.

Finally, several developers provide libraries and tools to help practitioners to develop ANN applications on GPUs [4].

6.5 Conclusions

Parallel EAs have been using traditional clusters and MPP for over two decades. However, in this decade other architectures like GPUs are becoming increasingly adopted for general purpose parallel processing. As the legacy parallel EAs were not designed mainly for data-parallel architectures, current implementations of parallel EAs, if ported with outchange, show minor performance gains even with the high throughput of GPUs.

In this chapter we have reviewed the use of GPUs to implement bioinspired algorithms to solve optimization problems. We have commented the GPU computing general approach, and given an overview of currently usable programming languages and software tools.

Most of the bio-inspired methods use the GPU mainly to speed up just the fitness evaluation (usually the most time-expensive process). In most of the EC approaches, competition and selection are performed by CPU, while fitness evaluation, mutation and reproduction are performed on GPU (which is a massively parallel machine with shared memory). GPU allows processors to communicate with any other processors directly, thus more flexible fine-grained algorithms can be implemented on GPU.

In general, approaches found in literature obtain speedups up to several thousands times higher on GPU compared to CPU sequential versions of the same algorithms.

However, as the programming tools improve, newer EC approaches run the whole optimization algorithm on the GPU side, with no need of CPU interaction.

Acknowledgements. This work has been supported in part by the CEI BioTIC GENIL (CEB09-0010) MICINN CEI Program (PYR-2010-13 and PYR-2010-29) project, UGR PR-PP2011-5, the Junta de Andalucía TIC-3903 and P08-TIC-03928 projects, and the Jaén University UJA-08-16-30 project.

The authors are very grateful to the anonymous referees whose comments and suggestions have contributed to improve this chapter.

References

1. Alba, E.: Parallel Metaheuristics: A New Class of Algorithms. Wiley (2005) ISBN: 978-0-471-67806-9, http://eu.wiley.com/WileyCDA/WileyTitle/productCd-0471678066.html
2. Alba, E., Nebro, A.J., Luna, F.: Advances in parallel heterogeneous genetic algorithms for continuous optimization. International Journal of Applied Mathematics and Computer Science 14(3), 101–117 (2004)
3. Bernhard, F., Keriven, R.: Spiking neurons on gpus. In: International Conference on Computational Science, Workshop General Purpose Computation on Graphics Hardware (GPGPU): Methods, Algorithms and Applications, Reading, UK (2006)
4. billconan, kavinguy: Ann libraries to develop on gpus (2011), http://www.codeproject.com/KB/graphics/GPUNN.aspx
5. Buck, I., Foley, T., Horn, D., Sugerman, J., Fatahalian, K., Houston, M., Hanrahan, P.: Brook for gpus: stream computing on graphics hardware. ACM Trans. Graph. 23, 777–786 (2004), Doi http://doi.acm.org/10.1145/1015706.1015800
6. Cantú-Paz, E.: A survey of parallel genetic algorithms. Calculateurs Paralleles, Reseaux et Systems Repartis 10 (1998)

7. Cotta, C., Talbi, E.-G., Alba, E.: Parallel hybrid metaheuristics. In: Parallel Metaheuristics, a New Class of Algorithms, pp. 347–370. John Wiley (2005)
8. Crainic, T.G., Gendreau, M.: Towards a taxonomy of parallel tabu search heuristics (1997)
9. Dorigo, M., Di Caro, G.: The ant colony optimization meta-heuristic, pp. 11–32. McGraw-Hill Ltd., UK (1999), http://dl.acm.org/citation.cfm?id=329055.329062
10. Fan, Z., Qiu, F., Kaufman, A., Yoakum-Stover, S.: Gpu cluster for high performance computing. In: SC Conference, p. 47 (2004), Doi http://doi.ieeecomputersociety.org/10.1109/SC.2004.26
11. Fernández, F., Tomassini, M., Vanneschi, L.: An empirical study of multipopulation genetic programming. Genetic Programming and Evolvable Machines 4, 21–51 (2003), http://dx.doi.org/10.1023/A:1021873026259, doi:10.1023/A:1021873026259
12. García-López, F., Melián-Batista, B., Moreno-Pérez, J.A., Moreno-Vega, J.M.: Parallelization of the scatter search for the p-median problem. Parallel Computing 29(5), 575–589 (2003), http://www.sciencedirect.com/science/article/pii/S01678191030%00437, doi:10.1016/S0167-8191(03)00043-7
13. García-López, F., Melián-batista, B., Moreno-pérez, J.A., Moreno-vega, J.M.: The parallel variable neighborhood search for the p-median problem. Journal of Heuristics 8, 200–222 (2004)
14. Genetic, D.B., Miki, M., Hiroyasu, T., Yoshida, T., Fushimi, T.: Parallel simulated annealing with adaptive temperature. In: Proceedings of IEEE International Conference on Systems, Man and Cybernetics 2002, pp. 1–6 (2002)
15. Glover, F., Laguna, M.: Tabu Search. Kluwer Academic Publishers, Norwell (1997)
16. Harding, S., Banzhaf, W.: Fast genetic programming and artificial developmental systems on gpus. In: 21st International Symposium on High Performance Computing Systems and Applications, HPCS 2007, p. 2 (2007)
17. Illinois, U.: The LLVM Compiler Infrastructure. University of Illinois at Urbana-Champaign (2011), http://llvm.org
18. Janson, S., Merkle, D., Middendorf, M.: Parallel ant colony algorithms, Tech. rep., Parallel Metaheuristics. Wiley Book Series on Parallel and Distributed Computing (2005)
19. Koomey, J.G., Berard, S., Sanchez, M., Wong, H.: Implications of historical trends in the electrical efficiency of computing. IEEE Annals of the History of Computing 33, 46–54 (2011), Doi http://doi.ieeecomputersociety.org/10.1109/MAHC.2010.28
20. Li, J., Wang, X., He, R., Chi, Z.: An efficient fine-grained parallel genetic algorithm based on GPU-Accelerated. In: International Conference on Network and Parallel Computing, pp. 855–862 (2007)
21. Li, J., Zhang, L., Liu, L.: A parallel immune algorithm based on fine-grained model with gpu-acceleration. In: Proceedings of the 2009 Fourth International Conference on Innovative Computing, Information and Control, ICICIC 2009, pp. 683–686 (2009), Doi http://dx.doi.org/10.1109/ICICIC.2009.44, http://dx.doi.org/10.1109/ICICIC.2009.44
22. Luo, Z., Liu, H.: Cellular genetic algorithms and local search for 3-SAT problem on graphic hardware. In: IEEE CEC 2006, pp. 2988–2992 (2006)

23. Luo, Z., Liu, H., Wu, X.: Artificial neural network computation on graphic process unit. In: Proceedings of the 2005 IEEE International Joint Conference on Neural Networks, vol. 1, pp. 622–626 (2005)
24. Luong, T.V., Melab, N., Talbi, E.G.: GPU-based Island Model for Evolutionary Algorithms. In: Genetic and Evolutionary Computation Conference (GECCO), Portland, USA (2010)
25. Madera, J., Alba, E., Ochoa, A.: A parallel island model for estimation of distribution algorithms. In: Lozano, J., Larrañaga, P., Inza, I., Bengoetxea, E. (eds.) Towards a New Evolutionary Computation. STUDFUZZ, vol. 192, pp. 159–186 (2006)
26. Maitre, O., Baumes, L.A., Lachiche, N., Corma, A., Collet, P.: Coarse grain parallelization of evolutionary algorithms on gpgpu cards with easea. In: Proceedings of the 11th Annual Conference on Genetic and Evolutionary Computation, GECCO 2009, pp. 1403–1410 (2009), http://doi.acm.org/10.1145/1569901.1570089, http://doi.acm.org/10.1145/1569901.1570089
27. Martínez-Zarzuela, M., Díaz Pernas, F.J., Díez Higuera, J.F., Rodríguez, M.A.: Fuzzy ART Neural Network Parallel Computing on the GPU. In: Sandoval, F., Prieto, A.G., Cabestany, J., Graña, M. (eds.) IWANN 2007. LNCS, vol. 4507, pp. 463–470. Springer, Heidelberg (2007)
28. Martinez-Zarzuela, M., Diaz-Pernas, F., Diez, J., Anton, M., Gonzalez, D., Boto, D., Lopez, F., DelaTorre, I.: Multi-scale neural texture classification using the gpu as a stream processing engine. Machine Vision and Applicactions (in press 2010)
29. Meuth, R.J., Wunsch, D.C.: A survey of neural computation on graphics processing hardware. In: IEEE 22nd International Symposium on Intelligent Control (ISIC 2007), pp. 524–527 (2007)
30. Munawar, A., Wahib, M., Munetomo, M., Akama, K.: Hybrid of genetic algorithm and local search to solve max-sat problem using nvidia cuda framework. Genetic Programming and Evolvable Machines 10, 391–415 (2009)
31. Nebro, A.J., Durillo, J.J., Luna, F., Dorronsoro, B., Alba, E.: Mocell: A cellular genetic algorithm for multiobjective optimization. International Journal of Intelligent Systems, 25–36 (2007)
32. Oh, K.S., Jung, K.: Gpu implementation of neural networks. Pattern Recognition 37(6), 1311–1314 (2004)
33. Parejo, J., Ruiz-Cortés, A., Lozano, S., Fernandez, P.: Metaheuristic optimization frameworks: a survey and benchmarking. In: Soft Computing - A Fusion of Foundations, Methodologies and Applications, pp. 1–35., http://dx.doi.org/10.1007/s00500-011-0754-8, doi:10.1007/s00500-011-0754-8
34. Pospichal, P., Jaros, J.: Gpu-based acceleration of the genetic algorithm, Tech. rep., GECOO competition (2009)
35. Pospichal, P., Jaros, J., Schwarz, J.: Parallel Genetic Algorithm on the CUDA Architecture. In: Di Chio, C., Cagnoni, S., Cotta, C., Ebner, M., Ekárt, A., Esparcia-Alcazar, A.I., Goh, C.-K., Merelo, J.J., Neri, F., Preuß, M., Togelius, J., Yannakakis, G.N. (eds.) EvoApplicatons 2010. LNCS, vol. 6024, pp. 442–451. Springer, Heidelberg (2010)
36. Pospichal, P., Schwarz, J., Jaros, J.: Parallel genetic algorithm solving 0/1 knapsack problem running on the gpu. In: 16th International Conference on Soft Computing Mendel 2010, pp. 64–70 (2010)

37. Resende, M.G.C., Ribeiro, C.C.: Parallel greedy randomized adaptive search procedures (2004)
38. Rudolph, G.: Parallel approaches to stochastic global optimization. In: Joosen, W., Milgrom, E. (eds.) Parallel Computing: From Theory to Sound Practice, pp. 256–267. IOS Press (1992)
39. Selman, B., Kautz, H.: Domain-independent extensions to gsat: Solving large structured satisfiability problems. In: Proc. IJCAI 1993, pp. 290–295 (1993)
40. Thompson, C.J., Hahn, S., Oskin, M.: Using modern graphics architectures for general-purpose computing: a framework and analysis. In: Proceedings of the 35th Annual ACM/IEEE International Symposium on Microarchitecture, MICRO 35, pp. 306–317 (2002), `http://portal.acm.org/citation.cfm?id=774861.774894`
41. Tsutsui, S., Fujimoto, N.: Solving quadratic assignment problems by genetic algorithms with gpu computation: a case study. In: GECCO 2009, pp. 2523–2530 (2009)
42. Tsutsui, S., Fujimoto, N.: Aco with tabu search on a gpu for solving qaps using move-cost adjusted thread assignment. In: Krasnogor, N., Lanzi, P.L., Engelbrecht, A., Pelta, D., Gershenson, C., Squillero, G., Freitas, A., Ritchie, M., Preuss, M., Gagne, C., Ong, Y.S., Raidl, G., Gallager, M., Lozano, J., Coello-Coello, C., Silva, D.L., Hansen, N., Meyer-Nieberg, S., Smith, J., Eiben, G., Bernado-Mansilla, E., Browne, W., Spector, L., Yu, T., Clune, J., Hornby, G., Wong, M.L., Collet, P., Gustafson, S., Watson, J.P., Sipper, M., Poulding, S., Ochoa, G., Schoenauer, M., Witt, C., Auger, A. (eds.) GECCO 2011: Proceedings of the 13th Annual Conference on Genetic and Evolutionary Computation, pp. 1547–1554 (2011)
43. Vidal, P., Alba, E.: Cellular Genetic Algorithm on Graphic Processing Units. In: González, J.R., Pelta, D.A., Cruz, C., Terrazas, G., Krasnogor, N. (eds.) NICSO 2010. SCI, vol. 284, pp. 223–232. Springer, Heidelberg (2010)
44. Wong, M., Wong, T.: Parallel hybrid genetic algorithms on Consumer-Level graphics hardware. In: IEEE CEC 2006, pp. 2973–2980 (2006)
45. Wong, M.L., Wong, T.T.: Implementation of Parallel Genetic Algorithms on Graphics Processing Units. In: Gen, M., Green, D., Katai, O., McKay, B., Namatame, A., Sarker, R.A., Zhang, B.-T. (eds.) Intelligent and Evolutionary Systems. SCI, vol. 187, pp. 197–216. Springer, Heidelberg (2009)
46. Wong, M., Wong, T., Fok, K.: Parallel evolutionary algorithms on graphics processing unit. In: IEEE CEC 2005, vol. 3, pp. 2286–2293 (2005)
47. Yu, Q., Chen, C., Pan, Z.: Parallel Genetic Algorithms on Programmable Graphics Hardware. In: Wang, L., Chen, K., S. Ong, Y. (eds.) ICNC 2005. LNCS, vol. 3612, pp. 1051–1059. Springer, Heidelberg (2005)
48. Zhang, S., He, Z.: Implementation of Parallel Genetic Algorithm Based on CUDA. In: Cai, Z., Li, Z., Kang, Z., Liu, Y. (eds.) ISICA 2009. LNCS, vol. 5821, pp. 24–30. Springer, Heidelberg (2009)

Part III
Nature-Inspired Solutions for Intelligent Networking

Chapter 7
Scalability Analysis: Reconfiguration of Overlay Networks Using Nature-Inspired Algorithms

Simone A. Ludwig

Abstract. Overlay networks are virtual networks of nodes and logical links built on top of the existing network infrastructure, with the purpose of contributing new functionality. There are many different solutions proposed to tackle a range of specific needs such as content distribution and caching, file sharing, improved routing, multicast and streaming, ordered message delivery, and enhanced security and privacy. In this chapter, the focus lies on the optimization of overlay networks in terms of cost, performance, and reliability. In particular, the main objective is the optimization of data mirroring. Three different optimization approaches are introduced. The first approach is based on a "related work" implementation using Genetic algorithms, the second makes use of artificial immune systems, and the third approach uses the Particle swarm optimization approach. The three algorithms are implemented and experiments are conducted to measure the overall performance, the behavior and feasibility of network and link failures, as well as a scalability analysis is performed.

7.1 Introduction

Many different kind of overlay networking technologies have emerged in the past years. Research and development of overlay systems have primarily focused on developing technologies that solve the challenges of reliability and efficient processing of networks by providing a higher-level network that is built on top of the normal network, the so-called overlay network. As the overlay network is built on top of an existing network, it relies on the underlay

Simone A. Ludwig
Department of Computer Science, North Dakota State University, Fargo, ND, USA
e-mail: `simone.ludwig@ndsu.edu`

J. Kołodziej et al. (Eds.): Advances in Intelligent Modelling and Simulation, SCI 422, pp. 137–154.
springerlink.com

network for basic networking function such as routing and forwarding. The nodes in an overlay network are connected via logical links and can span many physical links.

In particular, the huge number of Internet users, estimated to almost 2 billions to date [1], as well as the delivery of huge amounts of data and media has become commonplace. Multimedia content such as videos is posing an increasing challenge to the networks, as well as social collaboration and social media web sites, which use and distribute large amounts of data on a daily basis. These developments of the evolving web have a profound impact on the network requirements in terms of performance and reliability. Therefore, it is essential that overlay networks have to ensure an efficient and scalable service for Internet users.

An overlay network built on top of an existing network consists of a set of distributed nodes that are deployed on the Internet. The nodes of the overlay network are expected to meet the following requirements [2]:

- Provide the infrastructure to support the execution of one or more distributed applications.
- Support high-level routing and forwarding tasks necessary in a network. The overlay network should provide data-forwarding capabilities that are different from the ones that are part of the Internet.
- Deployment should be across the Internet to allow third parties to participate in the organization and operation of overlay networks.

Overlay networks have many advantages that can be listed as such [3]:

- Overlay networks are incrementally deployable –The overlay network does not require changes to the existing Internet infrastructure, only additional servers are necessary. Once nodes are added to an overlay network, the control of paths of data becomes possible with great precision.
- Overlay networks are adaptable – Even though the abstraction of an overlay network constrains the packets to flow over a constrained set of links, the set of links can be constantly optimized over metrics that the different applications require.
- Overlay networks are robust – Robustness is a result of the given increased control and adaptable nature of the overlay networks. With a sufficient number of nodes deployed, an overlay network should be able to route between any two nodes in two independent ways, i.e., overlay networks are able to route around faults occurring in the network.
- Overlay networks are customizable – Given that overlay nodes can be multipurpose computers, they can be easily equipped with whatever is necessary. To give an example, overlay networks make extensive use of disk space that allows overlay networks to provide savings in terms of bandwidth when the content is not consumed simultaneously in different parts of the network.

The main properties of overlay networks are adaptability and robustness. These two features are the major diving force behind the research of overlay networks. The aim of this paper is the optimization of an overlay network in terms of cost, performance, and reliability. In particular, the main focus of the optimization is the application of data mirroring. One "related work" approach, as well as two additional approaches are compared in terms of performance and scalability. Part of this research presented in this book chapter was published in [4].

The remainder of this chapter is as follows: Section 7.2 describes related work; Section 7.3 introduces the approaches used; in Section 7.4 the experimental setup and results are described; and in Section 7.5 the findings are summarized.

7.2 Related Work

There are many different types of overlays that were developed meeting a range of purposes, e.g. content distribution and caching [5], overlays used for file sharing [6], improved routing [7], multicast and streaming [8], ordered message delivery [9], and enhanced security and privacy [10].

In the next few paragraphs we list and describe in more detail particular overlay network implementations that are developed. A routing overlay network is an overlay that controls and/or modifies the path of data delivery through the network. Routing overlay networks [7] improve the performance and robustness of packet delivery. This improvement is achieved by delegating the task of selecting paths to users, who can choose among more reliable routes, less loaded routes, shorter routers, or higher bandwidth routes. Overlay networks do not require support from routers as other path selection methods do. The packets still traverse the underlying routing infrastructure, however, their path is defined by the logical edges of the overlay.

Another approach is called Overcast [11], which is an application-level multicast system that can be incrementally deployed making use of the Internet infrastructure. Basically, the implementation consists of a collection of nodes that are placed at strategic locations in an existing network, which in turn implement a network abstraction on top of the network provided by the underlying network. Overcast provides multicast that is scalable and reliable by using a simple protocol for building efficient data distribution trees that automatically adapt to changing network conditions. The simulations conducted indicate that the Overcast provides roughly 70-100% of the total bandwidth possible, at a cost of somewhat less than twice the network load.

Another well-known overlay network implementation is RON (Resilient Overlay Network) [12]. RON is an architecture that allows distributed Internet applications to detect path errors/outages and recover from them within seconds, thus improving wide-area routing protocols that take several minutes to recover. It is an application-layer overlay on top of the existing Internet routing substrate. RON monitors the quality of the Internet paths in order to decide whether a route change of the packets need to take place to improve the overall quality of the overlay network. RON was able to improve the loss rate, latency, or throughput perceived by data transfers (5% of the transfers doubled their TCP throughput, and 5% of the transfers had the data loss reduced by a factor of 0.05).

Research concerned in particular with the optimization of overlay networks is manifold. One special class of overlay networks is the Service Overlay Networks (SON) [13]. The aim is to design the overlay with reliability constraints in mind. The assumption is that a SON could enter an inadmissible state for two reasons; (1) due to insufficient resources to accommodate new connections, and (2) due to some hardware malfunctioning. Two sub-functions, which are Maximum Profit (MP) and Minimum Cost (MC) need to be optimized. Given the competing relationship between MP and MC, the aim is to ensure the systems' operability. Their approach is based on the Lagrange multipliers, which shows that MP and MC separately can achieve the same network designs. The paper's contribution is the provisioning of guidelines to the formulation of the design of a reliable and economically optimal SON.

Given that the optimization of an overlay network configuration is NP-hard, classic exact techniques are only applicable to solve very small instances of overlay networks. Among a broad set of modern heuristics and meta-heuristics methods for optimization, nature-inspired methods have emerged as promising techniques for solving for example network design problems since they are able to compute approximate solutions within acceptable range of execution times [14].

The research and developments of nature-inspired networking techniques have fostered new techniques in networking, in particular due to their dynamic nature, resource constraints and heterogeneity. In particular, an Ant Colony Optimization approach was used in the AntNet routing protocol [15]. Their protocol uses agents to concurrently explore the network and exchange collected information in the same way as ants explore the environment. The main idea used from the Ants is their indirect communication capability depositing pheromone for the routing optimization.

Multi-objective evolutionary algorithms, in particular, NSGA-II was used in [16] to optimize a multicast overlay network based on two criteria; the first is to optimize the total end to end delay of a multicast tree, and the second is to maximize the link utilization.

A swarm-intelligence based approach was used in a layered overlay multicast approach for routing web streams [17]. An architecture was adopted to improve service capabilities, for satisfying the request of multi-constrained Quality of Service (QoS) routing of large-scale multi-domain web streams. The approach is based on meeting the uncertainty of the network status description, in order to find the QoS-satisfied routes using an effective mathematical model.

A multi-swarm approach for neighbor selection in peer-to-peer overlay networks is described in [18]. Their approach is inspired by the commonalities of P2P systems and Particle Swarm in a dynamic environment. A multi-swarm interactive pattern was introduced to match the dynamic nature of P2P networks.

In [19], genetic algorithms are applied to the reconfiguration of the topology and link capacities of an operational network. It does this in order to adapt to changes of its operational conditions, in which nodes and links might become unavailable, the traffic patterns might change, and the quality of service requirement and priorities of different users and applications might change suddenly.

Another example of a genetic algorithm approach is the Genetic-Algorithm-Based Neighbor-Selection Strategy for Hybrid Peer-to-Peer Networks [20]. The strategy enhances the decision process applied to transfer coordination. An investigation of the strategy revealed that it affects the system throughput and distribution efficiency, as well as peer contribution, especially for low-connectivity peers.

Plato is a genetic algorithm approach to run-time reconfiguration in autonomic computing systems [21]. The genetic algorithm approach uses the evolutionary computation technique to automate the decision-making process of an autonomic system. It enables a system to dynamically evolve target reconfigurations at run time, and at the same time, it balances the trade-offs between functional and non-functional requirements to changes in the environmental requirements and conditions. In particular, their approach is applied to the reconfiguration of a collection of remote data mirrors demonstrating it to be a good optimization method for diffusing data, and minimizing operational costs; at the same time it maximizes data reliability and network performance.

This paper closely follows the Plato approach using genetic algorithms to optimize an overlay network based on cost, performance, and reliability. However, it varies in the following two aspects (1) the fitness functions are normalized in order to compare the three different measures, and (2) besides the use of genetic algorithms, an algorithm based on Artificial Immune Systems, and another based on Particle Swarm Optimization (PSO) are implemented and evaluated. As can be seen in the evaluation section, the two additionally implemented algorithms perform well, and in particular the PSO approach shows a better performance than the Genetic algorithm approach. Furthermore, we also address the issue of scalability of the three nature-inspired algorithms, and comment on the feasibility of the approaches for the automatic reconfiguration of overlay networks.

7.3 Approaches

This research addresses the dynamic reconfiguration of a collection of remote data mirrors. In remote data mirroring, data copies of critical data is stored at one or more secondary site(s), which prevents the protected data from failures that may affect the primary copy [22]. There are two important design criteria for remote data mirroring; the first is to choose the type of network link connection to the mirrors, and the second is to choose the remote mirroring protocol. Each link in the network has a cost associated, as well as throughput, latency and loss rate which determine the overall remote mirror design performance [22]. There are two types of remote mirroring protocols that are synchronous and asynchronous propagation, both affecting network performance and data reliability. In synchronous propagation the secondary site applies each write before the write completes at the primary site, and in asynchronous propagation updates get queued at the primary site and are periodically propagated to the secondary site in a batch-fashion.

The optimization design criteria are the same as for Plato [21]. The main goal is the construction and maintenance of an overlay network for the data to be distributed to all nodes while fulfilling the following requirements:

1. Overlay network must remain connected at all times;
2. Overlay network should never succeed the allocated monetary budget;
3. The data should be distributed as efficiently as possible, meaning the amount of bandwidth consumed when diffusing data should be minimized.

Three algorithms have been implemented in order to evaluate and compare their performance. The first algorithm is an implementation of Genetic Algorithms as in the Plato implementation [21], the second one is an implementation of the aiNet algorithm inspired by Artificial immune systems, and the third algorithm is an implementation of the discrete Particle Swarm Optimization approach.

Table 7.1. Link Propagation Methods

Time interval	Avg. data batch size in GB
0	0
1 min	0.0436
5 min	0.2036
1 hr	2.091
4 hr	6.595
12 hr	15.12
24 hr	27.388

The fitness function used as a measure for all three algorithms are slightly modified compared to the Plato approach [21]. The differences are that normalization of the overall fitness value is done in order to have an overall fitness value in the range of 0 and 1, as well as the sum of weights for each part of the fitness function is 1.

The fitness function consists of three parts (as in Plato); the first part evaluates the overlay network in terms of cost, the second in terms of performance, and the third part evaluates the reliability of the overlay network. The overall fitness function (Eq. (7.1)) is the weighted average of all three fitness portions. Please note that the sum of the weights needs to sum up to 1 (Eq. (7.2)).

$$F_{overall} = w_1 * F_{cost} + w_2 * F_{perf} + w_3 * F_{rel} \tag{7.1}$$

$$\sum_{i=1}^{3} w_i = 1 \tag{7.2}$$

Looking at the different fitness sub-functions, the fitness sub-function for cost is given as:

$$F_{cost} = 1 - \frac{cost}{budget} \tag{7.3}$$

where *cost* is the sum of operational expenses of all active links and *budget* is a user supplied value on the maximum amount of money for an operating overlay network. The sub-function for the performance consists of two parts: latency and bandwidth as given below:

$$F_{perf} = 0.5 * (1 - \frac{latency_{avg}}{latency_{wc}}) + 0.5 * (\frac{bandwidth_{sys} - bandwidth_{eff}}{bandwidth_{sys}} + bound) \tag{7.4}$$

where $latency_{avg}$ is the average latency over all active links, and $latency_{wc}$ is the largest latency value measure over all links in the underlying network; $bandwidth_{sys}$ is the total available bandwidth across the overlay network given the active links, and $bandwidth_{eff}$ is the total effective bandwidth across the overall network after data has been coalesced, and *bound* is a limit on the best value that can be achieved throughout the network.

The last fitness sub-function measures the overlay network in terms of reliability consisting of two parts as given below:

$$F_{rel} = 0.5 * (\frac{links_{used}}{links_{max}}) + 0.5 * (1 - \frac{dataloss_{pot}}{dataloss_{max}}) \tag{7.5}$$

where $links_{used}$ is the number of active links, and $links_{max}$ is the maximum number of possible links given the network structure; and $dataloss_{pot}$ is the total amount of data that could be lost during write coalescing using the propagation methods as given in Table 7.1, and $dataloss_{wc}$ is the amount of data that could be lost during write coalescing using the propagation method with the largest time window.

7.3.1 Genetic Algorithm Implementation

Genetic algorithms [23] are a class of stochastic search algorithms based on biological evolution. In particular, the principles of the evolution via natural selection are applied, employing a population of individuals that undergo selection, as well as variation-inducing operators such as mutation and crossover. A fitness function is used to evaluate individuals.

The genetic algorithm in short works as follows: a population is created with a group of individuals that are created randomly. The individuals in the population are then evaluated. The evaluation function, called fitness function, gives the individuals a score based on how well they perform at the given iteration. Two individuals are then selected based on their fitness, i.e., the higher the fitness, the higher the chance of being selected. These individuals then "reproduce" to create one or more offspring(s), and afterwards the offsprings are mutated randomly. This continues until a suitable solution has been found or a certain number of iterations have passed.

Table 7.2 lists the parameters used for the implementation, which is identical to the one proposed in Ramirez et al. [21] with the exception of the modified fitness function as explained above. Two-point crossover is employed, as well as Tournament selection is used.

Table 7.2. GA parameters

Parameter	Value
Population size	100
Crossover	Two-point
Crossover probability	0.1
Mutation probability	0.05
Selection method	Tournament (k=2)

7.3.2 Artificial Immune System Implementation

An Artificial Immune System (AIS) models the natural system's ability to detect foreign cells in the body. It is a new computational paradigm with the ability to perform pattern recognition that is mainly applied to anomaly detection.

There are different views on how natural immune systems have been developed. These models include the classical view of lymphocytes that are used to distinguish between self- and non-self, clonal selection theory, danger theory, etc. The implementation we have adopted and applied is based on the clonal selection theory. According to Burnet's clonal selection theory [24], the immune system undergoes a selection mechanism during the lifetime of the individual. The theory states that activation of lymphocytes occurs when the binding with a suitable antigen happens. Once activated, clones of the lymphocyte are produced expressing identical receptors to the original lymphocyte that encountered the antigen. This process ensures that only lymphocytes specific to an activating antigen are produced in large numbers.

Based on this biological background, the clonal algorithm was introduced by de Castro and Timmis [25], named *aiNet*. In order to present the pseudo code of the *aiNet* algorithm the following terms needs to be introduced:

- *Network cell*: individual of the population; each cell is a real-valued vector in an Euclidean shape-space (in our case it is an integer-valued vector);
- *Fitness*: fitness of a cell in relation to an objective function to be optimized (in our case the optimization is a maximization problem);
- *Clone*: offspring cells that are identical copies of their parent cell. The offspring will further suffer a somatic mutation so that they become variations of their parent.

The pseudo code of the slightly adapted (the concept of affinity was not used as not applicable) optimization version of the *aiNet* algorithm is summarized as follows:

Algorithm 4. AIS pseudocode

```
Randomly initialize a population of cells (the initial number of cells is not rele-
vant)
while stopping criterion is not met do
  Determine the fitness of each network cell.
  Generate a number of clones for each network cell.
  Mutate each clone proportionally to the fitness of its parent cell, but keep the
  parent cell.
  Determine the fitness of all individuals of the population.
  for each clone do
    Select the cell with highest fitness and calculate the average fitness of the
    selected population.
  end for
end while
```

The parameters used in the AIS implementation are stated in Table 7.3.

Table 7.3. AIS parameters

Parameter	Value
Number of cells	2
Number of clones	6

7.3.3 Particle Swarm Optimization Implementation

Particle Swarm Optimization (PSO) as introduced in [26], is a swarm based global optimization algorithm. The algorithm models the behavior of bird swarms searching for an optimal food source. The movement of a single particle is influenced by its last movement, its knowledge, and the swarm's knowledge. PSO's basic equations are:

$$x_i(t+1) = x_i(t) + v_{ij}(t+1) \tag{7.6}$$

$$\begin{aligned} v_{ij}(t+1) = w(t)v_{ij}(t) + c_1 r_{1j}(t)(xBest_{ij}(t) - x_{ij}(t))) \\ + c_2 r_{2j}(t)(xGBest_j(t) - x_{ij}(t))) \end{aligned} \tag{7.7}$$

where x represents a particle, i denotes the particle's number, j the dimension, t a point in time, and v is the particle's velocity. *xBest* is the best location the particle ever visited (the particle's knowledge), and *xGBest* is the best location any particle in the swarm ever visited (the swarm's knowledge). w is the inertia weight and used to weigh the last velocity, c_1 is a variable to weigh the particle's knowledge, and c_2 is a variable to weigh the swarm's knowledge. r_1 and r_2 are uniformly distributed random numbers between zero and one.

PSO is usually used on continuous and not discrete problems. In order to solve the discrete overlay network assignment using the PSO approach, several operations and entities have to be defined. The implementation follows in part the implementation for solving the traveling salesman problem as described in [27]. First, a swarm of particles is required. A single particle represents one overlay network, i.e., every particle's position in the search space must correspond to a possible overlay network. Velocities are implemented as lists of changes that can be applied to a particle (its vector) and will move the particle to a new position (a new overlay network). Changes are exchanges of values of the overlay network. Further, minus between two matches (particles), multiplication of a velocity with a real number, and the addition of velocities have to be defined. Minus is implemented as a function of particles. This function returns the velocity containing all changes that have to be applied to move from one particle to another in the search space. Multiplication randomly deletes single changes from the velocity vector, if the multiplied real number is smaller than one. If the real number is one, no changes are applied. For a real number larger than one, random changes are added to the velocity vector.

The PSO implemented uses guaranteed convergence, which means that the best particle is guaranteed to search within a certain radius, implying that the global best particle will not get trapped in local optima. Table 7.4 shows the specific parameters chosen for the implementation.

Table 7.4. PSO parameters

Parameter	Value
Number of particles	100
Inertia weight	0.001
Weight of local knowledge	0.5
Weight of global knowledge	0.5
Radius	2
Neighborhood size	4

7.4 Experiments and Results

The three algorithms were tuned with the parameter values given in the previous section. Since the GA and PSO implementation have probabilities involved, all parameters of the algorithms were set, so that the number of iterations needed were kept constant, but at the same time the number of function evaluations is kept as equal as possible. Furthermore, all experiments were conducted 25 times in order to account for statistical variations.

The first set of experiments was performed measuring the accuracy of all approaches given different settings of the weights, comparing the different effects on the accuracy. The second set investigates single link failures, and the third set evaluates complete network failures. The last set of experiments explores the effect of increasing overlay network sizes, and the impact on the accuracy, as well as the performance in terms of actual execution times.

7.4.1 Overall Comparison of Approaches

Table 7.5 shows the results of the fitness scores of all approaches for the configuration of a 25-node network using different weight combinations, thereby favoring cost, performance or reliability. The values were taken at iteration 500. It can be seen that the highest fitness score can be achieved with F2 and F3, which only consider the cost and performance fitness sub-functions respectively (a smaller number of links achieves a higher fitness score). The worst fitness score is observed for F4, when optimizing the overlay network based on reliability given the trade-off between the number of links and the potential data loss. Overall, PSO outscores the other approaches for all weight settings.

Figure 7.1 shows the fitness values plotted for a 25-node network for increasing iterations. It can be observed that the PSO approach reaches the maximum fitness score of 0.8333 at iteration 270, which is much earlier than the GA and AIS approaches. The GA approach takes 610 iterations, and the

Table 7.5. Fitness function and fitness values (500 iterations)

Fitness function	w_1	w_2	w_3	Best fitness GA	Best fitness AIS	Best fitness PSO
F1	$0.\overline{3}$	$0.\overline{3}$	$0.\overline{3}$	0.8233	0.8069	0.8333
F2	1	0	0	0.8910	0.9021	0.9129
F3	0	0	0	0.8922	0.8965	0.9087
F4	0	1	1	0.6037	0.6053	0.6086
F5	0.5	0	0	0.8978	0.8796	0.8999
F6	0	0.5	0.5	0.7394	0.7337	0.7500
F7	0.5	0.5	0.5	0.7301	0.7297	0.7499

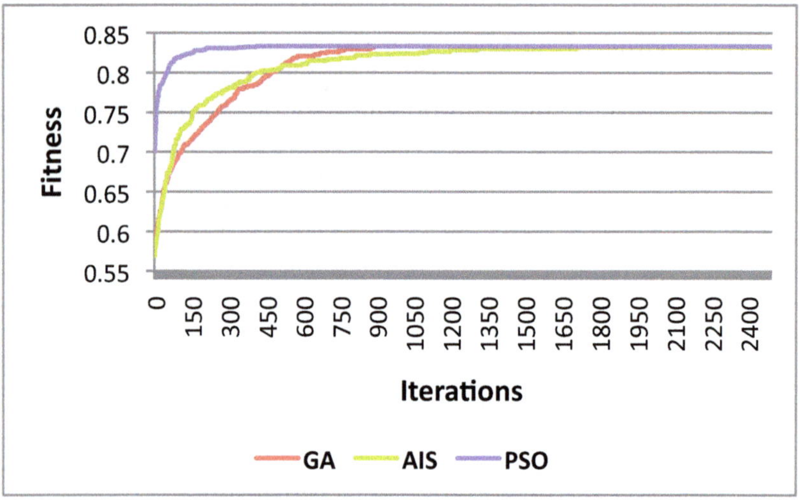

Fig. 7.1. Fitness of all approaches

AIS approach requires 1420 iterations to reach the maximum fitness score. The fitness function F1 with an equal weight distribution was used.

7.4.2 Investigation of Network and Link Failures

Figure 7.2shows three complete reconfigurations of a 25-node overlay network. The simulation runs the optimization every 2,500 iterations, due to an artificially induced breakdown of the network. As can be seen in Figure 7.2, all three approaches, GA, AIS and PSO, can reconfigure overlay networks and achieve the maximum fitness score after around 1,200, 300 and 1,600 iterations respectively.

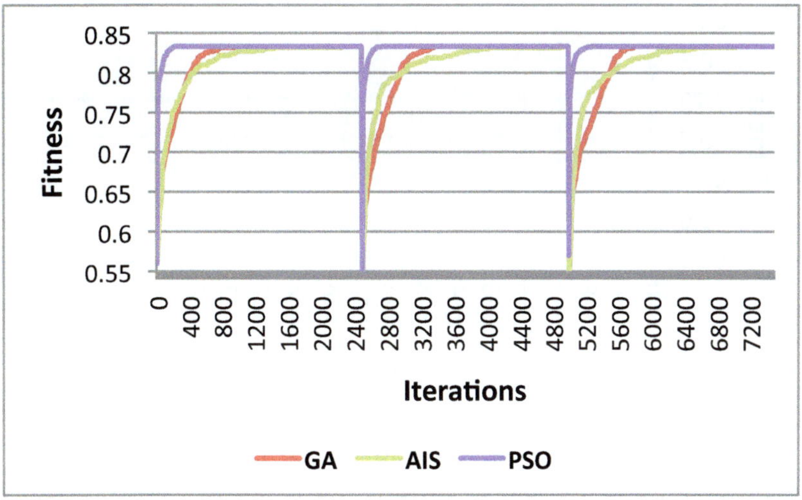

Fig. 7.2. Reconfiguration of complete overlay network with 25 nodes

Figure 7.3 shows the reconfiguration of a 25-node network. The first evolution shows the fitness trend when F1 is used (weights are all $0.\overline{3}$), after a failure occurs the network switches the weights to w_1=w_2=0.25 and w_3=0.5 in order to stress more on reliability rather than an equal contribution of all three fitness sub-functions. Even though the fitness score is lower with the second weight setting, however, the network configuration is more stable to protect against future failures.

Figure 7.4 shows the fitness curves for successive link failures without reconfiguration. It can be seen that the generation of new overlay network configurations work fine until 84% of link failures occur, however, at the cost of reducing fitness scores. Decreasing fitness scores can be observed with the lowest of 0.34 when 80% of link failures occur; after 84% an overlay network cannot be constructed anymore, and therefore the reconfiguration needs to be started.

7.4.3 Scalability Analysis

The scalability analysis investigates the performance impact with increasing network sizes. We successively increase the number of networks starting with 10 nodes all the way up to 60 nodes.

Figure 7.5 shows the fitness scores of the approaches for a network of size 60. What can be seen, compared to Figure 7.1, is that even though PSO achieves larger fitness improvements at the beginning, however, after around 2,700 iterations GA has already achieved the maximum fitness score, whereas it takes the PSO approach many more iterations to reach the maximum fitness

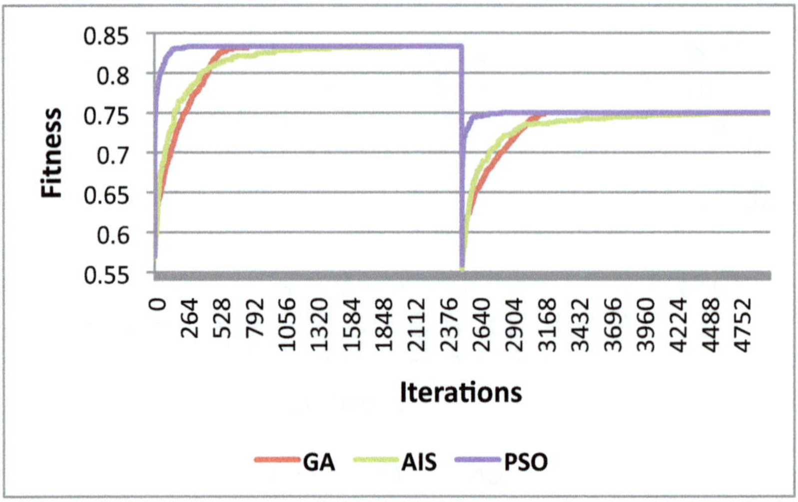

Fig. 7.3. Reconfiguration of 25-node network with varying weight distributions of fitness function

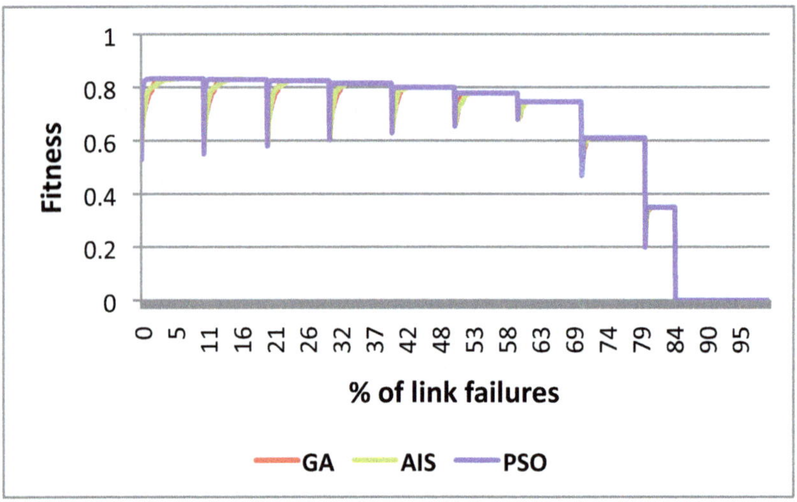

Fig. 7.4. Fitness scores with increasing percentage of link failures

score (89426 iterations on average). The AIS approach does not scale very well. Even though it achieves the maximum fitness score eventually; many more iterations are necessary compared to the PSO and GA approaches.

Figure 7.6 shows the number of iterations it takes to achieve the optimal fitness scores for all approaches. It can be seen that the number of iterations necessary to achieve the optimum is first lower for PSO, but after a network size of around 40 nodes, GA reaches the maximum fitness values faster than

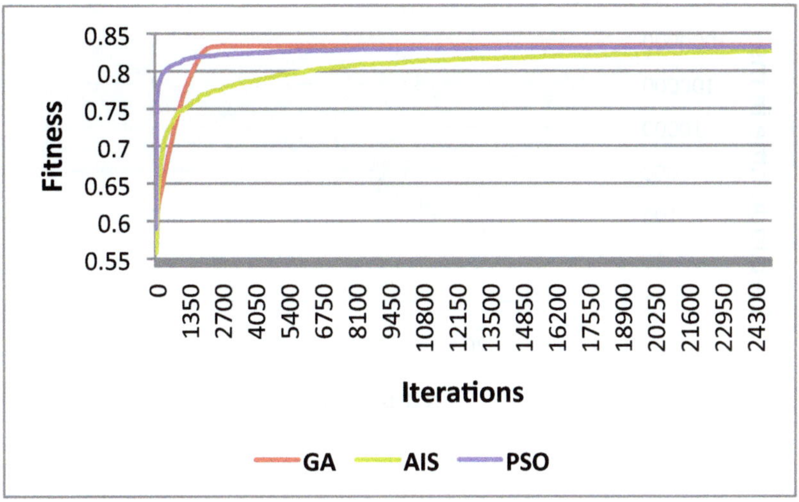

Fig. 7.5. Fitness of all approaches for network size of 60 nodes

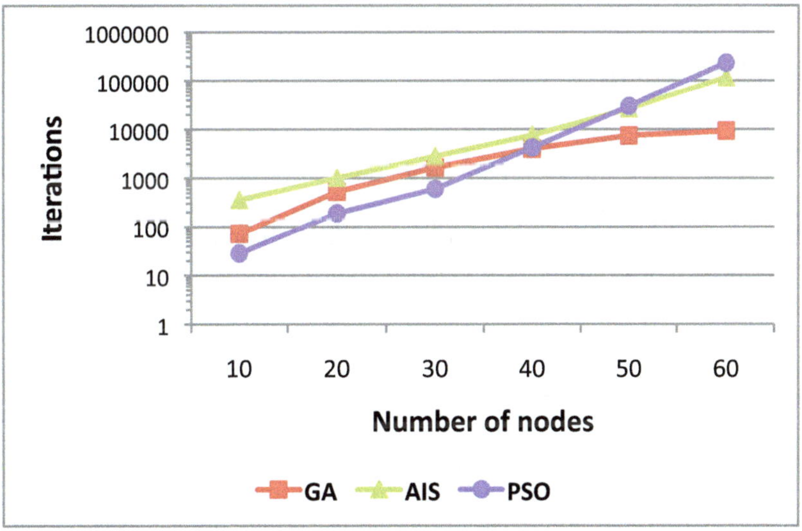

Fig. 7.6. Maximum fitness scores for different network sizes

PSO. For larger networks (above 50 nodes) AIS shows a slightly better fitness score than PSO. Please note that the y-axis is in logarithmic scale.

Figure 7.7 shows the same trend as Figure 6 in terms of execution time. However, the two lines for PSO and GA cross approximately at 43 nodes, whereas for AIS and PSO the cross happens after 50 nodes. Please also note that the y-axis is in logarithmic scale.

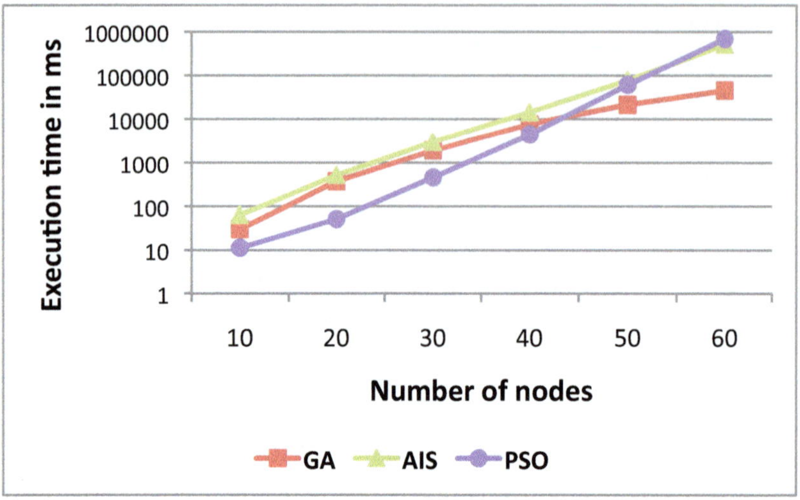

Fig. 7.7. Execution times in milliseconds of maximum fitness scores achieved (Figure 7.6)

7.5 Conclusion

This paper investigated three approaches for the reconfiguration of overlay networks and could be used as a guide for overlay network construction and configuration. The first approach was based on genetic algorithms used in literature; the second was based on Artificial Immune Systems, and the third approach was based on a discrete implementation of Particle Swarm Optimization.

In summary, the experiments conducted show that for networks up to 40 nodes the PSO approach achieves the maximum fitness score much faster than the GA and the AIS approaches. PSO usually takes 270 iterations to reach the maximum fitness value, whereas GA needs 610 iterations and AIS needs 1420 iterations to achieve the same for a 25-node network.

Furthermore, the experiments regarding link and complete network failures showed that the approaches are able to reconfigure the networks in a reasonable amount of time. It takes a network of 20 nodes to be reconfigured by GA on average 376 ms, AIS 1035 ms, and PSO 51 ms. Again, PSO outperformed GA and AIS. For the reconfiguration of a 60-node network, GA takes 12 min, AIS 18 min, and PSO needs 21 min. Everything less than a minute is unacceptable for an automatic reconfiguration method. Of course, if sub-optimal reconfigurations are sufficient, then for example the GA method can achieve 90% after approximately 245 iterations which equates to an execution time of roughly 5.5 s; 95% can be achieved in 32 s. Therefore, the automatic reconfiguration can be used if sub-optimal configurations are sufficient.

In addition, the scalability analysis revealed that even though the performance of PSO is better than GA for smaller networks (up to 40 nodes), however, networks with a larger number of nodes are optimized faster with the GA approach as shown in Figures 6 and 7, even though better sub-optimal fitness scores can be achieved by PSO for a small number of iterations.

In conclusion, if the execution time is of essence then the PSO algorithm is the better choice achieving overlay network reconfigurations within a shorter period of time, as well as when larger network sizes are used and sub-optimal scores are acceptable. However, if the fitness is paramount, then PSO should be used for the reconfiguration of networks up to 40 nodes, and for larger networks the GA approach should be used.

Acknowledgements. This research is based on work supported by North Dakota EPSCoR and National Science Foundation Grant EPS-0814442.

References

1. Internet Usage Statistics, World Internet Users and Population Stats (2011), http://www.internetworldstats.com/stats.htm (last retrieved in December 2011)
2. Tarkoma, S.: Overlay Networks: Toward Information Networking. CRC Press, Auerbach Publications (2010) ISBN: 978-1-4398-1371-3
3. Awerbuch B., Terzis, A.: A Robust Routing Algorithm for Overlay Networks. Technical Report (2004), http://www.cs.jhu.edu/~terzis/reprouting.pdf (last retrieved in December 2011)
4. Ludwig, S.A.: Nature-Inspired Reconfiguration of Overlay Networks. In: Proc. of the Third World Congress on Nature and Biologically Inspired Computing (NaBIC), Salamanca, Spain (2011)
5. Kostic, D., Rodriguez, A., Albrecht, J., Vahdat, A.: Bullet: High bandwidth data dissemination using an overlay mesh. In: Proc. of the Nineteenth ACM Symposium on Operating Systems Principles (SOSP), Bolton Landing, NY, USA (2003)
6. Cohen, B.: Incentives build robustness in Bit Torrent. In: Proc. of the 1st Workshop on Economics of Peer-to-Peer Systems, Berkley, CA, USA (2003)
7. Gummadi, K.P., Madhyastha, H., Gribble, S.D., Levy, H.M., Wetherall, D.J.: Improving the reliability of internet paths with one-hop source routing. In: Proc. of OSDI 2004 (2004)
8. Chu, Y.H., Rao, S.G., Seshan, S., Zhang, H.: A case for end system multicast. IEEE Journal of Selected Areas in Communication 20(8) (2002)
9. Lumezanu, C., Spring, N., Bhattacharjee, B.: Decentralized message ordering for publish/subscribe systems. In: Proc. of the ACM/IFIP/USENIX 2006 International Conference on Middleware, Melbourne, Australia (2006)
10. Syverson, P., Goldschlag, D., Reed, M.: Anonymous connections and onion routing. Proceedings of Security and Privacy, IEEE Journal on Selected Areas in Communications 16(4), 482–494 (1997)

11. Jannotti, J., Gifford, D.K., Johnson, K.L., Kaashoek, M.F., O'Toole, J.W.: Overcast: reliable multicasting with on overlay network. In: Proceedings of 4th Conference on Symposium on Operating System Design & Implementation, Berkeley, CA, USA, vol. 4 (2000)
12. Andersen, D., Balakrishnan, H., Kaashoek, F., Morris, R.: Resilient overlay networks. In: Proceedings of the Eighteenth ACM Symposium on Operating Systems Principles (SOSP 2001), USA (2001)
13. Lam, N., Dziong, Z., Mason, L.G.: Service Overlay Network Design with Reliability Constraints. In: Proceedings of IEEE 7th International Workshop on the Design of Reliable Communication Networks, Washington, D.C. (2009)
14. Baeck, T., Fogel, D., Michalewicz, Z. (eds.): Handbook of Evolutionary Computation. IOP Publ. Ltd., Bristol (1997)
15. Caro, G.D., Dorigo, M.: AntNet: distributed stigmergetic control for communications networks. J. Artif. Int. Res. 9(1), 317–365 (1998)
16. Montoya, J., Donoso, Y., Montoya, E., Echeverri, D.: Multiobjective model for multicast overlay networks over IP/MPLS using MOEA. In: Proceedings of International Conference on Optical Network Design and Modeling, pp. 1–6 (2008)
17. Zhao, Y., Wang, J., An, Y., Xia, F.: A Layered Overlay Multicast Algorithm with PSO for Routing Web Streams. In: Liu, W., Luo, X., Wang, F.L., Lei, J. (eds.) WISM 2009. LNCS, vol. 5854, pp. 205–213. Springer, Heidelberg (2009)
18. Abraham, A., Liu, H., Badr, Y., Grosan, C.: A multi-swarm approach for neighbor selection in peer-to-peer networks. In: Proceedings of the 5th International Conference on Soft Computing as Transdisciplinary Science and Technology (CSTST 2008), pp. 178–184. ACM, New York (2008)
19. Montana, D., Hussain, T., Saxena, T.: Adaptive reconfiguration of data networks using genetic algorithms. In: Proceedings of the Genetic and Evolutionary Computation Conference, pp. 1141–1149 (2002)
20. Koo, S.G.M., Kannan, K., Lee, C.S.G.: On neighbor-selection strategy in hybrid peer-to-peer networks. Journal of Future Generation Comp. Syst., 732–741 (2006)
21. Ramirez, A.J., Knoester, D.B., Cheng, B.H.C., McKinley, P.K.: Plato: A Genetic Algorithm Approach to Run-Time Reconfiguration in Autonomic Computing Systems. Journal of Cluster Computing (2010)
22. Keeton, K., Santos, C., Beyer, D., Chase, J., Wilkes, J.: Designing for disasters. In: Proceedings of the 3rd USENIX Conference on File and Storage Technologies, USA, pp. 59–62 (2004)
23. Holland, J.H.: Adaptation in Natural and Artificial Systems. University of Michigan Press, Ann Arbor (1975)
24. Burnet, F.M.: The Clonal selection theory of acquired immunity. Vanderbilt Univ. Press, Nashville (1959)
25. de Castro, L.N., Timmis, J.: An Artificial Immune Network for Multimodal Function Optimisation. In: Proceedings of the IEEE Congress on Evolutionary Computation, Honolulu, Hawaii, USA (2002)
26. Kennedy, J., Eberhart, R.: Particle swarm optimization. In: Proceedings of IEEE International Conference on Neural Networks, Perth, Western Australia (1995)
27. Clerc, M.: Discrete particle swarm optimization - illustrated by the traveling salesman problem. In: New Optimization Techniques in Engineering. STUDFUZZ. Springer (2004)

Chapter 8
Analysis of Emergent Behavior for GA-based Topology Control Mechanism for Self-Spreading Nodes in MANETs*

Stephen Gundry, Jianmin Zou, Elkin Urrea, Cem Safak Sahin, Janusz Kusyk, and M. Umit Uyar

Abstract. We introduce a genetic algorithm based MANET topology control mechanism to be used in decision making process of adaptive and autonomic systems at run time. A mobile node adapts its speed and direction using limited information collected from local neighbors operating in an unknown geographical terrain. We represent the genetic operators (i.e., selection, crossover and mutation) as a *dynamical system model* to describe the behavior of a single node's decision mechanism. In this dynamical system model each mobile node is viewed as a stochastic variable. We build a homogeneous Markov chain to study the convergent nature of multiple mobile nodes running our algorithm, called FGA. Each state in our chain represents a configuration of

Stephen Gundry · Jianmin Zou · M. Umit Uyar
Department of Electrical Engineering, The City College of New York, NY 10031 USA
e-mail: {sgundry00,jzou,umit}@ccny.cuny.edu

Elkin Urrea
Lehman College of the City University of New York, Bronx, NY 10468 USA
e-mail: eurrea@gc.cuny.edu

Cem Safak Sahin
BAE Systems - AIT, Burlington, MA 01803 USA
e-mail: csafaksahin@gmail.com

Janusz Kusyk · M. Umit Uyar
The Graduate Center of the City University of New York, New York, NY 10016 USA
e-mail: jkusyk@gc.cuny.edu

* This research was initiated with support from collaborative participation in the Communications Networks Consortium sponsored by the U.S. Army Research Lab under the Collaborative Technology Alliance Program, Cooperative Agreement DAAD19-01-2-0011, the National Science Foundation grants ECS-0421159 and CNS-0619577 and US Army, Fort Monmouth, New Jersey, Army Communications-Electronics RD&E Center Grant W15P7T-09-CS021.

J. Kołodziej et al. (Eds.): Advances in Intelligent Modelling and Simulation, SCI 422, pp. 155–183.
springerlink.com

the nodes in a MANET for a given instant. The homogeneous Markov chain model of our FGA is shown to be ergodic; its convergence is demonstrated using Dobrushin's contraction coefficients. We also observe that the nodes with longer communication ranges utilize more information about their neighborhood to make better decisions, require less movement and converge faster, whereas smaller communication ranges utilize limited information, take more time to escape local optima, and, hence, consume more energy.

8.1 Introduction

Scalability is a crucial requirement for future computing, communication, wireless sensor networks (WSNs), mobile ad hoc networks (MANETs), and multi-robot systems. Decentralization of WSNs and MANETs makes them suitable for different types of applications (e.g., military missions, search and rescue missions, etc.) where centralized control is not practical or possible. Furthermore, decentralized and autonomous mobile nodes improve the scalability of MANETs with respect to various design parameters including the required bandwidth, time delay to exchange control messages, and the total number of mobile nodes. Scalability and robustness will improve when each mobile node decides its own next movement direction and speed. Autonomous systems are also self-sufficient and capable of adapting their behavior to rapidly changing and unfamiliar environments.

In this chapter, we use a genetic algorithm based deployment and topology control mechanism, called force-based genetic algorithm (FGA) [24], in order to determine a mobile node's movement decision in terms of speed and direction. Each mobile node independently runs its own GA-based software to spread over an unknown terrain while maintaining a uniform distribution without synchronization among the nodes nor through a centralized control unit. Furthermore, autonomous mobile nodes only require their local neighborhood information for making relocation decisions. It is important to note that having these nodes to make deployment and topology control decisions reduces communication overhead, increases adaptivity of MANETs over unknown deployment terrains, and improves the robustness of the network. In this work, without loss of generality, all mobile nodes are considered to be identical; each node broadcasts periodic light-weight messages to its one-hop neighbors, where each message includes the node's location, direction, speed, and number of neighbors. One can envision many applications for our GA-based spread and topology control mechanism ranging from military to commercial applications, for example locating humans trapped in rubble after an earthquake and spreading military assets under harsh and bandwidth-limited conditions. In these applications a large number of mobile nodes can gather information from multiple viewpoints simultaneously, allowing them to share information and adapt to the environment quickly and comprehensively. A common objective among these applications is the uniform distribution of autonomous mobile nodes operating on areas without a priori geographical terrain knowledge.

Applying Vose's *dynamical system model* [29], we analyze the convergence of our FGA. One of the most important characteristics of GA-based approaches is that fitter chromosomes have a higher probability of survival in future generations. As a population grows, the probability that the best individual will survive to the next generation increases. The dynamical system model provides the trajectory of the expected next population and, therefore, provides an indication of the FGA behavior. We view each autonomous mobile node running FGA as a stochastic variable and show that FGA generates candidate solutions, representing speed and directional choices, approaching towards a uniform distribution of the nodes.

Using the dynamical system model results, we build an ergodic homogeneous Markov chain for our FGA. In this Markov chain model, each state represents a different node configuration. Using Dobrushin's contraction coefficients in conjunction with the total fitness of the final stationary distribution, we prove that the mobile nodes under the guidance of FGA converge to an acceptable uniform distribution.

In Sect. 8.2 of this chapter we review existing research on applications of GAs, schema theory and Markov chains. In Sect. 8.3, we outline standard GA operators and introduce our FGA. The dynamical system model of our FGA is given in Sect. 8.4. The ergodic homogeneous Markov model of our FGA and its convergence analysis are in Sect. 8.5. Concluding remarks and future work are discussed in Sect. 8.6.

8.2 Related Work

Our FGA is inspired by the force based distribution in physics where each particle reacts to intermolecular forces to achieve a minimal energy distribution. FGA is run by each mobile node as a stand-alone topology control mechanism to uniformly distribute mobile nodes over an unknown terrain [24]. Our initial work on the dynamical system model was introduced in [22, 28, 27] and the formal analysis of our GA-based topology control framework in [10].

Distribution of MANET nodes has been studied in different contexts. In [14] and [30], mobile nodes deploy into a bounded area. In [7], bio-inspired algorithms are shown to be effective to optimize the path that a robot takes to reach its assigned target. GA has also been proven as an efficient approach in different decentralized robotic applications. For example, in [4], a GA is presented with a distance-safety criteria for a mobile robot motion. In [25], the goal of autonomous robots is to move in a highway without collision. An adaptive GA is presented in [8] to identify targets while avoiding obstacles where mobile robots collect environmental information with their video cameras and light sensors to arrive a target.

Schema theory, Markov chains, and dynamic system models are among the widely used tools to study the efficiency of GAs. Schema theory [13, 9] describes how *schemata*, templates specifying groups of similar chromosomes, are expected to propagate from one generation to the next. In [13], a theorem is formulated to give a lower bound on the expected fraction of a population in a schema after one generation of a simple GA. Finite and discrete-time Markov chain models of GAs have been used successfully to understand the complex dynamics of a simple GA. For example, in [18], a Markov chain, whose states are populations, is given for a simple GA. In [29], this work was extended to a *random heuristic search* model of GAs where each individual of the next generation is selected from a probability distribution over individuals in the search space. In [26], a GA is analyzed using an elitist strategy on a Markov chain by assuming that the genetic operation in the generation changes is restricted to selection, crossover, and mutation, and by evaluating the eigenvalues of the transition matrix of the Markov chain. Convergence of canonical GAs is studied in [19]. Markov chain analysis of a GA that uses various selection mating schemes is presented in [15]. An analysis of the progression of GAs in high noise environments limiting fitness functions to include noise conditions can be found in [17]. Other works provide insight into analysis of Markov chains: definition of two common Markov chain Monte Carlo methods including the Gibb's sampler and the Metropolis sampler [31], a topology control algorithm for autonomous swarms that makes use of the Gibbsian sampler simulated annealing process [1], an intuitive model of MANETs using a Markov chain representation [3], and definitions of fundamental properties of finite Markov chains,with graph theory for (non)recurrent and (non)transient examples [16] .

8.3 Force-based Genetic Algorithm (FGA)

The concept of *survival of the fittest* is the main principle behind GAs, which are typically applied to problems where deterministic or heuristic methods are not present or cannot provide satisfactory results. GAs are especially suitable for problems where the search space is large, complex, or poorly understood. GAs use a population of abstract representations for candidate solutions to optimization problems that evolve toward better solutions based on mimicking the genetic operators of selection, mutation, and crossover. The group of abstract representations are called *chromosomes* where each solution corresponds to an *individual* in the population of possible results. In general, there is a randomly selected initial population, which is an important feature of GAs since it eliminates the need for extensive knowledge about the problem space. Various self-learning and assisted-learning methods exist to determine the fitness of candidate solutions. The fitness function evaluates each solution (i.e., chromosome) depending on how close it is to the the desired result.

Chromosomes with a better fitness value are more likely to reproduce offspring that share their characteristics. In this process, called *crossing over*, genes from the *parents* are exchanged. A typical stopping criteria for GA is either when the new generation contains a solution close enough to the desired solution. Otherwise the process continues until a maximum number of iterations is reached.

Let us now present our FGA [24, 20, 23] where the fitness function is inspired by the molecular distribution of forces in physics [12]. In this context, MANET nodes, each running FGA independently, distribute themselves uniformly over an unknown geographical terrain. Each node calculates the virtual forces applied to itself by its near neighbors located within its communication range (R_{com}). This virtual force sums to zero when all near neighbors are optimally distributed around the mobile node; if the sum is not equal to zero, the mobile node runs FGA to find a new location such that its virtual total force is reduced (if possible). The objective of our FGA is to obtain a uniform node distribution over an unknown geographical area without any centralized control unit or global knowledge of the network.

8.3.1 Chromosomes in FGA

In the mobility model described in this chapter, each mobile node can move into one of six hexagonal directions in a given terrain. As an example, suppose mobile nodes can move at two different speeds (e.g., immobile and mobile). Therefore, six different directions with two different speeds can be coded into a 4-bit chromosome $\langle b_3b_2b_1b_0\rangle$. The three low order bits $\langle Xb_2b_1b_0\rangle$ represent hexagonal movement directions where $\langle 000\rangle$ is north, $\langle 001\rangle$ northeast, $\langle 010\rangle$ southeast, $\langle 011\rangle$ south, $\langle 100\rangle$ southwest, and $\langle 101\rangle$ northwest. The high order bit of the chromosome $\langle b_3XXX\rangle$ is used for defining different speeds (e.g., $\langle 0\rangle$ for immobile and $\langle 1\rangle$ for mobile). For example, a chromosome of $\langle 1001\rangle$ means that the mobile node should move one position to the northeast.

8.3.2 Fitness Function for FGA

A fitness function is used to measure the quality of the chromosomes within a solution space Ω, where Ω represents all possible speeds and directions defined in our mobility model. Note that a binary chromosomes is a result of mapping from a continuous space in the node's R_{com} to the possible locations that it can move into (i.e., $R^n \mapsto Z^n$).

FGA's fitness function is based on the virtual forces applied to a node by its neighboring nodes. The virtual force inflicted on a node by its neighbor is inversely proportional to the distance between them. The force from a closer neighbor is greater than a farther one. The force exerted on node N_i by its neighboring node N_j is calculated as:

$$F_{ij} = \begin{cases} (d_{th} - d_{ij}) & \text{if } 0 < d_{ij} < d_{th} \\ 0 & \text{if } d_{th} \le d_{ij} \le R_{com} \end{cases} \tag{8.1}$$

where, d_{ij} is the Euclidean distance between nodes N_i and N_j, d_{th} is the threshold to define the local neighborhood. The virtual force value is used as a part of each node's fitness calculation in our FGA. The fitness function f_i is given as the sum of all the partial forces that are exerted on node N_i by its k neighboring nodes:

$$f_i = \sum_{j=1}^{k} F_{ij} = \sum_{j=1}^{k} (d_{th} - d_{ij}) \tag{8.2}$$

Notice that the objective of FGA is to find a set of parameter values (i.e., chromosomes) such as speed and direction that minimize the fitness function f_i in Eqs. (8.1) and (8.2). The best fitness value (i.e., the lowest value of force) between node i and j is obtained when the two nodes are d_{th} units apart from each other. Similarly, the worst fitness value corresponds to two nodes that are next to each other (i.e., $d_{ij} \approx 0$).

8.4 Dynamical System Model of FGA

We apply Vose's *dynamical system model* [29] to study the convergence of our decentralized topology control mechanism where FGA is run by each autonomous mobile node to achieve a uniform spread over an unknown geographical area. The dynamical system model calculates the cumulative effects of GA operators of selection, mutation, and crossover as a population evolves through generations. This model provides expected population distribution after each generation by observing the changes in the chromosomes. Also, using this model, it is possible to analyze the effect of the initial population on the final convergence of our FGA.

Dynamical system model uses the information in the past populations to create the state of the current population. Due to the nondeterministic nature of the genetic operators, we cannot exactly predict what the next population will be. However, we can calculate the probability distribution over the set of possible populations defined by the genetic information of the past populations (or *states*) to estimate the expected next population.

As the size of a population grows, the probability that the next state of a population will be the targeted one approaches to one since the chance that copies of the best chromosome will be included in the population increases. In this case, the GA behavior becomes less stochastic but more deterministic, and, hence, the trajectory of expected next population gives a safe indication of the actual FGA behavior.

8.4.1 Population Representation

Let Ω denote the search space, and n the cardinality of Ω. Then, using a fixed-length binary string representation, $\Omega = \{0,1\}^{\ell}$, where ℓ is the string length. We will identify the elements of Ω with the integers in the range of $[0,n)$.

A population P can be represented as an incidence vector: $P = \langle P_0, P_1, \ldots, P_{n-1}\rangle$, where P_k is the number of copies of individual $k \in \Omega$ in the population, hence $\sum_{k=0}^{n-1} P_k = \mathcal{N}$, where $\mathcal{N}$ is the population size. To obtain a more general representation, we can also describe the population as a *population vector* $p = \langle p_0, p_1, \ldots, p_{n-1}\rangle$, in which p_k is the proportion that an individual $k \in \Omega$ appears in the population.

For example, suppose a population consists of $\{00,00,01,10,11,10,00,10,10,11\}$ for $\ell = 2$, therefore, $\mathcal{N} = 10$ and $p_0 = 0.3$ since the individual 00 appears three times in the population. For this example, $p = \langle 0.3, 0.1, 0.4, 0.2\rangle$. Properties of population vectors include that p is an element of the vector space R^n (addition and/or multiplication by scalar produce other vectors within R^n), each entry p_k must lie in the range $[0,1]$, and that all entries of p sum to 1 ($\sum_{k=0}^{n-1} p_k = 1$).

The set of all vectors in R^n satisfying these properties is called a *simplex*, denoted by Λ. A simplex is a set of the set of probability distributions over Ω. In general,

$$\Lambda = \left\{ p \in R^n : p_i \geq 0 \text{ and } \sum_{i=0}^{n-1} p_i = 1 \right\} \tag{8.3}$$

We can now view the actions of our FGA on a population as a trajectory of vectors $p \in R^n$. When $n = 2$, the simplex is a straight line segment in the plane R^2, running from $\langle 1,0\rangle$ to $\langle 0,1\rangle$. All real populations correspond to points within the simplex. However, since components in the corresponding population vectors must be rational numbers not all points in the simplex correspond to finite populations. Therefore, only those rational points with common denominator N correspond to valid population vectors. For example, vertex $\langle 1,0\rangle$ represents populations with copies of individual 1 only. The points in between are populations containing a mixture of individuals. When $n = 3$ and $n = 4$, the simplex is a triangle embedded in R^3 and a tetrahedron embedded in R^4.

8.4.2 Heuristic Functions

FGA can be described through a heuristic function $\mathcal{G}(p) : \Lambda \to \Lambda$, where $\mathcal{G}(p)$ contains all the details of selection , crossover, and mutation operators represented by $\mathcal{F}$, C, and $\mathcal{U}$ heuristic functions, respectively. In other words, $\mathcal{G}(p) = \mathcal{U}(C(\mathcal{F}(p))) = \mathcal{U} \circ C \circ \mathcal{F}(p)$ where $\mathcal{G}(p)$ is a discrete-time dynamical system that has the following interpretations: *(i)* $\mathcal{G}(p)$ is the *expected* next generation population, *(ii)* $\mathcal{G}(p)$ is the *limiting next population* as the population size goes to infinity, and *(iii)* $\mathcal{G}(p)_k$ is the *probability* that $k \in \Omega$ is selected to be in the next generation Later sections describe each of the three heuristic functions in more detail.

8.4.3 Selection

Selection is the first operator in FGA. In order to construct our model, we first define *selection heuristic function* $\mathcal{F}(p) : \Lambda \to \Lambda$. Let $p = (p_0, p_1, ..., p_{n-1})$ be our current population and the probability $P_r(k)$ that any individual k will be selected for the next population is given by:

$$Pr(k) = \frac{f_k p_k}{\sum_{i=0}^{n-1} f_i p_i} \tag{8.4}$$

where f_i represents the fitness value of individual $i \in \Omega$. Then the fitness proportional selection operator $\mathcal{F}(p)$ expressed in terms of matrices as

$$\mathcal{F}(p) = \frac{diag(f)p}{f^T p} \tag{8.5}$$

where f is the fitness function expressed as a vector ($f_k = f(k)$), $diag(f)$ is the diagonal matrix with entries from vector f along its diagonal, and $f^T p$ is the average fitness of the population p (inner product of vectors f and p).

We now give an example which we will continue through the remaining sections.

Let $\ell = 2$, $\mathcal{N} = 10$, initial population $P = \langle 3, 1, 4, 2 \rangle$, and fitness function $f(x) = 4x$. This gives $p = \left\langle \frac{3}{10}, \frac{1}{10}, \frac{2}{5}, \frac{1}{5} \right\rangle$, $f_0 = 0$, $f_1 = 4$, $f_2 = 8$, and $f_3 = 12$. Using Eq. (8.5), we calculate $F(p)$ as follows:

$$\mathcal{F}(p) = \frac{\begin{bmatrix} 0 & 0 & 0 & 0 \\ 0 & 4 & 0 & 0 \\ 0 & 0 & 8 & 0 \\ 0 & 0 & 0 & 12 \end{bmatrix} \begin{bmatrix} 0.3 \\ 0.1 \\ 0.4 \\ 0.2 \end{bmatrix}}{\begin{bmatrix} 0 & 4 & 8 & 12 \end{bmatrix} \begin{bmatrix} 0.3 \\ 0.1 \\ 0.4 \\ 0.2 \end{bmatrix}} = \frac{\begin{bmatrix} 0 \\ 0.4 \\ 3.2 \\ 2.4 \end{bmatrix}}{\begin{bmatrix} 6.0 \end{bmatrix}} = \begin{bmatrix} 0 \\ 0.0667 \\ 0.5333 \\ 0.4 \end{bmatrix} \tag{8.6}$$

If proportional selection is the only genetic operator, then $\mathcal{G} = \mathcal{F}$ and we now have the evolutionary equation for $\mathcal{G}$.

8.4.4 Crossover

Crossover operator never changes the number of zeros or ones in a given chromosome pair, but only shuffles their bits. In this example, we use a *one-point crossover* (1X) where the *crossover point* is chosen at random from the numbers $1,\ldots,\ell-1$, and a new chromosome is produced by combining the pieces of the original parent chromosomes. For instance, if we combine parent chromosomes $a_3a_2a_1a_0$ and $b_3b_2b_1b_0$ with a crossover point of 2, the offspring chromosomes would be $a_3a_2b_1b_0$ and $b_3b_2a_1a_0$.

Crossover can be easily written as a binary string or *mask* when implementing crossover operator in a computer algorithm. Therefore, any linear crossover operator is represented by a vector $m \in \Omega$ so that the offspring of parents a and b are $a\otimes m \oplus b \otimes \overline{m}$ and $a \otimes \overline{m} \oplus b \otimes m$, where $\overline{m}$ is the complement of m, and $\otimes, \oplus$ denote component-wise multiplication and addition, respectively. For example, if we have parents (1011) and (0110), the mask (1100) indicates that we take the first two bits of the first parent and the last two from the second parent to produce (1010).

The effects of applying crossover can be represented with the *crossover heuristic function* $C(p): \Lambda \to \Lambda$ such that the kth component of $C(p)$ is the probability that individual $k \in \Omega$ results from applying crossover to population p.

The probability that a given chromosome k is created by applying the crossover operator to a population p is found by summing over all the possible ways this can happen. Hence, $C(p)$ is defined as:

$$C(p)_k = \sum_{i,j} p_i p_j r(i,j,k) \tag{8.7}$$

where individuals $i, j \in \Omega$, and $r(i,j,k)$ is the probability of creating chromosome k from parent chromosomes i and j.

One-point crossover is not symmetric, so that $r(i,j,k) \neq r(j,i,k)$, however we can define symmetric matrices M_k having $(i,j)^{\text{th}}$ entries as:

$$M_k = \frac{1}{2}(r(i,j,k) + r(j,i,k)) \tag{8.8}$$

Therefore,

$$C(p)_k = p^T M_k p \tag{8.9}$$

To continue the example started in the previous section, there are two valid masks $m \in \Omega$ for $\ell = 2$ that represent one-point crossover. They are $m = \mathbf{2} = (10)$ and $m = \mathbf{3} = (11)$ with probabilities μ_c and $(1-\mu_c)$, respectively. We now calculate $r(i,j,0)$, the probability that individuals i and j cross to form individual $\mathbf{0} = (00)$ with a crossover rate of μ_c using $m = \mathbf{2} = (10)$ and $m = \mathbf{3} = (11)$:

$$r(i,j,\mathbf{0}) = \begin{bmatrix} \mu_c+(1-\mu_c) & \mu_c & 0 & 0 \\ 1-\mu_c & 0 & 0 & 0 \\ \mu_c+(1-\mu_c) & \mu_c & 0 & 0 \\ 1-\mu_c & 0 & 0 & 0 \end{bmatrix} \tag{8.10}$$

For example, to produce $\mathbf{0} = (00)$ from parents $\mathbf{0} = (00)$ and $\mathbf{2} = (10)$, we can use either mask (10) in view of $(00)\otimes(10)\oplus(10)\otimes(01) = (00)$, or mask (11) in view of $(00)\otimes(11)\oplus(10)\otimes(00) = (00)$. The probability is clearly $\mu_c + (1-\mu_c)$, since either the first bit must come from $\mathbf{0}$ and the other from $\mathbf{2}$ with probability μ_c, or both bits come from $\mathbf{0}$ with probability $(1-\mu_c)$.

We now proceed to calculate $r(j,i,0)$, the probability that individuals j and i cross to form individual $\mathbf{0} = (00)$ using $\overline{m} = \mathbf{1} = (01)$ and $\overline{m} = \mathbf{0} = (00)$:

$$r(j,i,\mathbf{0}) = \begin{bmatrix} \mu_c+(1-\mu_c) & 1-\mu_c & \mu_c+(1-\mu_c) & 1-\mu_c \\ \mu_c & 0 & \mu_c & 0 \\ 0 & 0 & 0 & 0 \\ 0 & 0 & 0 & 0 \end{bmatrix} \tag{8.11}$$

From the matrix above, we can see that the probability of producing $\mathbf{0} = (00)$ from parents $\mathbf{0} = (00)$ and $\mathbf{2} = (10)$ is zero since offspring $(00)\otimes(01)\oplus(10)\otimes(10) = (10)$ and $(00)\otimes(00)\oplus(10)\otimes(11) = (10)$ do not result in individual $\mathbf{0} = (00)$.

We now combine matrices $r(i,j,0)$ and $r(j,i,0)$ to obtain symmetric matrix M_0 using Eq.(8.8):

$$M_0 = \frac{1}{2}(r(i,j,0) + r(j,i,0)) \tag{8.12}$$

$$M_0 = \begin{bmatrix} 1 & \frac{1}{2} & \frac{1}{2} & \frac{(1-\mu_c)}{2} \\ \frac{1}{2} & 0 & \frac{\mu_c}{2} & 0 \\ \frac{1}{2} & \frac{\mu_c}{2} & 0 & 0 \\ \frac{(1-\mu_c)}{2} & 0 & 0 & 0 \end{bmatrix} \tag{8.13}$$

And doing this for each individual $k \in \Omega$, we have

$$C(p) = \left(p^T M_0 p, p^T M_1 p, \ldots, p^T M_{n-1} p\right) \tag{8.14}$$

In the same manner as M_0 was concluded, M_1 is calculated as

$$M_1 = \frac{1}{2}(r(i,j,1) + r(j,i,1)) \tag{8.15}$$

$$M_1 = \begin{bmatrix} 0 & \frac{1}{2} & 0 & \frac{\mu_c}{2} \\ \frac{1}{2} & 1 & \frac{(1-\mu_c)}{2} & \frac{1}{2} \\ 0 & \frac{(1-\mu_c)}{2} & 0 & 0 \\ \frac{\mu_c}{2} & \frac{1}{2} & 0 & 0 \end{bmatrix} \tag{8.16}$$

The probabilities in each matrix M_k are exactly those that are in M_0, but shuffled around according to the permutation matrix α_k. Therefore, we only have to calculate M_0 (*mixing matrix*) to capture the whole effect of crossover. This means that

$$M_k = \alpha_k M_0 \alpha_k^T \text{ for each } k \in \Omega \tag{8.17}$$

In order to calculate permutation matrix α_k, we consider each element within Ω as a group under bitwise addition (modulo 2). In our example, the group table for addition is:

$$\begin{array}{c|cccc} & 00 & 01 & 10 & 11 \\ \hline 00 & 00 & 01 & 10 & 11 \\ 01 & 01 & 00 & 11 & 10 \\ 10 & 10 & 11 & 00 & 01 \\ 11 & 11 & 10 & 01 & 00 \end{array} \tag{8.18}$$

The element $\mathbf{0} = (00)$ is the identity in this group. The corresponding permutation matrices are

$$\alpha_0 = \begin{bmatrix} 1&0&0&0 \\ 0&1&0&0 \\ 0&0&1&0 \\ 0&0&0&1 \end{bmatrix} \quad \alpha_1 = \begin{bmatrix} 0&1&0&0 \\ 1&0&0&0 \\ 0&0&0&1 \\ 0&0&1&0 \end{bmatrix}$$
$$\alpha_2 = \begin{bmatrix} 0&0&1&0 \\ 0&0&0&1 \\ 1&0&0&0 \\ 0&1&0&0 \end{bmatrix} \quad \alpha_3 = \begin{bmatrix} 0&0&0&1 \\ 0&0&1&0 \\ 0&1&0&0 \\ 1&0&0&0 \end{bmatrix} \tag{8.19}$$

In general, we can observe that for any given α_k, a 1 is set where entry $(i, j)^{\text{th}}$ in Table 8.18 is equal to k. For instance, for α_2, a 1 is placed where entry $(i, j)^{\text{th}}$ is equal to $i \oplus j = \mathbf{2} = (10)$, otherwise a 0 is assigned.

To continue with our example, let us assume $\mu_c = 0.7$, hence the complete set of matrices for this crossover operator using Eq.(8.17) are:

$$M_0 = \begin{bmatrix} 1 & 0.5 & 0.5 & 0.15 \\ 0.5 & 0 & 0.35 & 0 \\ 0.5 & 0.35 & 0 & 0 \\ 0.15 & 0 & 0 & 0 \end{bmatrix} \quad M_1 = \begin{bmatrix} 0 & 0.5 & 0 & 0.35 \\ 0.5 & 1 & 0.15 & 0.5 \\ 0 & 0.15 & 0 & 0 \\ 0.35 & 0.5 & 0 & 0 \end{bmatrix}$$
$$M_2 = \begin{bmatrix} 0 & 0 & 0.5 & 0.35 \\ 0 & 0 & 0.15 & 0 \\ 0.5 & 0.15 & 1 & 0.5 \\ 0.35 & 0 & 0.5 & 0 \end{bmatrix} \quad M_3 = \begin{bmatrix} 0 & 0 & 0 & 0.15 \\ 0 & 0 & 0.35 & 0.5 \\ 0 & 0.35 & 0 & 0.5 \\ 0.15 & 0.5 & 0.5 & 1 \end{bmatrix} \tag{8.20}$$

Now, we calculate $C(p)$ using Eq.(8.9)

$$C(p)_k = \mathcal{F}(p)^T M_k \mathcal{F}(p) = \begin{bmatrix} 0.0249 \\ 0.0418 \\ 0.5089 \\ 0.4249 \end{bmatrix} \tag{8.21}$$

8.4.5 Mutation

The mutation operator randomly changes a bit with usually a small probability in order to provide diversity to a population and protect against getting stuck in a local optima. Mutation can be defined by means of mutation masks. If $j \in \Omega$, then the result of mutating j using a mutation mask $m \in \Omega$ is $j \oplus m$. The mutation heuristic is defined by giving a probability distribution $\mu \in \Lambda$ over mutation masks. In other words, μ_m is the probability that $m \in \Omega$ is used. Given a population $p \in \Lambda$, the *mutation heuristic* $\mathcal{U}(p) : \Lambda \to \Lambda$ is defined by

$$\mathcal{U}(p)_k = \sum_{j \in \Omega} \mu_{j \oplus k} p_j \tag{8.22}$$

where $\mu_{j\oplus k}$ is the probability that an individual j mutates to an individual k using a mutation mask $m \in \Omega$. The contribution of mutation can also be described in terms of a $n \times n$ matrix U (*mutation matrix*) that directly gives the effect of mutation on a population vector. Hence, Eq.(8.22) can simply be shown as

$$\mathcal{U}(p)_k = Up \tag{8.23}$$

We continue the numerical example. Let us assume $\mu_m = 0.2$, then the mutation matrix U is

$$U = \begin{bmatrix} (1-\mu_m)^2 & (1-\mu_m)\mu_m & \mu_m(1-\mu_m) & \mu_m^2 \\ (1-\mu_m)\mu_m & (1-\mu_m)^2 & \mu_m^2 & \mu_m(1-\mu_m) \\ \mu_m(1-\mu_m) & \mu_m^2 & (1-\mu_m)^2 & (1-\mu_m)\mu_m \\ \mu_m^2 & \mu_m(1-\mu_m) & (1-\mu_m)\mu_m & (1-\mu_m)^2 \end{bmatrix} \tag{8.24}$$

$$U = \begin{bmatrix} 0.64 & 0.16 & 0.16 & 0.04 \\ 0.16 & 0.64 & 0.04 & 0.16 \\ 0.16 & 0.04 & 0.64 & 0.16 \\ 0.04 & 0.16 & 0.16 & 0.64 \end{bmatrix} \tag{8.25}$$

Now, we calculate $\mathcal{U}(p)$ using Eq.(8.23)

$$\mathcal{U}(p) = UC(p) \tag{8.26}$$

$$\mathcal{U}(p) = \begin{bmatrix} 0.64 & 0.16 & 0.16 & 0.04 \\ 0.16 & 0.64 & 0.04 & 0.16 \\ 0.16 & 0.04 & 0.64 & 0.16 \\ 0.04 & 0.16 & 0.16 & 0.64 \end{bmatrix} \begin{bmatrix} 0.0249 \\ 0.0418 \\ 0.5084 \\ 0.4249 \end{bmatrix} = \begin{bmatrix} 0.1210 \\ 0.1190 \\ 0.3990 \\ 0.3610 \end{bmatrix} \tag{8.27}$$

Fig. 8.1 shows the expected next generation population for all four chromosomes of length $\ell = 2$ in our numerical example. Chromosome $\mathbf{3} = (11)$ has the highest probability to survive in the entire solution space with a probability of 0.4916 after 10 generations. Chromosomes $\mathbf{0}, \mathbf{1}$ and $\mathbf{2}$ have probabilities of 0.0878, 0.1621, and 0.2635 respectively.

As we can see from the example, the dynamical system model takes into account the construction and destruction of strings (chromosomes) by considering all possible ways of performing crossover and mutation operators in a GA. In the most general case (e.g. one point crossover) every string in the search space Ω can potentially be constructed. This effect can be captured by a matrix to represent the effects of recombination, where the strings or vectors are indexed from 0 to $(2^\ell - 1)$. The entry $(i, j, k)th$ of this array gives the probability that vectors i and j will recombine to produce vector k. The effect of mutation is similar to represent, but since only one string is involved, it is simpler.

8.4.6 Estimating FGA Behavior

We now apply the dynamical system model to predict the behavior of FGA in a typical scenario. The dynamical system model will help us to calculate the cumulative effects of GA operators as a population evolves through

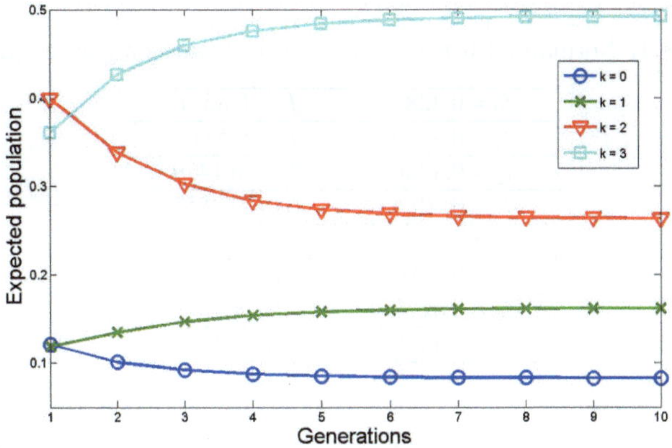

Fig. 8.1. Expected population for 10 generations where k is the chromosome number in the example.

generations. We define our geographical area as a two dimensional $d_h \times d_w$ area divided into logical hexagonal cells, where a unique Cartesian coordinate pair (x,y) is assigned to each one of them. Fig. 8.2 shows an area with 25 logical cells and five mobile nodes. A mobile node may move into six different directions (i.e., D_0 through D_5) and is not allowed to move beyond the area boundaries. The direction and speed of movement for all nodes in the network are determined by our FGA topology control algorithm running on each node. In this example we assume that all nodes remain in the same position except node N_i at location (2, 2) which runs FGA in order to optimize its position.

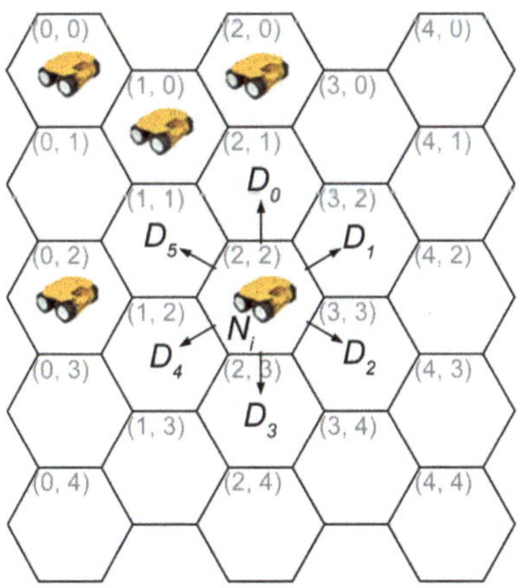

Fig. 8.2. Example: mobile nodes at t with $R_{com} = 3$.

Table 8.1. Normalized fitness values for a chromosome of length $\ell = 4$

$f_0 = 0.4286$	$f_8 = 0.8571$
$f_1 = 0.4286$	$f_9 = 0.5714$
$f_2 = 0.4286$	$f_{10} = 0.1429$
$f_3 = 0.4286$	$f_{11} = 0.2857$
$f_4 = 0.4286$	$f_{12} = 0.4286$
$f_5 = 0.4286$	$f_{13} = 0.8571$
$f_6 = 0.4286$	$f_{14} = 1.0$
$f_7 = 0.4286$	$f_{15} = 1.0$

As mentioned in Sect. 8.3.1, each chromosome is encoded using a bit string of length ℓ representing different combinations of the node's speed and direction. In the hexagonal grid, N_i at the location (2,2) can move in one of 6 different directions, therefore the node's speed and direction are encoded using 1 bit and 3 bits, respectively. Therefore, the chromosome length is set to $\ell = 4$, which requires a solution space Ω with a size of $2^\ell = 16$.

$$\mathcal{F}(p)_0 = \begin{bmatrix} 0.1739 \\ 0 \\ 0 \\ 0.1739 \\ 0.1739 \\ 0 \\ 0 \\ 0 \\ 0.0870 \\ 0 \\ 0 \\ 0.2174 \\ 0.1739 \\ 0 \\ 0 \\ 0 \end{bmatrix} \quad \mathcal{C}(p)_0 = \begin{bmatrix} 0.1590 \\ 0.0105 \\ 0.0250 \\ 0.1479 \\ 0.1544 \\ 0.0116 \\ 0.0019 \\ 0.0114 \\ 0.1032 \\ 0.0121 \\ 0.0193 \\ 0.1726 \\ 0.1419 \\ 0.0139 \\ 0.0017 \\ 0.0141 \end{bmatrix} \quad \mathcal{U}(p)_0 = \begin{bmatrix} 0.1696 \\ 0.0034 \\ 0.0034 \\ 0.1692 \\ 0.1704 \\ 0.0017 \\ 0.0017 \\ 0.0017 \\ 0.0869 \\ 0.0030 \\ 0.0030 \\ 0.2105 \\ 0.1696 \\ 0.0017 \\ 0.0017 \\ 0.0021 \end{bmatrix}$$

$(a) \qquad (b) \qquad (c)$

Fig. 8.3. Expected population distribution after (a) $\mathcal{F}$, (b) C, and (c) $\mathcal{U}$ operators for population vector p after one generation.

Table 8.1 shows the normalized fitness values for chromosomes of length $\ell = 4$. For instance, chromosome **13** = (1101) represents the node moving one hexagonal grid northwest. Notice in Table 8.1 that chromosome **13** with a high fitness value of $f_{13} = 0.8571$ among the other candidate solutions; since

$$\mathcal{F}(p)_{40} = \begin{bmatrix} 0.0004 \\ 0 \\ 0.0194 \\ 0.0015 \\ 0 \\ 0 \\ 0.0003 \\ 0 \\ 0.0019 \\ 0.0006 \\ 0.9254 \\ 0.0502 \\ 0.0002 \\ 0 \\ 0 \\ 0 \end{bmatrix} \qquad \mathcal{C}(p)_{40} = \begin{bmatrix} 0.0006 \\ 0.0001 \\ 0.0280 \\ 0.0022 \\ 0 \\ 0 \\ 0.0004 \\ 0 \\ 0.0110 \\ 0.0011 \\ 0.8892 \\ 0.0573 \\ 0.0003 \\ 0 \\ 0.0092 \\ 0.0006 \end{bmatrix} \qquad \mathcal{U}(p)_{40} = \begin{bmatrix} 0.0007 \\ 0.0001 \\ 0.0277 \\ 0.0023 \\ 0 \\ 0 \\ 0.0006 \\ 0.0001 \\ 0.0108 \\ 0.0011 \\ 0.8896 \\ 0.0572 \\ 0.0003 \\ 0 \\ 0.0090 \\ 0.0006 \end{bmatrix}$$

$(a) \qquad (b) \qquad (c)$

Fig. 8.4. Expected population distribution after (a) $\mathcal{F}$, (b) C, and (c) $\mathcal{U}$ operators for population p after 40 generations.

FGA favors smaller fitness values, chromosome **13** has little expectations that copies of itself will be included in forthcoming populations compared to, say, chromosome **11** with a fitness of 0.2857. Let us start with a random initial population P consisting of chromosomes **0**, **3**, **4**, **11**, and **12**, two copies of chromosome **8**, and three copies of chromosome **14**. Using Eq. (8.5), we can calculate $F(p)$ for FGA as shown in Fig. 8.3(a). After applying the selection operator to population P, chromosome **11**, with only one copy in the initial population, has the highest probability (0.2174) of being selected for crossover. On the other hand, chromosome **14** with three copies in population P, has zero probability that copies of itself will be selected due to its high fitness value of $f_{14} = 1$. Recall that the goal of FGA is to find candidate solutions that minimize the fitness function shown in Eq. (8.2). Using Eq. (8.9) with a crossover probability of $c = 0.5$, we obtain $C(p)$ as shown in Fig. 8.3(b). We observe from Fig. 8.3(b) that chromosome **11** still remains the individual with the highest probability to survive at the next generation after crossover operation. Also notice that most of the 16 chromosomes have non-zero probabilities compared with the values of $F(p)$ in Fig. 8.3(a), which implies that FGA developed a non-zero possibility of survival for many individuals after crossover. For a mutation probability $\mu = 0.01$, $\mathcal{U}(p)$ is calculated using Eq. (8.22) as shown in Fig. 8.3(c). Chromosome **11** has the highest probability of surviving and be a part of the next generation population, with probability of 0.2105.

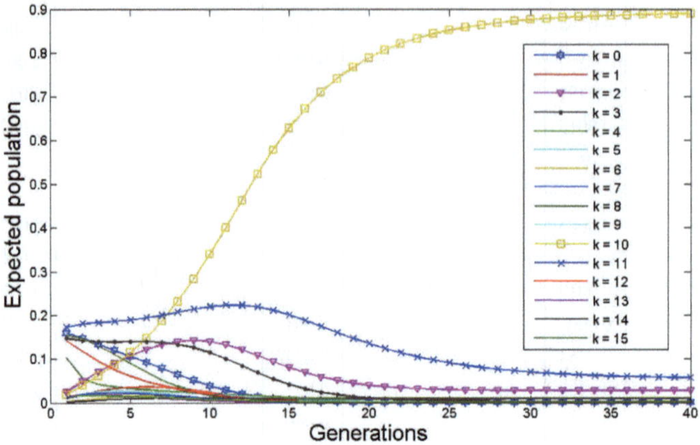

Fig. 8.5. Expected population distribution for chromosomes of length $\ell = 4$ with $\mu = 0.01$ and $c = 0.5$, after 40 generations. For clarity of presentation, we inserted symbols over the lines depicting only the most important chromosomes.

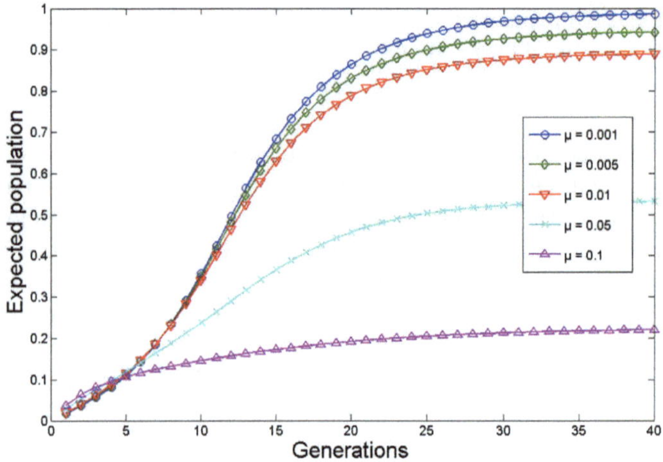

Fig. 8.6. Expected population distribution of chromosome **10** = (1010) for different mutation rates of μ after 40 generations.

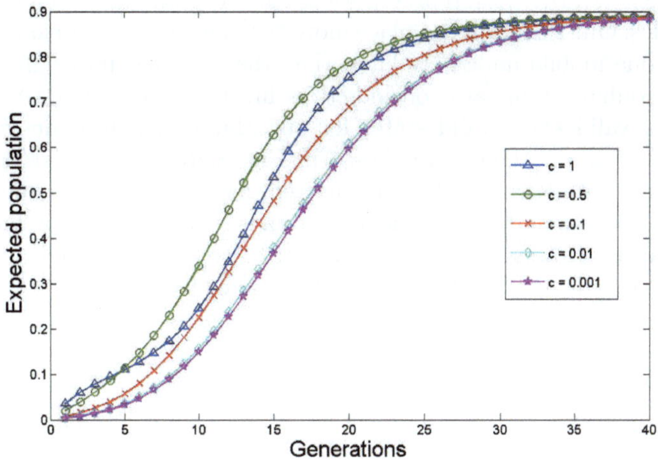

Fig. 8.7. Expected population distribution of chromosome **10** = (1010) for different crossover rates c after 40 generations.

Fig. 8.4 shows the expected population distribution after 40 generations. Chromosome **10** = (1011) has the highest probability to survive in the entire solution space with a probability of 0.8892 after 40 generations. We observe the similar result from Fig. 8.5 where the expected generation population for all 16 chromosomes of length $\ell = 4$ during 40 generations are displayed. The probabilities of crossover (c) and mutation (μ) are set to 0.5 and 0.01, respectively. In Figs. 8.4 and 8.5 chromosome **10** represents N_i at the location (2,2) moving one grid southeast after running FGA (Fig. 8.2).

Figures 8.6 and 8.7 examine the effect of mutation and crossover in FGA. They show the expected population distributions of chromosome **10** = (1010), the fittest chromosome in our example, for different mutation and crossover rates. Small changes in the mutation rate μ cause a significant impact in FGA's trajectory (Fig. 8.6). As mutation increases, chromosome **10** gets weaker, and it is more likely that it (and any offspring) will die out. On the other hand, Fig. 8.7 shows that changes in the crossover rate of c do not have a major impact on the expected population distributions of chromosome **11**.

8.5 Markov Chain Model for FGA

In this section, we introduce a Markov chain model to represent the different node configurations that a MANET with N nodes can occupy. In this model, a state $s_i \in S = \{s_1, s_2, \ldots, s_r\}$, represents a particular geometric configuration of mobile nodes. Our FGA will not place more than one node at a given location. If at least one mobile node is isolated (i.e., the node has no neighbors) the geometric configurations are considered as invalid states. Figure 8.8 shows examples of valid and invalid states for our Markov chain model for three nodes with $R_{com} = 2$. The state shown in Fig. 8.8(a) is considered a valid state because the network is fully connected, meaning that all nodes in the network are reachable by other nodes through either one-hop or multi-hop communication. However, the state in Fig. 8.8(b) is considered an invalid state since there is one isolated node in this configuration.

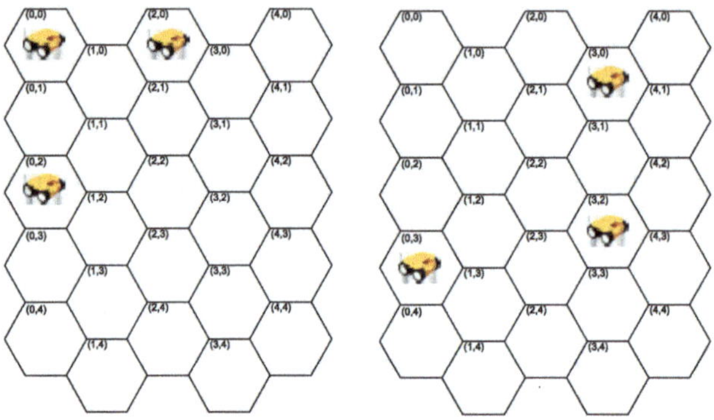

Fig. 8.8. Examples of our Markov chain model for $R_{com} = 2$ and $N = 3$: (a) is a valid state, while (b) is an invalid state.

The number of valid states, S_{val}, for a given MANET can be calculated as the number of permutations of a set without repetition minus the number of invalid states, S_{inv}:

$$S_{val} = \frac{(d_h \times d_w)!}{N! \cdot (d_h \times d_w - N)!} - S_{inv} \tag{8.28}$$

where $(d_h \times d_w)$ is the size of the hexagonal grid. Since S_{val} grows factorially, it is infeasible to build models to analyze the convergence properties of realistic networks. However, the size of S_{val} does not represent the computational complexity of FGA, where the nodes base their decisions only on the local neighborhood information with a near linear complexity.

Examples of the number of valid states are given in Tables 8.2 and 8.3. In Table 8.2, we list S_{val} for the population of 3 nodes with varying R_{com} and the size of the deployment terrain. Table 8.3 demonstrates the changes in S_{val} when number of nodes and the R_{com} change, with fixed deployment terrain.

We can see in Tables 8.2 and 8.3 that when R_{com} increases, S_{val} gets bigger, as there are less invalid states in a given topology (i.e., the nodes can relocate further without partitioning the network). Also, when the deployment terrain becomes more saturated by the nodes, there are fewer configuration permutations.

Table 8.2. Number of valid states for 3 nodes with varying R_{com} and hexagonal grid sizes.

	Hexagonal Grid Sizes						
R_{com}	3×3	4×4	5×5	6×6	7×7	8×8	9×9
1	32	82	154	248	364	502	662
2	81	358	857	1554	2449	3542	4833
3	84	540	1752	3856	6819	10574	15121
4	84	560	2227	5916	12033	20614	31539
5	84	560	2298	6932	15995	30362	50309

Table 8.3. Number of valid states in 5×5 hexagonal grid with varying R_{com} and number of nodes.

	Number of Nodes						
R_{com}	3	4	5	6	7	8	9
1	154	453	1364	4026	11339	29888	72656
2	857	4943	24301	97980	318124	827843	1737443
3	1752	10709	48864	170792	474014	1076233	2039653
4	2227	12559	53075	177084	480698	1081575	2039653
5	2298	12650	53130	177100	480700	1081575	2039653

A *transition matrix* Q is a numerical representation of a Markov model specifying the probability that a system will transition from one state s_i to another state s_j in a single time unit. A *right stochastic matrix* has an equal number of rows and columns and each of the rows has a sum of one, thus signifying that there are no *hidden* states [31]. Given the locations of a set of mobile nodes at time t, the probability that they will move from a given configuration (i.e., a *state*) to any other configuration at time $(t+1)$ can be calculated using the dynamical system model. Thus, for a given MANET, a one-step memoryless transition matrix for FGA can be generated. It is important to note that we are not concerned with the identity of the node that has moved from one location to another, merely, the change in the geometric configuration of the mobile nodes in a given transition.

8.5.1 Homogeneous Finite Markov Chains

A homogeneous Markov chain on a finite space S having an initial distribution $\nu(s)$ (i.e., the probability of being in a particular state) has a transition matrix that is the same at every instant of time (i.e., $Q_i = Q$, where $i = 1,2,\cdots$). Given this definition, the system is considered *memoryless*. The distribution of states $s \in S$ at times $t \geq 0$ is given by $\mu_t(s_0,\cdots,s_n) = \nu(s_0,\cdots,s_n)Q_1\cdots Q_t$. Likewise, the iterated transition matrix $Q^t = Q_1Q_2\cdots Q_t$ demonstrating the likelihood of transitioning from one state to another in t steps. A useful subset of Markov chains includes those that are *ergodic* because it has been proved that they will eventually converge to a stationary distribution [31, 5, 2]. For a transition matrix to be ergodic, it must be both *irreducible* and *aperiodic.*

A Markov chain is considered irreducible if all states can be reached from all others. In a Markov chain $(X_0, X_1,\ldots)$ with state space $S = s_1,\ldots,s_k$ and transition matrix Q, a state s_i communicates with another state s_j , writing $s_i \to s_j$, if the chain has positive probability of ever moving to s_j starting from s_i. If $s_i \to s_j$ and $s_j \to s_i$, we say that the states s_i and s_j intercommunicate, and write $s_i \leftrightarrow s_j$.

Definition 8.1. A Markov chain $(X_0, X_1,\ldots)$ with state space $S = \{s_1,\ldots,s_k\}$ and transition matrix Q is **irreducible** if for all $s_i, s_j \in S$ there exists $s_i \leftrightarrow s_j$.

Definition 8.2. A state s_i of a Markov chain is called **absorbing** if it is impossible to leave it (i.e., $Q_{s_is_i} = 1$); a Markov chain is absorbing if it has at least one absorbing state, and if from every state it is possible to go to an absorbing state (not necessarily in one step).

For instance, there is no absorbing state in the Markov chain shown in Fig. 8.9 for our FGA topology control algorithm, since $Q_{s_is_i} \neq 1$ for all $s_i \in S$.

The following lemma states that our topology control model using FGA is irreducible [20].

Lemma 8.1. *A Markov chain model for our FGA topology control algorithm is irreducible if and only if* $Q(T_{s_j} < \infty | s_0 = s_i) > 0$ *for all* $s_i, s_j \in S$ *assuming* $Q^0(s_0 = s_i)$ *and* T_{s_j} *is the shortest number of steps from* s_i *to* s_j *(i.e., the probability of moving from a given state to another in finite time is non-zero).*

Proof. Suppose there exists integers n, m such that $Q^n_{s_is_k} > 0$ and $Q^m_{s_ks_j} > 0$. If $l = n+m$, then $Q^l_{s_is_j} > Q^n_{s_is_k}Q^m_{s_ks_j} > 0$ by using the Chapman-Kolmogorov equations [11]. Therefore, we conclude that the Markov chain can go from s_i to s_j by first going from s_i to s_k in n steps, and then (independent of the past) going from s_k to s_j in additional m steps. Moreover the state space S can be only partitioned in one communicating class C (that is, if every state is accessible from every other), hence $S = C$ and the chain (and its transition matrix Q) is said to be irreducible. ■

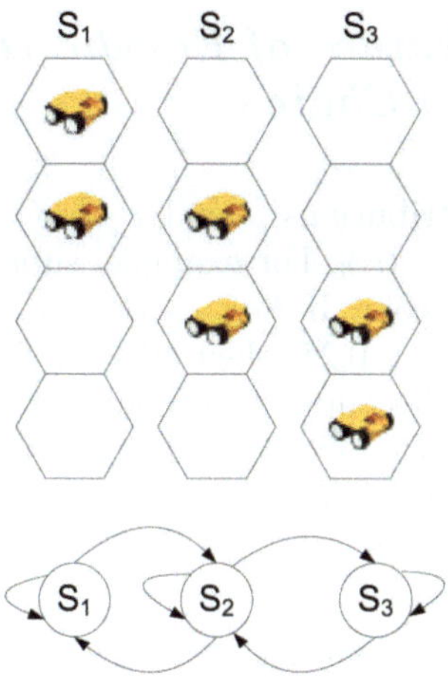

Fig. 8.9. Markov chain model for $N = 2$, $R_{com} = 1$, and 4×1 hexagonal grid.

If we consider the chain in Fig. 8.9(b) for $N = 2$, $R_{com} = 1$ and a 4×1 hexagonal grid, we see that the state space $S = \{1,2,3\}$ cannot be broken into disjoint subsets. Hence, there is only one communicating class $C = S$ where the set of states $C = \{1,2,3\}$ all communicate with one another.

For a finite or infinite set $\{a_1, a_2, \ldots\}$ of positive integers, $gcd\{a_1, a_2, \ldots\}$ is the greatest common divisor of $a_1, a_2, \ldots$. The **period** $d(s_i)$ of a state $s_i \in S$ is defined as $d(s_i) = gcd\{n \geq 1 : (Q^n)_{i,i} > 0\}$. In other words, the period of s_i is the greatest common divisor of the set of times that the chain can return to s_i, given that we start with $s_0 = s_i$. If $d(s_i) = 1$, then the state s_i is **aperiodic**.

Definition 8.3. A Markov chain is said to be **aperiodic** if all its states are aperiodic.

We show in the following lemmas that FGA is aperiodic, irreducible and hence ergodic [20].

Lemma 8.2. *In a Markov chain for our FGA, $d(s_i) = 1$ for all states $s_i \in S$, hence the Markov chain is aperiodic.*

Proof. Using our FGA, a node may not move if FGA did not find a fitter position. Therefore, the 1-step transition probability is $Q_{s_i s_i} > 0$ for each state $s_i \in S$. This implies that any state is accessible from itself (i.e., a self-loop transition exists). Therefore the chain is aperiodic (See Fig. 8.9). ■

Lemma 8.3. *The analytical configuration-based Markov chain representation of our FGA topology control algorithm is both irreducible and aperiodic, and hence ergodic.*

Proof. Lemma 8.1 demonstrates that every state is reachable in finite time from every other state. Lemma 8.2 shows that *self-loop* transitions exist for every valid state. Therefore, our FGA is ergodic [31]. ■

8.5.2 Convergent Nature of Ergodic Homogeneous Finite Markov Chains

For a finite set S with distributions μ and ν on S, the *total variation* is defined as $\|\mu - \nu\| = \sum_n |\mu(s_i) - \nu(s_i)|$. For example, suppose we are comparing a *fair* coin with an *unfair* coin. If $Pr_{fair} = [Pr_{heads} = 0.5, Pr_{tails} = 0.5]$ and $Pr_{unfair} = [Pr_{heads} = 0.75, Pr_{tails} = 0.25]$ then $\left\|Pr_{fair} - Pr_{unfair}\right\| = |(0.5 - 0.75)| + |(0.5 - 0.25)| = 0.5$. Dobrushin's contraction coefficient [6] extends this notion and provides a rough measure of orthogonality between the distributions in a Markov kernel (i.e., transition matrix Q). Contraction coefficient c for transition matrix Q is defined as:

$$c(Q) = \frac{1}{2} \cdot \underbrace{max}_{s_i, s_j} \left|Q(s_i, \cdot) - Q(s_j, \cdot)\right| \tag{8.29}$$

The coefficient c(Q) represents half the largest total variation between any two rows in the transition matrix. When $c(Q) = 1$, any two rows (i.e. distributions) of the transition matrix are disjoint. When $c(Q) = 0$, all of the rows in the transition matrix $Q(s_i, \cdot)$ are equal. The application of these measures lead to the following statements (proofs are in Winkler [31]:

Lemma 8.4. *(from [31]) Let Q_1 and Q_2 be transition matrices and let μ and ν be probability distributions:* $|\mu Q_1 - \nu Q_1| \leq c(Q_1)|\mu - \nu|$, $c(Q_1 Q_2) \leq c(Q_1)c(Q_2) \Rightarrow |\mu Q_1 - \nu Q_1| \leq |\mu - \nu|$, $|\mu Q_1 - \nu Q_1| \leq 2 \cdot c(Q_1)$.

Lemma 8.5. *(from [31]) For each transition matrix Q the sequence $(c(Q^t))t \geq 0$ decreases.*

Lemma 8.6. *(from [31]) If Q is ergodic then the sequence decreases to* 0.

Theorem 8.1. *(from [31]) For an ergodic homogeneous transition matrix Q on a finite space with a stationary distribution μ, then uniformly for all distributions ν, $\nu Q^t \to \mu$ as $t \to \infty$.*

Winkler demonstrates in Lemma 8.4 that the interaction of various distributions within a set with an ergodic system (transition matrix) reduces the orthogonality (i.e. the total variation) between them (assuming they are not disjoint). Lemmas 8.5 and 8.6 state that as a distribution cycles through an ergodic system it converges to a stationary distribution. Theorem 8.1 extends this result to include any initial distribution of the set. Using Theorem 8.1 we can state that the analytical model of our FGA topology control mechanism will converge to a stationary behavior [20]:

Theorem 8.2. *If the transition matrix Q for a Markov chain of our* FGA *topology control algorithm is ergodic, then Q will converge to a stationary distribution.*

Proof. As stated in Lemma 8.3, the transition matrix Q for a Markov chain of our FGA topology control algorithm is ergodic. Therefore, using Theorem 8.1, Q will converge to a stationary distribution. ■

In the following sections, we analytically find the convergence measure of various system configurations to support the statement given by Theorem 8.2. We also study the fitness of the final stationary distributions.

8.5.3 Convergence of FGA Analytical Model

In the analysis of the Markov model of our FGA, the likelihood that the mobile nodes will move from one configuration to another is based on their current location. This transition is memoryless with regard to previous movements. For simplicity and without loss of generality, we present two case studies. First, the number of mobile nodes in the network was set to 3 for each analysis and the size of the hexagonal grid was varied between 3×3, 4×4 and 5×5 (larger values have an exponentially growing number of states, as shown in Table 8.2). Likewise, the communication range R_{com} was varied as 1 and 3 units. In the second case, the hexagonal grid was set to 5×5 and the number of nodes was varied as 3, 4 and 5. Again, the communication range R_{com} was varied as 1 and 3 units.

Figure 8.10 shows the Dobrushin's contraction coefficients iterated over time ($c(Q^t)$ as $t \to \infty$) for $R_{com} = 1$ and 3, and hexagonal grids of size 3×3, 4×4 and 5×5. The contraction coefficient gives a rough measure of the orthogonality (i.e., uniqueness) between the distributions (i.e., rows) in the transition matrix. For all parameter sets the contraction coefficients go to 0 as the system converges towards a stationary distribution. In Fig. 8.10, our FGA topology control algorithm converges slower for larger hexagonal grids since larger spaces require more time steps to evolve into different configurations (i.e., states). If a state is not reachable in a single time step, it will take several steps for the iterated transition matrix Q^t to have a non-zero value. Smaller hexagonal grids can accomplish this transition in fewer iterations and therefore converges to a stationary distribution faster. Figure 8.11 also shows the Dobrushin's contraction coefficients iterated over time for $R_{com} = 1$ and 3 and the number of nodes varied as $N = 3$, 4 and 5. Here we observe that larger numbers of nodes take longer to disperse themselves in their environment as they negotiate around each other. In both Figs. 8.10 and 8.11 we observe that larger communication ranges converge faster by having access to extended knowledge about environment. These measures demonstrate that the system does in fact converge to a stationary distribution. Note that these time steps represent an upper bound for convergence times; the actual convergence occurs much faster based on initial configuration [21].

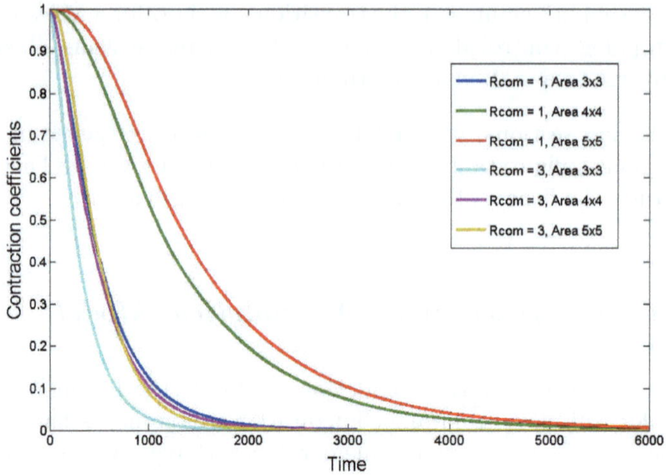

Fig. 8.10. Contraction coefficients for analytical model of FGA topology control algorithm when $N = 3$ for various R_{com} and hexagonal grid sizes.

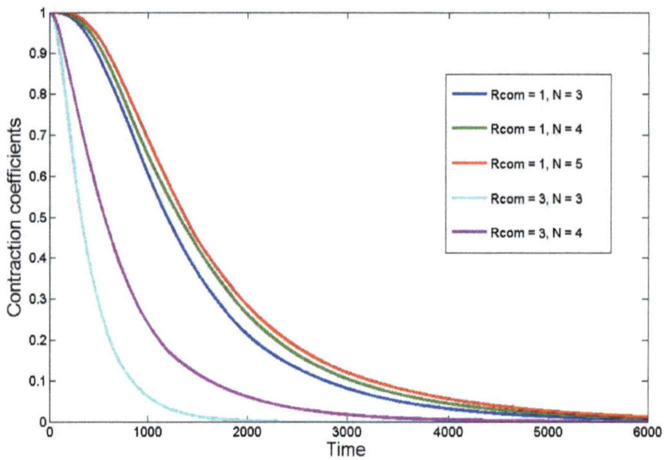

Fig. 8.11. Contraction coefficients for analytical model of FGA topology control algorithm when the hexagonal grid is 5×5 for various R_{com} and numbers of nodes.

8.5.4 Fitness Analysis for Stationary Distribution

We demonstrated analytically in Sect. 8.5.3 that our FGA converges to a stationary distribution. Let us now analyze the fitness of the final stationary distribution to determine if the FGA topology control algorithm will converge to a desired configuration. The total fitness for each state is determined by adding the individual fitness of each of the nodes in a state. For the two cases of Sect. 8.5.3, a total fitness of 0 is denoted as a *high* fitness state, a fitness between 0 and $N/3$ is denoted as a *medium* fitness state, and any value above $N/3$ is considered as a *low* fitness state. Figure 8.12 shows examples of high, medium and low fitness states for $R_{com} = 2$, $N = 3$ in a 5×5 hexagonal grid.

Figure 8.13(a) shows the total fitness of the stationary distribution when $R_{com} = 1$, $N = 3$ with different grid sizes.

The first group of columns in Fig. 8.13(a) represents the aggregated fitness values for $R_{com} = 1$ and $N = 3$ in a 3×3 grid. Out of the 32 possible states available, there is a 97% chance of being in one of the 8 high fitness states when the system converges to a stationary distribution. There is a 3% chance that the system will be in a medium fitness configuration (representing 24/32 states) and a 0% chance of being in a low fitness state. When $R_{com} = 1$, it is impossible to have a low fitness state without becoming disconnected. This demonstrates that the FGA is pushing the system to a desirable configuration. For each of the three configurations in Fig. 8.13(a), the probability of the system stabilizing in a high fitness state when $R_{com} = 1$ for 3×3, 4×4 and 5×5 hexagonal grids it is 97%, 93%, and 93%, respectively.

The first three columns in Fig. 8.13(b) represents the aggregated fitness for the system with $R_{com} = 3$ and $N = 3$ in a 3×3 hexagonal grid. We observe that the probability of being in a medium total fitness state is very high. This observation is expected since $R_{com} = 3$ in a 3×3 represents a densely populated space. The high probability of medium total fitness states corresponds to configurations that are optimal, but penalized for proximity to the border. Larger grid spaces for a similar communication range demonstrate that when crowding is not a problem, the nodes will configure themselves in an optimal configuration and have low fitness states with a probability around 10^{-4}%.. As before, Fig. 8.13(b) shows that when the system converges to a stationary distribution in 4×4 and 5×5 hexagonal grid spaces, there will be a 84% and 77%chance that the system will be in a state with perfect fitness.

In Fig. 8.14 we study the output distributions of the cases for the hexagonal space of 5x5 where the number of nodes was varied as $N = 3$, 4 and 5 for communication ranges of $R_{com} = 1$ and 3. The communication range R_{com} was varied as 1 and 3 units. In Fig. 8.14(a) we observe similar results as in Fig. 8.13(a) where there is a high probability of being in a high fitness state. Likewise, in Fig. 8.14(a) we observe that larger numbers of nodes with a larger R_{com} of 3 corresponds to a higher likelihood of being penalized for being near the border. Still, we observe that FGA has a very high probability of evolving the nodes to an optimal configuration.

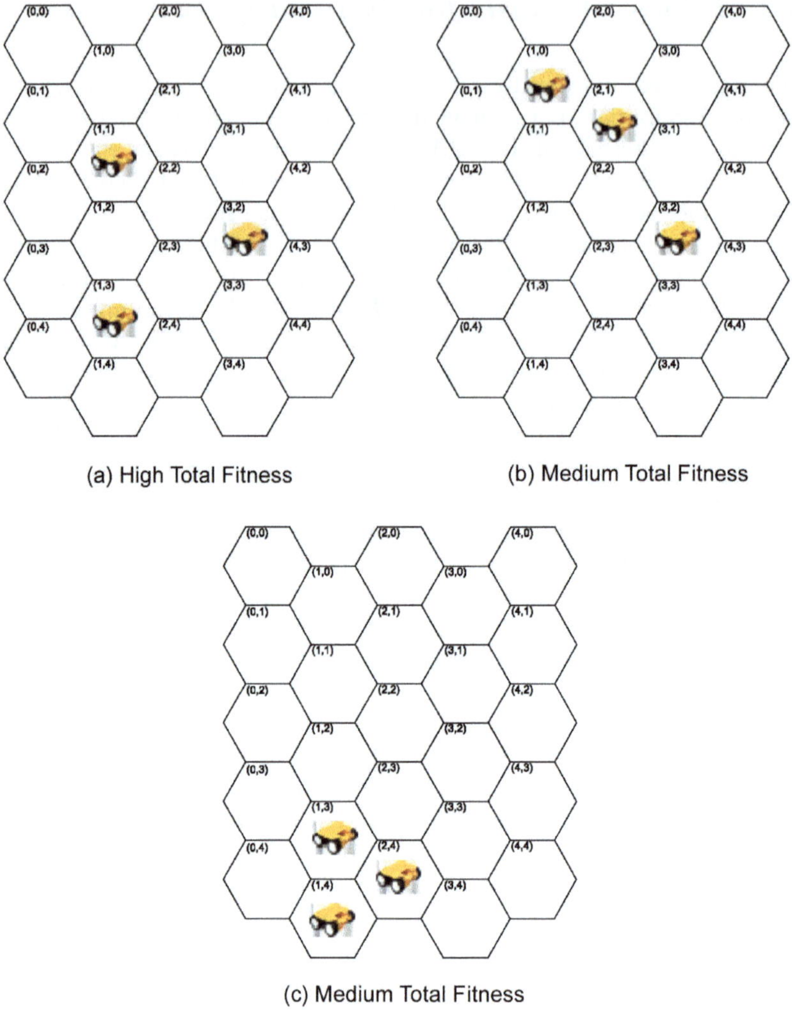

Fig. 8.12. Example of (a) high, (b) medium and (c) low fitness state for $R_{com} = 2$, $N = 3$ and a 5×5 hexagonal grid.

8.6 Conclusions and Future Work

In this chapter, we analyze the convergence properties of our genetic algorithm based topology control mechanism for a uniformly separated node distribution. Here, a mobile node adapts its speed and direction using limited information collected from local neighbors operating in an unknown geographical terrain. We introduce a dynamical system model representing the behavior of a single node's decision mechanism in the network. Based on our

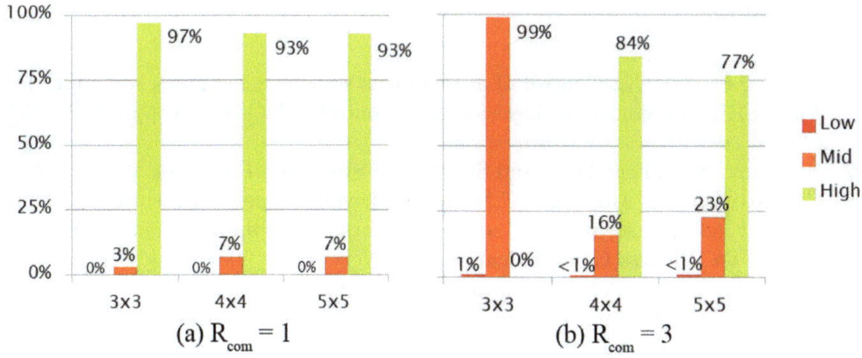

Fig. 8.13. Aggregate total fitness of stationary distribution when $N = 3$ with 3×3, 4×4, and 5×5 hexagonal grids for (a) $R_{com} = 1$ and (b) $R_{com} = 3$.

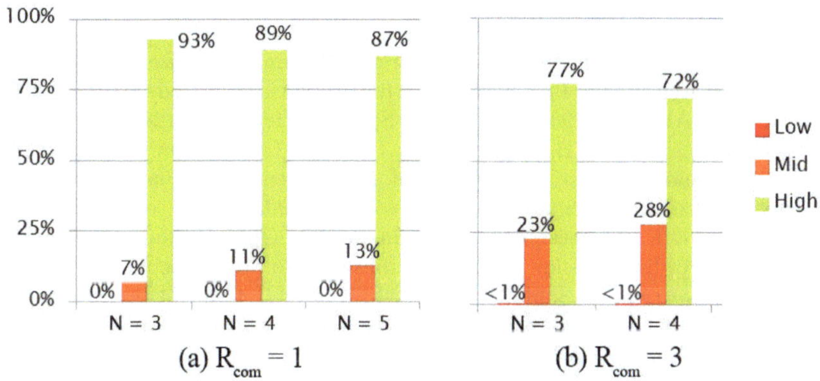

Fig. 8.14. Aggregate total fitness of stationary distribution when the hexagonal grid is 5×5, for (a) $R_{com} = 1$ and (b) $R_{com} = 3$ where $N = 3$, $N = 4$, and $N = 5$.

dynamical model, we build a homogeneous and ergodic Markov chain to analyze the convergence of FGA using Dobrushin's contraction coefficients. We show that FGA can provide acceptable levels of node dispersion over unknown geographical terrains. One very interesting observation is that the nodes using longer communication ranges have more information to make decisions based on a larger neighborhood set; as a result they require less movement and converge faster. Shorter communication ranges make decisions based on a smaller neighborhood set, spend more time in local minima, and consume more energy. Future work will include a more detailed convergence analysis of our FGA and comparison of our analytical model results with measurements from realistic case studies.

References

1. Baras, J.S., Tan, X.: Control of autonomous swarms using gibbs sampling. In: 43rd IEEE Conference on Decision and Control, CDC, vol. 5, pp. 4752–4757 (2004)
2. Borovkov, A.A.: Ergodicity and Stability of Stochastic Processes. Wiley (1998)
3. Campos, C.A.V., Moraes, L.F.M.: A markovian model representation of individual mobility scenarios in ad hoc networks and its evaluation. EURASIP Journal on Wireless Communications and Networking 2007(35946), 14 (2007)
4. Chen, M., Zalzala, A.: Safety considirections in the optimization of the paths for mobile robots using genetic algorithms. In: Proc. of First Int. Conference on Genetic Algorithms in Engineering Systems: Innovations and Applications (1995)
5. Ching, W.K., Ng, M.K., Ching, W.: Markov Chains: Models, Algorithms and Applications. International Series in Operations Research & Management Science. Springer (2005)
6. Dobrushin, R.: Central limit theorem for nonstationary Markov chains. Teor. Veroyatnost Primenen, 365–425 (1956)
7. Garro, B.A., Sossa, H., Vazquez, R.A.: Path Planning Optimization Using Bio-Inspirited Algorithms. In: Gelbukh, A., Reyes-Garcia, C.A. (eds.) MICAI 2006. LNCS (LNAI), vol. 4293, pp. 319–330. Springer, Heidelberg (2006)
8. Gesu, V.D., Lenzitti, B., Bosco, G.L., Tegolo, D.: A distributed architecture for autonomous navigation of robots. In: International Workshop on Computer Architectures for Machine Perception, vol. 0, p. 190 (2000)
9. Goldberg, D.E.: Genetic Algorithms in Search, Optimization and Machine Learning. Addison-Wesley Longman Publishing Co., Inc., Boston (1989)
10. Gundry, S., Urrea, E., Sahin, C.S., Zou, J., Kusyk, J., Uyar, M.U.: Formal convergence analysis for bio-inspired topology control in manets. In: IEEE Sarnoff Symposium, pp. 1–5 (2011)
11. Hachigian, J.: Collapsed Markov chains and the Chapman-Kolmogorov equation. The Annals of Mathematical Statistics 34(1), 233–237 (1963)
12. Heo, N., Varshney, P.: A distributed self spreading algorithm for mobile wireless sensor networks. IEEE Wireless Communications and Networking (WCNC) 3(1), 1597–1602 (2003)
13. Holland, J.H.: Adaptation in Natural and Artificial Systems. University of Michigan, New York (1975)
14. Howard, A., Mataric, M., Sukhatme, G.: Mobile sensor network deployment using potential fields: A distributed, scalable solution to the area coverage problem. In: Proc. Int. Conf. on Distributed Autonomous Robotic Systems (2002)
15. Huang, C.F.: A Markov chain analysis of fitness proportional mate selection schemes in genetic algorithm. In: GECCO 2002: Proceedings of the Genetic and Evolutionary Computation Conference, p. 682. Morgan Kaufmann Publishers Inc., San Francisco (2002)
16. Jarvis, J., Shier, D.: Graph-theoretic analysis of finite Markov chains. In: Shier, D.R., Wallenius, K.T. (eds.) Applied Mathematical Modeling: A Multidisciplinary Approach. CRC Press, Cambridge (1999)
17. Nakama, T.: Markov chain analysis of genetic algorithms in a wide variety of noisy environments. In: GECCO 2009: Proceedings of the 11th Annual Conference on Genetic and Evolutionary Computation, pp. 827–834 (2009)

18. Nix, A.E., Vose, M.D.: Modeling genetic algorithms with markov chains. Annals of Math. and Artificial Intelligence 5(1), 79–88 (1992)
19. Rudolph, G.: Convergence analysis of canonical genetic algorithms. IEEE Transactions on Neural Networks 5 (1994)
20. Sahin, C.S., Gundry, S., Urrea, E., Uyar, M.U.: Optimizing topology in manets using evolutionary algorithms. In: Chiong, R. (ed.) Variants of Evolutionary Algorithms for Real-World Applications (2011)
21. Sahin, C.S., Gundry, S., Urrea, E., Uyar, M.U., Conner, M., Bertoli, G., Pizzo, C.: Markov chain models for genetic algorithm based topology control in manets. In: Applications of Evolutionary Computation: EvoApplications 2010, pp. 41–50 (2010)
22. Sahin, C.S., Urrea, E., Uyar, M.U.: Bio-inspired techniques for topology control of mobile nodes. In: Formal and Practical Aspects of Autonomic Computing and Networking: Specification, Development, and Verification, pp. 217–240 (2011)
23. Sahin, C.S., Urrea, E., Uyar, M.U., Conner, M., Bertoli, G., Pizzo, C.: Design of genetic algorithms for topology control of unmanned vehicles. International Journal of Applied Decision Sciences (IJADS), Special Issue on "Decision Support Systems for Unmanned Vehicles" 3(3) (2010)
24. Sahin, C.S., Urrea, U., Uyar, M.U., Conner, M., Hökelek, I., Bertoli, G., Pizzo, C.: Genetic algorithms for self-spreading nodes in manets. In: GECCO 2008: Proceedings of the 10th Annual Conference on Genetic and Evolutionary Computation, pp. 1115–1116 (2008)
25. Shinchi, T., Tabuse, M., Kitazoe, T., Todaka, A.: Khepera robots applied to highway autonomous mobiles. Artificial Life and Robotics 7, 118–123 (2000)
26. Suzuki, J.: A Markov chain analysis on simple genetic algorithms. IEEE Transaction on Systems, Man, and Cybernetics 25(4) (2005)
27. Urrea, E.: Knowledge sharing agents using genetic algorithms in mobile ad hoc networks. Ph.D. Thesis, The Graduate Center of the City University of New York (2010)
28. Urrea, E., Sahin, C.S., Uyar, M.U., Conner, M., Bertoli, G., Pizzo, C.: Estimating behavior of a ga-based topology control mechanism for self-spreading nodes in manets. In: IEEE Intl. Conf. on Military Communications (MILCOM), pp. 1275–1280 (2010)
29. Vose, M.D.: The Simple Genetic Algorithm. MIT Press, Cambridge (1998)
30. Winfield, A.F.: Distributed sensing and data collection via broken ad hoc wireless connected networks of mobile robots. In: Distributed Autonomous Robotic Systems, vol. 4, pp. 273–282 (2000)
31. Winkler, G.: Image Analysis, Random Fields and Markov Chains Monte Carlo Methods. Springer, Heidelberg (2006)

Chapter 9
Evolutionary P2P Networking for Realizing Adaptive Networks

Kei Ohnishi, Mario Köppen, Kaori Yoshida, and Yuji Oie

Abstract. Recently, a peer-to-peer (P2P) network model became very popular. This model is different from a conventional client-server network model. While a conventional client-server network model explicitly distinguishes nodes providing services (servers) from nodes receiving services (clients), a P2P network model does not assign fixed roles to nodes. One type of P2P networks are unstructured P2P networks that do not include any mechanism to manage data locations and which consist only of nodes that communicate with each other through direct connections. A direct connection between two nodes in unstructured P2P networks is represented by a logical network link, and therefore, a structure formed by logical links and nodes, that is, a P2P network topology, can be formed freely. Thus, unstructured P2P networks are flexible. However, they require some mechanisms to realize quick, accurate, and reliable searches. A free-form P2P network topology can be a control object for realizing such searches. One method that can adaptively modify a free-form topology of a running unstructured P2P network for quick, accurate, and reliable searches is an evolutionary algorithms (EA) inspired by biological genetics and evolution. Although EA is not the only way for adjusting the entire network topology of a running P2P network, EA would be suitable for this modification because EA can hold several solution candidates (i.e. future options for network topologies) simultaneously in a running P2P network. An evolutionary P2P networking technique (EP2P) and a parallel evolutionary P2P networking technique (P-EP2P) are both a fusion technique of an EA and an unstructured P2P network, and optimize the manner in which the nodes belong to different P2P network topologies in realtime. We

Kei Ohnishi · Mario Köppen · Kaori Yoshida · Yuji Oie
Kyushu Institute of Technology, 680-4 Kawazu, Iizuka, Fukuoka 820-8502, Japan
e-mail: `ohnishi@cse.kyutech.ac.jp`, `mario@ndrc.kyutech.ac.jp`, `kaori@ai.kyutech.ac.jp`, `oie@cse.kyutech.ac.jp`

J. Kołodziej et al. (Eds.): Advances in Intelligent Modelling and Simulation, SCI 422, pp. 185–213.
springerlink.com

will introduce EP2P and P-EP2P and show simulation results that validate both techniques.

9.1 Introduction

Recently, peer-to-peer (P2P) network models have attracted a great deal of attention. The concept of the P2P network model is completely different from that of a conventional client-server network model. While a conventional client-server network model explicitly distinguishes hosts providing services (servers) from hosts receiving services (clients), a P2P network model does not assign fixed roles to hosts. Hosts composing P2P networks, referred to as *peers*, can be both servers and clients, so that P2P networks could be used to facilitate autonomic and decentralized service management. In a client-server network, all data required by clients are held only by a server. Meanwhile, in a P2P network, data are held by each peer. So, no matter what type of P2P network is used, a mechanism to locate peers holding data requested by some peer is needed in P2P networks. We use the general term of "*node*" for representing "*peer*" hereinafter.

P2P networks are in fact roughly classified into three types. The first type are hybrid P2P networks. Hybrid P2P networks consist of an index server and nodes. The index server does not hold any data, but manages what data each node holds. Therefore, a node asks the index server when it desires some data, and from this it knows which node holds the desired data, and then it acquires the desired data from that node through direct communication, that is peer-to-peer communication. One of this type of real P2P file sharing networks is Napster [2], which first operated in 1999 and handles music files. Napster is actually the first real-world application of P2P network.

The second type are pure P2P networks. A pure P2P network consists of nodes that communicate with each other through direct connections, and does not include an index server. For example, nodes in P2P file sharing networks [15] provide files to each other through direct connections among the nodes. A direct connection between two nodes is represented by a logical network link, and therefore, a structure formed by logical links and nodes, that is, a P2P network topology, can be formed freely. Actually, according to [15], there are several forms in pure P2P networks. One form are structured P2P networks that have a mechanism to manage data locations in a network, This form forces nodes to form a specific network topology to reliably locate requested resources. The other form are unstructured P2P networks that do not have such a mechanism for reliably locating resources. One of the real unstructured P2P file sharing networks is Gnutella, which first operated in 2000. There are several Gnutella based networks such as LimeWire [1] and Phex [3]. In this book chapter, we will focus on such unstructured P2P networks.

The third type are super-node based P2P networks. This type are sort of hybrid P2P networks. However, super-node based P2P networks are not using a fixed index server but so-called super nodes. The super nodes are basically the same as the fixed index server in terms of functionality, but they are dynamically selected from among general nodes according to the current situation. In fact, hybrid P2P networks have the problem of single point failure because an index server is fixed. However, super-node based P2P networks are more flexible because they include a mechanism to vary super nodes equivalent to index servers in a dynamic fashion. One of the real applications of super-node based P2P networks is Skype [4], which is an application that enables voice and video calls and chats over the Internet and was first developed in 2003.

As mentioned above, here we focus on unstructured P2P networks. Unstructured P2P networks are flexible and robust because they do not rely on servers or super nodes at all. So, they are expected to have good scalability in terms of the number of nodes. However, since searching unstructured P2P networks for desired data is blind, unstructured P2P networks need some mechanisms to enhance search performance. To enhance search performance in unstructured P2P networks, query forwarding methods and data replication methods [7], [8], [16], [14], [24] have been investigated. The basic strategy to enhance search performance in unstructured P2P networks is that a data replication method places more data at easy reachable nodes by a given query forwarding method or that a query forwarding method forwards queries to nodes holding more data. However, enhancing search performance means that particular nodes are more frequently accessed, thus causing a load bias to particular nodes. This problem is actually similar to the problem of single point failure in client-server network services. So, it is important to consider a trade-off between search performance and load balancing in unstructured P2P networks.

Furthermore, a free-form P2P network topology can also sometimes be a control object for enhancing the quality or efficiency of P2P services. For example, a P2P network topology can be modified dynamically and adaptively in order to realize quick, accurate, and reliable searches. We here focus on topologies of unstructured P2P networks as a control object to realize quick, accurate, and reliable searches in the networks. To achieve that, we need a method for adaptively adjusting topologies of unstructured P2P networks.

One method that can adaptively optimize system parameters in general are is evolutionary algorithms (EAs), which were inspired by biological genetics and evolution [5]. One characteristic of EAs is that several solution candidates are held at any moment during an optimization process while a better solution is sought using these candidates. If we intend to use an EA as a method that adaptively optimizes a topology of a running P2P network, it is necessary to make several solution candidates coexistent, that is, several P2P network topologies, at any moment. Since, as mentioned above, a P2P network topology is a structure that is formed by logical links, it is possible to make several P2P network topologies coexisting at any given time. Therefore, it is also possible to adaptively change the topologies of the running P2P network by an EA. Although an EA is not the only way for adjusting the entire network topology of a running P2P network, an EA would be suitable for this case because an EA can hold several options (i.e. possible future network topologies) simultaneously in a running P2P network, which is a highly dynamic environment. EA has been used to optimize the parameter values of a P2P network using fitnesses obtained from a simulation model of the P2P network [17], [25], but this EA is not an online approach to optimize the parameters of P2P networks. Other computational intelligence techniques such as artificial immune systems, neural networks, and memetic algorithms have also been applied to the resource discovery problem in P2P networks in an offline manner [10], [9], [19]. Besides P2P networks, EAs have been applied to structure optimization of communication networks in an offline manner [26], [6], [12] and also to on-line optimization of communication networks, such as on-line optimization of routing tables of routers in the Internet [18] and of protocol stacks [11]. Furthermore, there are a number of methods for local topology reconstruction in a sole P2P network topology based on the observation of local network states [13], [23], [22]. However, these methods cannot adjust the entire network topology for a given purpose.

In the present chapter, we introduce an evolutionary P2P networking technique [20] (referred to as **EP2P** hereinafter) whereby the topologies of a running P2P network are dynamically and adaptively modified by EA and show results of simulations to examine whether EP2P can provide reliable search capability in dynamic P2P environments. In simulations, we assume dynamic P2P environments in which nodes leave and join the network with their own probability and in which search objects vary with time. This technique is a fusion of an evolutionary algorithm (EA) and a P2P network, which optimizes in real time several P2P network topologies that every node belongs to at a given time under a given fitness function.

EP2P mentioned above, however, does not consider a large size of networks as seen in real P2P networks. For example, LimeWire [1], which is one of the real P2P file sharing networks was reported to hold several millions of users simultaneously. As will be mentioned later, EP2P needs a special node that plays a role of collecting fitnesses from nodes and executing EA to adaptively change the P2P network topologies, which is also called a super node but different from that in the super-node based P2P networks in terms of functionality. It can happen that the super node is overloaded as the number of nodes increases and then the P2P network stops working due to that. Unlike

EP2P, a parallel evolutionary P2P networking technique, which is referred to as **P-EP2P** [21], aims at realizing adaptive large-scale networks. P-EP2P first divides an entire network into several smaller networks to avoid the overload of a super node and then applies EP2P to each of the small networks to make the entire network adaptive. In the present chapter, we also introduce P-EP2P and show results of simulations to examine whether P-EP2P can provide reliable search capability for large-scale P2P networks.

The remainder of this chapter is organized as follows. In Section 9.2, we describe EP2P and Section 9.3 shows simulation results of EP2P. In Section 9.4, we describe P-EP2P and Section 9.5 shows simulation results of P-EP2P. Finally, Section 9.6 presents conclusions and describes areas for future research.

9.2 Evolutionary P2P Networking

The evolutionary P2P networking technique (**EP2P**) is a technique that evolutionarily reconstructs a set of P2P network topologies for a given purpose (see Figure 9.1). Generally, in evolutionary methods, the next generation of individuals is created from individuals in the present generation which are better adapted to the environment. In EP2P, P2P network topologies correspond to the individuals described above, and all of the nodes are included in the P2P network topologies. In addition, in EP2P, the fitness function for the EA is a P2P network that includes several P2P network topologies that are optimized by the EA, and the mechanism of topology reconstruction for the P2P network is the EA. Moreover, in EP2P, the set of P2P network topologies is not determined by the EA through simulations in advance. Rather, the fitness of each P2P network topology is obtained from nodes in the running P2P network, and a set of P2P network topologies is reconstructed by applying evolutionary operators to individuals that encode the topologies based on the fitnesses obtained while running the P2P network.

Next, we explain the details of the EP2P used herein.

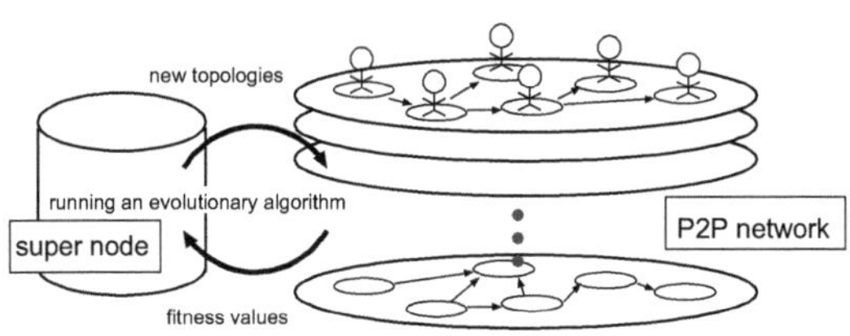

Fig. 9.1. Overview of the evolutionary P2P networking technique [20].

9.2.1 Network Composition

As shown in Figure 9.1, a network including EP2P is composed of a P2P network that includes several network topologies, in which all of the nodes are included at the same time, and a super node, in which the EA is used to optimize the topologies.

The actual role of the super node is (1) to determine links for a node that joins the network for the first time, (2) to reconstruct the network topologies by executing the EA, and (3) to monitor which nodes join the network at each moment. We will describe how each network topology is assigned a fitness and how the network topologies are reconstructed later herein. Note that here the super node does not manage which services each node can provide to other nodes. For example, in a P2P file-sharing network, the super node does not manage which files each node holds.

9.2.2 Joining and Leaving Nodes

The P2P nodes communicate their joining and leaving of the network to the super node. Thus, the super node can determine which P2P nodes have joined the network and whether these nodes are still in the network.

When a node joins the P2P network for the first time, the super node randomly determines the nodes to which the joining node will link from among all of the P2P nodes present in the network at that moment. Since there are several network topologies, the super node determines the links for all of the topologies for the joining node. Next, when the joining node leaves the network, the joining node first informs the super node that it will leave the network and then informs the super node when it rejoins the network. However, the target nodes linked to the node are the same as those to which the node was linked before leaving the network, although the target nodes may no longer be in the network.

9.2.3 Fitnesses Assigned by Nodes

A P2P node uses all network topologies including this node for a time period T and then assigns a fitness to each of the topologies. The fitness of each network topology is initially set to zero and again after every time interval T. Otherwise, each network topology basically increases the fitness by being used by the nodes.

When a P2P node searches the P2P network for P2P nodes that can provide the desired service, this P2P node uses all of the P2P network topologies in which it is included for the search. Therefore, it is possible that within a given allowed number of hops, H_{max}, the P2P node can find the desired data or service in some topologies while not being able to find the service in other topologies. If the desired data or service is found in a certain topology, the fitness of the topology is increased by one. Otherwise, the fitness does not change. Although disconnected network topologies are usually not useful for the search, such network topologies are unlikely to be selected by the EA.

If the above-mentioned search and assignment of fitnesses are conducted for a period of time T, each topology will be assigned a certain fitness. Then, the topologies with larger fitnesses are regarded as better in the EA used herein.

9.2.4 Representations of Network Topologies

In the EA, a solution candidate for an optimization problem is represented in an alternative form. This alternative form is designed by the person who is attempting to solve the problem using the EA and is referred to as a genotype or an individual. Meanwhile, a solution candidate itself is referred to as a phenotype in the EA. In EP2P, the P2P network topology is an object of optimization and an individual is an alternative form of a P2P network topology.

Suppose that a P2P network consists of L nodes. The P2P network topology assumed herein is generated by having each of the L nodes make N_C directed links to other nodes. Therefore, an individual is an internal representation of this network topology in the EA. The EA individual used here is shown in Figure 9.2. As shown in Figures 9.2(a) and 9.2(b), in which N_C is 1 and 2, respectively, the individual is a one-dimensional vector with $L \times N_C$ elements. An element in an individual is generally referred to as a gene in the EA.

Each node is assigned a serial number as its identifier, and the identifier corresponds to the index of the vector representing the individual. An element value of the individual represents an identifier of the node to which a focus node makes a directed link. A direction represented by a directed link indicates that a search query can be forwarded only in that direction. Thus, if flooding is used as a query forwarding method, a search query generated at some node is forwarded node by node in the direction represented by the directed links, and the paths for forwarding the query (flooding tree) are then determined accordingly. However, when a data, such as a file, is found during this search, the node having the object transmits the data to the node making the query by means of a direct communication.

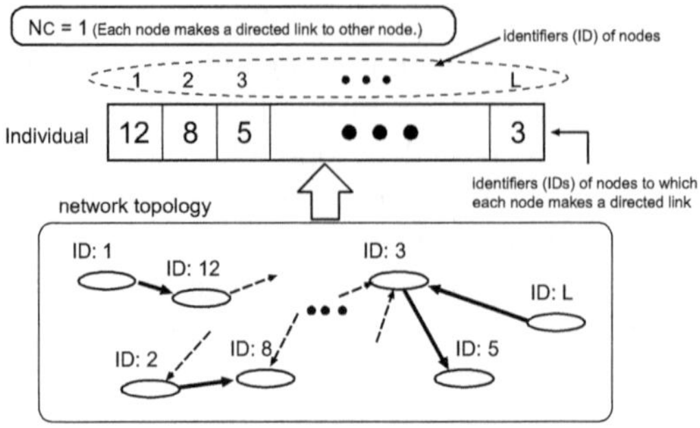

(a) $N_C = 1$.

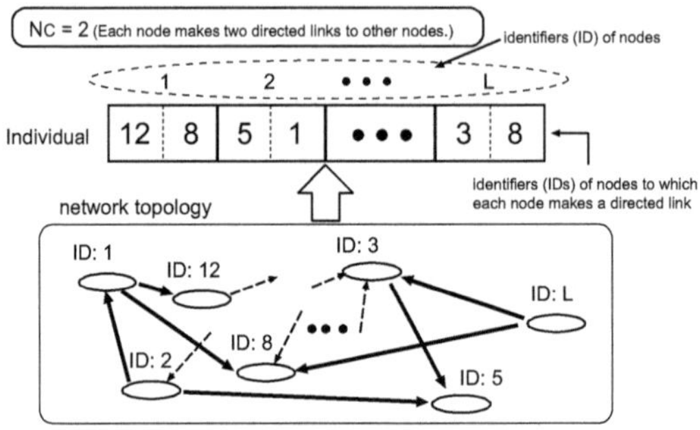

(b) $N_C = 2$.

Fig. 9.2. Representation of a P2P network topology in the EA (EA individual) [20].

9.2.5 Evolutionary Operators

Evolutionary operators are applied to the set of individuals mentioned above, which is referred to as a population, in order to generate a new set of individuals, which is referred to as the new population. The number of individuals held in the EA, i.e., the population size, is N. Evolutionary operators generally include a selection operator, which is inspired by natural selection in Darwinism, a recombination or crossover operator, which models genetic recombination, and a mutation operator, which models gene mutation. The evolutionary operators used in EP2P are explained below.

9.2.5.1 Selection

The selection operator used herein is a tournament selection with a tournament size of K. The tournament selection randomly selects K individuals from the EA population and selects the individual with the best fitness among the

K individuals. This selection procedure is repeated until N individuals have been selected. N is a population size and an even number.

9.2.5.2 Crossover

The crossover operator used in the present study is hereinafter referred to as *node linkage crossover* (NLX). This operator is applied to an EA population as follows. Other type of crossover operators can also be used and conventional crossover operators will be used for comparison in Section 9.3.

1. N individuals selected by the selection operator are divided into $N/2$ pairs of individuals. The selected individuals become parent individuals in this generation.
2. The crossover operator is applied to each pair of parent individuals with probability p_c. Child individuals generated from each pair of parent individuals are identical to the parent individuals before the crossover operator is applied. Each parent individual has a corresponding child individual.
3. For each pair of parent individuals to which the crossover operator is applied, one element is randomly selected from among the L elements of the individual. Recombination is conducted for the selected element with probability p_e.
4. For the element to which the recombination is to be applied, which child individual corresponding to one parent individual receives the element values of the other parent individual to be copied on itself is decided randomly.
5. After deciding which parent individual provides the element values for recombination, the node (element value) linkage generated by directed links between nodes is copied to the target child individual.

   For example, suppose that $N_C = 1$, and the fifth element has been selected as the initial element of the linkage. Initially, NLX refers to the value of the fifth element of the parent individual as a copy source. If the reference value is 10, then NLX refers to the value of the tenth element. Furthermore, if the value of the tenth element is 2, then NLX refers to the value of the second element. By referencing the element values N_L times, NLX generates N_L element values and then copies them to the child individual corresponding to the other parent individual. In this example, N_L is 3, and the values of the reference elements are 5, 10, and 2, in that order. Nodes corresponding to the values of the reference element are linked by directed links. An example of this form of recombination is illustrated in Figure 9.3(a).

   Figure 9.3(b) shows an example of NLX with $N_C = 2$. In Figure 9.3(b), the third node has been selected as the initial node of the linkage.

However, since each node makes two directed links, the third node has two elements that can be referred to by NLX, which, in this example, are 10 and 1. Then, NLX randomly chooses one of the two possible elements and refers to the value of the selected element, which is 10. Next, since the second node of the linkage, which is the tenth node, also has two elements, NLX randomly chooses one of the elements and refers to the value of the selected element, which is 2. In this way, the node linkage is formed. Generally, when $N_C \geqq 2$, NLX is performed in the same manner.

6. Repeat Stpdf (3) through (5) $N_C \times L$ times.

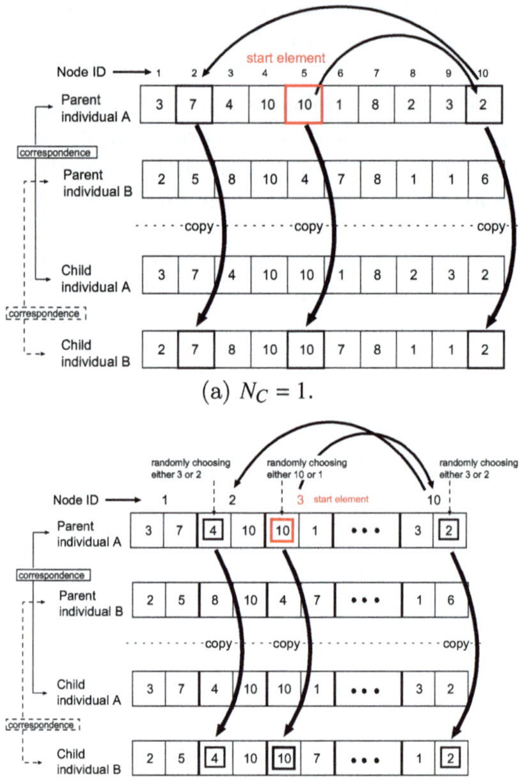

(a) $N_C = 1$.

(b) $N_C = 2$.

Fig. 9.3. Example of node linkage crossover (NLX) [20].

9.2.5.3 Mutation

The mutation operator used herein is such that the value at each position (the gene) on the N individuals obtained after the node linkage crossover (NLX) is randomly changed to some other possible value with probability p_m, which is referred to as the mutation rate. This mutation operator is introduced mainly for bringing novel genes that did not appear in the initial population. In addition, if we set the mutation rate to be higher, the EP2P approaches to a random method.

9.2.6 Timing for Topology Generation

The EA population obtained after applying the evolutionary operators is transformed into a new set of P2P network topologies on the super node mentioned above, and the nodes to which each node must make directed links are then communicated to each node in the network. The nodes then make directed links to other nodes according to this information. Nodes that are not present in the network at this moment obtain information on nodes to which they must link upon joining the network.

9.2.7 Procedure and Parameters

First, N individuals, which are N network topologies, are randomly generated as the initial EA population. The generated individuals are used in performing the object search for time T and then proceed to the fitness evaluation phase. The fitness of each individual is the number of successful searches using the individual in time T. Next, the evolutionary operators are applied to the present EA population based on their fitnesses in order to generate a new EA population. The segment procedure, starting from the use of the individuals (topologies) and ending with the application of the evolutionary operators, is repeated. In EP2P, the procedure is as follows:

1. The super node randomly initializes the EA population.
2. The super node transforms the EA population into P2P network topologies and then informs each P2P node of the topology information.
3. Each P2P node makes directed links to other nodes.
4. Each P2P node uses all of the network topologies for time T and then assigns a fitness to each individual corresponding to a P2P network topology.
5. The super node applies the evolutionary operators to the EA population.
6. Back to (2).

9.3 Simulations of EP2P

In this section, we examine whether the EP2P presented in Section 9.2 can contribute to reliable search in dynamic P2P environments through simulations.

The focus of the simulations here is to examine in what situation an evolutionary way works well for adaptation of network topologies. Actually, it is possible to consider a method for adaptively constructing single P2P network topology using fitness from all of the nodes, which is not an evolutionary approach using multiple network topologies. However, such a proposal is beyond the focus of the present contribution. Instead of proposing such a method, we prepare a random topology construction method that is realized in the framework of the evolutionary approach by making the mutation rate higher.

9.3.1 Dynamic P2P Environmental Model

In the P2P simulation model used herein, P2P nodes that join the network vary with time. Each of the L nodes decides whether it will join the network at each time unit according to its given probability. This probability is hereinafter referred to as the *participation probability*. The participation probability of each node is determined as a uniform random real number in $(0, 1]$. Each node joins the network with its participation probability. If a node does not join the network, then the node is in the state whereby the node leaves the network. After all of the nodes make a decision with regard to participation, each of the nodes conducts one search. A time unit is regarded as the period of time required for all of the nodes to make this decision and complete one search.

The parameter values of the EP2P used in the simulations are listed in Table 9.1. Some of the parameters take different values in the different evaluation scenarios that are described in detail later. A set of the parameter values shown in Table 9.1 is one of the possible sets. Other sets of the parameter values may yield different results from those obtained here. However, the focus of the simulations here is to examine in what situation an evolutionary way works well for adaptation of network topologies. For this purpose, most parameter values are fixed but only the mutation rate is changed to realize a random method for comparison of the evolutionary way, which does not rely on the previous good individuals. Thus, the crossover operator is considered to be the main driving force of evolution here. The effect of other types of crossover operators will be examined in Section 9.3.4.4.

The parameter setting used herein brings the effect that on average, approximately half of the nodes (approximately 1,000 nodes) join the network at any time.

Table 9.1. Parameter values of the EP2P used in the simulations.

Parameter	Description	Value
L	number of nodes (genes)	2,000
N	number of P2P network topologies (individuals)	50, 100 (for two evaluation scenarios)
T	time period for which generated topologies are used	50
H_{max}	maximum allowed number of hops for one search	3, 6 (for two evaluation scenarios)
N_C	number of directed links generated by a node	1
K	tournament size for the tournament selection	2
p_c	crossover rate	1.0
p_e	probability with which recombination is conducted for a selected element in NLX	0.1
N_L	length of node linkage in NLX	5
p_m	mutation rate	0,0.05,0.1,0.2,0.5,0.8,1.0

9.3.2 Evaluation Scenarios

We prepare two types of evaluation scenarios that include different dynamic P2P environments. For any evaluation scenario, node departure and participation occurs as explained in the previous section. The time period of the simulation, which is 1 to 5,000 time units, is the same for the two evaluation scenarios.

- Evaluation Scenario 1
  In the first evaluation scenario, replicas of the search objects are created in the network as unstructured file sharing networks. The replication method used here is such that a search object that was retrieved by the node making the query is stored in the storage of the node and will thereafter be shared among all nodes. This method is referred to as owner replication.

  The storage capacity for each node is determined as a uniform random integer in [50, 300]. Any one search object consumes one unit of storage capacity.

  There are 2,000 types of search objects, which is the same as the value of L shown in Table 9.1. Initially, each node has one type of search object. However, the objects for which nodes actually search vary with time. As mentioned above, the total simulation time period is 5,000 time units. From $1000(k-1)+1$ to $1000k$ time units, objects possessed by the $(400(k-1))$-th through $400k$-th nodes become search objects, where $k = 1,2,\cdots,5$. The search object that each node will search for is determined randomly from among the 400 current search objects that each node does not possess.

In the first evaluation scenario, the number of coexisting network topologies is 50, and the number of hops allowed for one search is 3. Therefore, it is possible to forward a search query to nodes a maximum of $150(=50\times3)$ times during one search.

- Evaluation Scenario 2
  In the second evaluation scenario, a replication method is not used. The search objects are P2P nodes, so storage capacity need not be considered. From 1 to 2,500 time units, the search objects are two of the 1st through 2,500th nodes. The two nodes as search objects have a participation probability that is greater than or equal to 0.95. One of the two nodes searches for the other node as a search object. Other nodes search for one of the two nodes selected randomly.

  Although the P2P network applications to which this evaluation scenario corresponds differs from the first evaluation scenario in the following aspects: there is no replication, and there are few search objects at any given moment. In the second evaluation scenario, the number of coexisting network topologies is 100, and the number of hops allowed for one search is six. Therefore, it is possible to forward a search query to nodes a maximum of $600(=100\times6)$ times during one search.

9.3.3 Observations

As mentioned above, the fitness of each network topology to which P2P nodes are assigned is the number of successful searches using the network topology. Therefore, it is expected that EP2P will evolve network topologies that have a low rate of search failures.

Thus, we consider the search failure rate of the present network topologies during a period of use, which is $T=50$, as an observation item. Search failure occurs when a requested object cannot be located using all of the coexisting network topologies. We observe the number of search failures in all of the searches conducted during 50 time units and then calculate the search failure rate every 50 time units based on the observed number of search failures.

In addition, we observe the average number of hops for only successful searches in the network topologies during a period of use ($T=50$). The average number of hops is also obtained every 50 time units.

9.3.4 Results

Simulation results for the first and second evaluation scenarios are shown in Figures 9.4 and 9.5, respectively. The results are averages over 10 independent simulation runs. Figures 9.4(a) and 9.5(a) show the time-varying search failure rates for the first and second evaluation scenarios, respectively, which assume dynamic P2P environments, and show the results obtained for seven mutation rates (p_m) and the results for the case in which the evolutionary operators are not applied (labeled "nothing"). For comparison, Figures 9.4(b) and 9.5(b) show the time-varying search failure rates for the first and second evaluation scenarios, respectively, that exclude node participation and departure. In other words, all of the nodes are in the network at any time. In addition, Figures 9.4(c) and 9.5(c) show the time-varying average number of hops only for successful searches for the first and second evaluation scenarios, respectively.

In addition, the average search failure rate over ten independent simulation runs and the standard deviation of the ten search failure rates at the end of simulation run (at 5,000 unit times) are shown in Table 9.2. The average and the standard deviation of the search failure rates are obtained for both of the first and the second evaluation scenarios including node participation and departure. Since Table 9.2 shows that the standard deviation is quite small both in the first and the second evaluation scenarios, we will discuss the performance of the EP2P using the average search failure rate below.

Table 9.2. The average search failure rate over ten independent runs and the standard deviation of the ten search failure rates at the end of the simulation run (at 5,000 time units) in the first and the second evaluation scenarios including node participation and departure. The label of "nothing" in the table represents no evolutionary operators.

The first evaluation scenario								
Value of p_m	nothing	0.0	0.05	0.1	0.2	0.5	0.8	1.0
Average search failure rate	0.717	0.999	0.148	0.052	0.028	0.013	0.013	0.012
Standard deviation	0.006	0.0002	0.011	0.010	0.013	0.011	0.006	0.004
The second evaluation scenario								
Value of p_m	nothing	0.0	0.05	0.1	0.2	0.5	0.8	1.0
Average search failure rate	0.758	0.988	0.496	0.513	0.666	0.857	0.891	0.912
Standard deviation	0.016	0.016	0.044	0.061	0.044	0.013	0.010	0.009

Figures 9.4(a) and 9.5(a) show that in all of the evaluation scenarios including different dynamic P2P environments, EP2P lowers the search failure rate with time just after the search objects are varied depending on the values of the mutation rate.

Furthermore, Figures 9.4(c) and 9.5(c) show that for the situation in which the search failure rate decreases with time, the average number of hops for successful searches increases, because the variance of the number of hops for successful searches becomes larger in processes in which the network topologies become reliable for more nodes. The number of the network topologies would be too small to meet demands of many nodes at once.

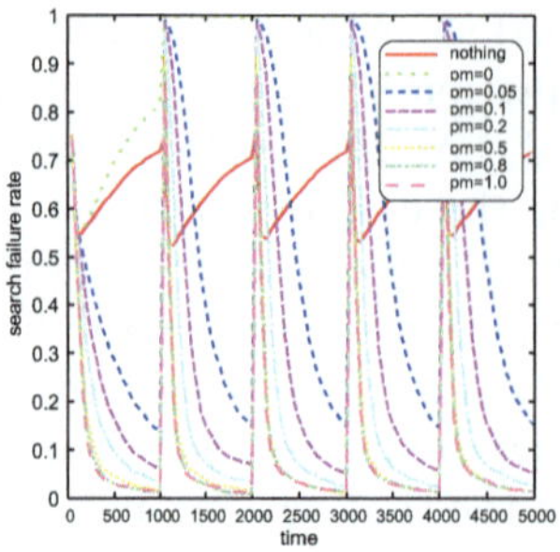

(a) Time-varying search failure rate.

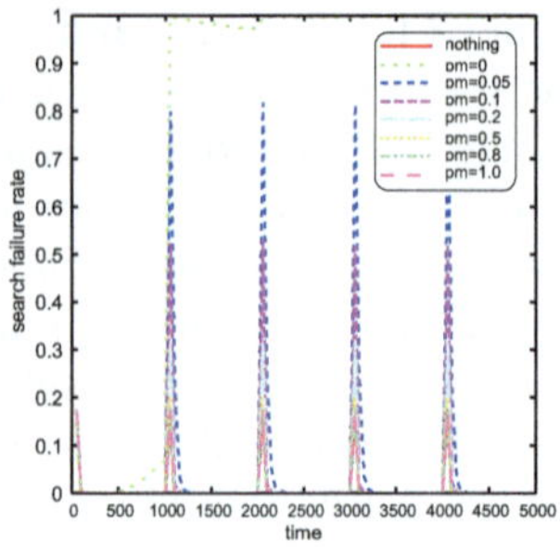

(b) Time-varying search failure rate when no node participation or departure occurs.

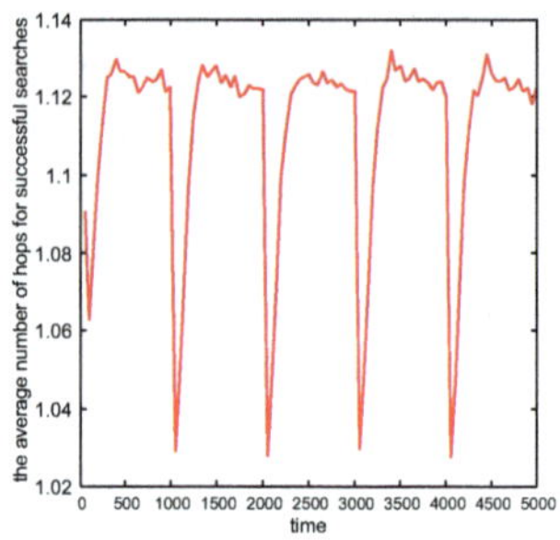

(c) Time-varying average number of hops for successful searches when the mutation rate, p_m, is 1.0.

Fig. 9.4. Results of the first evaluation scenario, in which there are several types of search objects and a replication method [20].

We discuss the results for each evaluation scenario below.

9.3.4.1 First Evaluation Scenario

Figure 9.4(a) shows that the higher the mutation rate, the lower the search failure rate. In addition, when the mutation rate, p_m, is 1.0, that is, when the network topologies are fully randomly reconstructed every 50 time units, the search failure rate is quickly minimized. This result means that a continuous large change in network topologies is superior to changing network topologies through an evolutionary method, which yields new individuals from the better individuals of the present generation, in terms of providing reliable search capability.

The reason for this result is as follows. Suppose that a node could find a requested object using the present network topologies at some moment. Then,

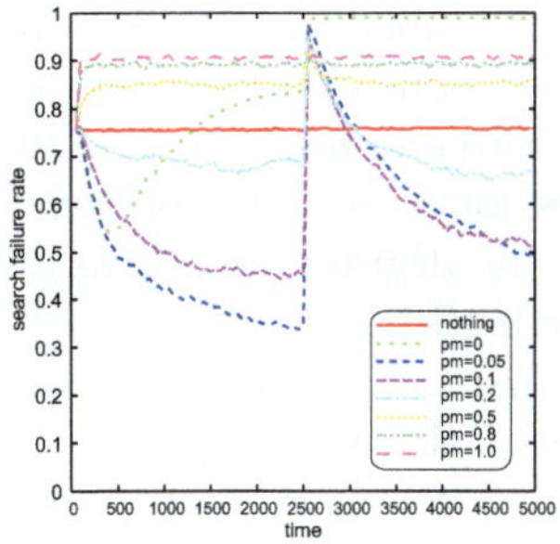

(a) Time-varying search failure rate.

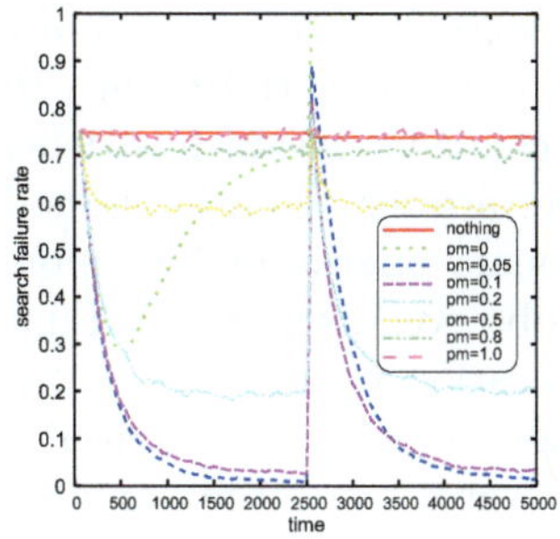

(b) Time-varying search failure rate when no node participation or departure occurs.

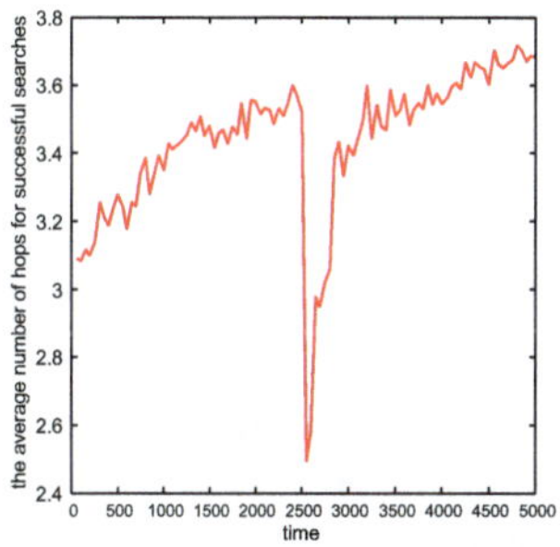

(c) Time-varying average number of hops for successful searches when the mutation rate, p_m, is 0.05.

Fig. 9.5. Results of the second evaluation scenario, in which there are few types of search objects and there is no replication method [20].

the node will no longer search for that requested object. Therefore, there is no guarantee that the network topologies that contributed to a successful search for a certain object will be useful for finding other objects later. In addition, the possibility exists that the locations of objects in the network change due to replication of the objects. From the viewpoint of the EA, the fitness landscape, which represents the distribution of the fitnesses on the search space, changes dynamically. In such an unstable situation in terms of the fitnesses of the EA individuals, it is difficult for the EA to evolutionarily identify better individuals and adapt them to the problem.

In the first evaluation scenario, if the evolutionary operators are not used, the topologies of the P2P network are not changed over time. In such a situation, the replication brings improvement of the search failure rate, but once most nodes have had search objects that are located among their reachable nodes, the search failure rate becomes worse with time. Therefore, we can observe in Figure 9.4(a) that the graph of the search failure rate for "nothing", which represents not using the evolutionary operators at all, is first improved with time and then degraded with the time needed until the search objects are totally changed. Meanwhile, when no node participation and departure occurs, there are more reachable nodes for each node even when not using the evolutionary operators. Therefore, the degradation of the search failure rate is not observed as shown in Figure 9.4(b).

If the mutation rate is zero in the first evaluation scenario, novel genes that were not supplied in the initial population never appear. Therefore, the selection operator forces the population to converge to one individual. Once such convergence of the network topologies occurs, the search failure rate is hardly improved. Therefore, we can observe in Figures 9.4(a) and 9.4(b) that the graph of the search failure rate for $p_m = 0$ is first improved and then degraded over the simulation time.

9.3.4.2 Second Evaluation Scenario

Figure 9.5(a) reveals that when the mutation rate is lower, but greater than zero, the search failure rate becomes lower. The results indicate that an evolutionary method, which produces the individuals for the next-generation from the better individuals of the present generation, is effective in this evaluation scenario. In this scenario, the locations of search objects in the network, i.e., the locations of nodes, do not vary because there is no replication, so that the fitness landscape in the EA is static. In addition, there are only two search objects, so that it is possible to somehow evolve network topologies that can lower the search failure rate, although the crossover operator might produce useless network topologies by recombining network topologies that are specialized for different search objects. In order to evolve network topologies more quickly that provide a more reliable search, it is necessary to evolve several coexisting network topologies specific for different search objects. In other words, diverse network topologies should be maintained in a single population.

If the mutation rate is zero in this second evaluation scenario, the selection operator forces the population to converge to one individual as in the first evaluation scenario. Therefore, we can observe in Figures 9.5(a) and 9.5(b) that the graph of the search failure rate for $p_m = 0$ is first improved and then degraded over the simulation time.

In this scenario, the quickest and most reliable search can be achieved using network topologies in which each node makes directed links to nodes as search objects. However, it is difficult for the EA considered herein to generate such network topologies from randomly initialized network topologies.

In present state there is no P2P application that fits the second evaluation scenario well. Especially, it is not realistic to assume that only few search objects exist in P2P networks at any moment. Meanwhile, no replication is possible in P2P applications whose search objects are nodes (users).

Therefore, in future work, we need to seek a way to maintain diverse network topologies that are specialized for many sorts of search objects in a single population, to apply the EP2P to P2P applications in which search objects are nodes (users), there are many search objects at any moment, and replication of search objects is not conducted.

9.3.4.3 EP2P in Dynamic and Static P2P Environments

Figures 9.4(a) and 9.4(b) show that although the rate at which the search failure rate decreases for the EP2P with the same parameter values differs for the dynamic and static P2P environments in terms of node participation and departure, the tendency of convergence of the search failure rate is roughly the same for the dynamic and static P2P environments for the first evaluation scenario. In addition, Figures 9.5(a) and 9.5(b) indicate that for the second evaluation scenario, the tendency of convergence of the search failure rate is roughly the same between the EP2Ps with the same parameter values in the dynamic and static P2P environments.

9.3.4.4 Other Types of Crossover Operators

We use other crossover operators in the EP2P and examine if those crossover operators can improve the results of NLX presented in this paper. The crossover operators used herein are one-point crossover and uniform crossover. Both of them are conventional and common in the evolutionary computation research field.

Both of the two crossover operators are applied with probability p_c to each of $N/2$ pairs of individuals that are formed from the selected N individuals. The one-point crossover first randomly selects a position on two parent individuals (vectors) and then exchange the values (genes) beyond that randomly selected position. The uniform crossover exchanges two values at each position on two parent individuals between the two individuals with probability of 50 %.

While NLX presented in this paper is meant to increase good node linkage in the population, the one-point crossover and the uniform crossover are not. Therefore, we can expect that NLX considering good node linkage yields better results than those two crossover operators. We apply the EP2P with the one-point crossover and the uniform crossover to the evaluation scenario 2 mentioned above, in which the EP2P with NLX and low mutation rate worked better than the periodic random topology reconstruction. We examine if the one-point crossover or the uniform crossover improves the result of the EP2P with NLX. The seven different values of mutation rate are also used here.

The simulation results are shown in Figure 9.6. Figure 9.6 does not include the result for the case in which the evolutionary operators are not applied.

Comparing Figure 9.6 with Figure 9.5 that represents the result of using NLX, the tendency of the results is so similar among the EP2P with the one-point crossover, the uniform crossover, and NLX, but the search failure rate for the case of using small mutation rate as $p_m = 0.05, 0.1$ is different. The EP2P is shown to be the best when using $p_m = 0.05, 0.1$. Since a main driving force to obtain better individuals is a crossover operator when the mutation rate is quite low, it is suggested that NLX considering the node linkage can produce better individuals than the two conventional crossover operators that do not consider the node linkage.

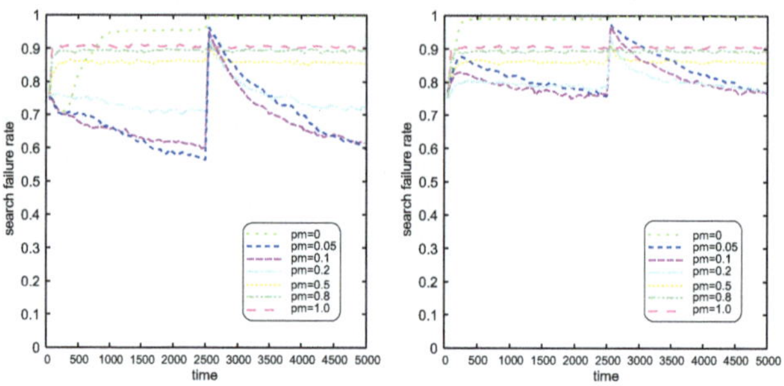

(a) The result for one-point crossover. (b) The result for uniform crossover.

Fig. 9.6. Time-varying search failure rate when using EP2P using other crossover operators [20].

Although NLX is shown to be better than the two conventional crossover operators above in EP2P, the performance of NLX may depend on values of N_L. We will investigate the effect of N_L value on the performance in future work when considering guidelines on the parameter values settings.

9.3.5 Using a Different Type of Network Topology

The simulation results presented above indicate that EP2P can provide reliable search capability in the second scenario. In other words, it is shown that topology reconstruction by EP2P is better than random topology reconstruction when only a few types of search objects are present in the network at any moment and these search objects are not replicated. In the EP2P used above, the number of directed links that each node makes to other nodes in a single network topology, N_C, is just one. As mentioned in Section 9.2, it is possible for every node to hold any number of directed links (N_C) in a network topology. In this section, we examine whether the EP2P that allows nodes to hold several directed links in a network topology can improve the results of the second scenario shown in Section 9.3, in which the EP2P was better than periodical random topology reconstruction.

Concretely, we compare two different EP2P with different configurations. In one EP2P, the number of directed links of each node, N_C, is set to be two, and the depth of the flooding tree, H_{max}, is three. In the other EP2P, the number of directed links of each node is set to be one, and the depth of the flooding tree is eight. In both of the EP2Ps, the number of times query forwarding is performed in a network topology in one search is eight. The application of the node linkage crossover (NLX) to the individuals was described in Section 9.2. The EP2P with $N_C = 2$ and the EP2P with $N_C = 1$ use different values for p_e and N_L to obtain the same expected numbers of element values that are copied by NLX between the two EP2Ps. Specifically, the EP2P with $N_C = 2$ takes $p_e = 0.2$ and $N_L = 3$, and the EP2P with $N_C = 1$ takes $p_e = 0.1$ and $N_L = 6$. In addition, the mutation rate for the two EP2Ps is the same as $p_m = 0.05$, which yielded the best result in the second scenario.

Figure 9.7 shows the time-varying search failure rate for the two different EP2Ps. The obtained results are the averages over 10 independent simulation runs. In addition, we show the results for the two EP2Ps, which exclude node participation and departure, in Figure 9.7.

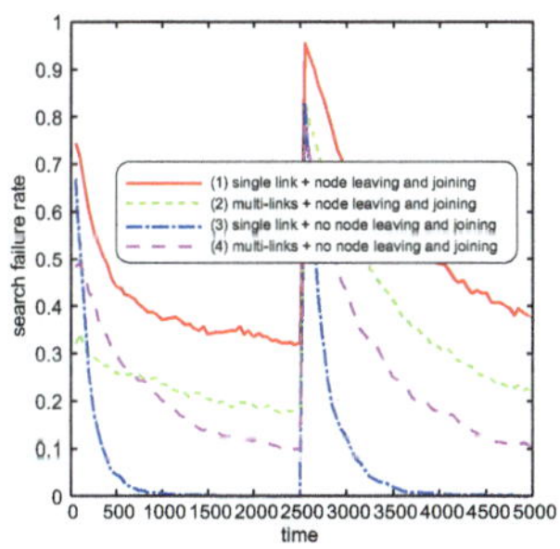

Fig. 9.7. Time-varying search failure rate when using EP2P with $N_C = 1$ and $N_C = 2$ [20].

Figure 9.7 shows that, in the second evaluation scenario including node joining and leaving, the EP2P with $N_C = 2$ achieves a lower search failure rate than the EP2P with $N_C = 1$. This suggests that the EP2P that allows nodes to hold several directed links is more suitable for dynamic P2P environments. In contrast, for the case in which node joining and leaving does not occur, the EP2P with $N_C = 1$ achieves a lower search failure rate than the EP2P with $N_C = 2$. These results indicate that the number of directed links that achieves more reliable search depends on how often node joining and leaving occurs.

9.4 Parallel Evolutionary P2P Networking

In the previous section, EP2P is shown to be able to adaptively reconstruct network topologies of a running P2P network. As mentioned already, EP2P needs a special node called a super node that plays the role of collecting fitness values from nodes and executing EA to adaptively change the P2P network topologies, but due to that, the super node may be overloaded as the number of nodes increases and then the P2P network may stop working. That is, EP2P has a scalability issue in terms of the number of nodes.

So, in this section, we introduce a parallel evolutionary P2P networking technique (P-EP2P) for realizing adaptive large-scale networks. The overview of P-EP2P is shown in Figure 9.8. P-EP2P first divides an entire network into several smaller networks to avoid the overload of a super node and then applies EP2P to each small network in parallel to make the entire network adaptive. Let N_G be the number of node groups, which are obtained by dividing the entire network, and G_k be the number of nodes in the k-th node group ($k = 1,2,\ldots,N_G$). In this contribution, we do not, however, discuss a process in which all of nodes are divided into multiple node groups, but assume that they are divided into multiple groups in advance. In addition, we do not discuss how to select a super node in each node group. However, we can easily find ways to obtain multiple node groups, such that the maximum number of nodes that each node group can hold is decided in advance and a node joining the network is assigned to one of available node groups.

P-EP2P, as mentioned before, divides all of nodes into multiple node groups and then applies EP2P to each group. We will explain P-EP2P by describing the difference between EP2P and P-EP2P. As mentioned above, we assume that all of the nodes are in advance divided into multiple groups. Then, the differences are the following two points: (1) how multiple super nodes gather fitnesses and (2) the evolutionary operators used. We here assume that the multiple super nodes are able to know only which nodes are in the network at any moment by exchanging information on the present nodes among them and also that every super node knows only links created by nodes in its node group.

9.4.1 Gathering Fitnesses by Super Nodes

In EP2P, one super node gathers fitnesses from all of the nodes. Meanwhile, in P-EP2P, one super node is assigned to each of the N_G node groups and gathers fitnesses only from G_k ($k = 1,2,\ldots,N_G$) nodes that belong to its node group. All L nodes simultaneously belong to N network topologies and are divided into N_G node groups. Therefore, it can happen that the node groups assign different fitnesses to identical network topologies.

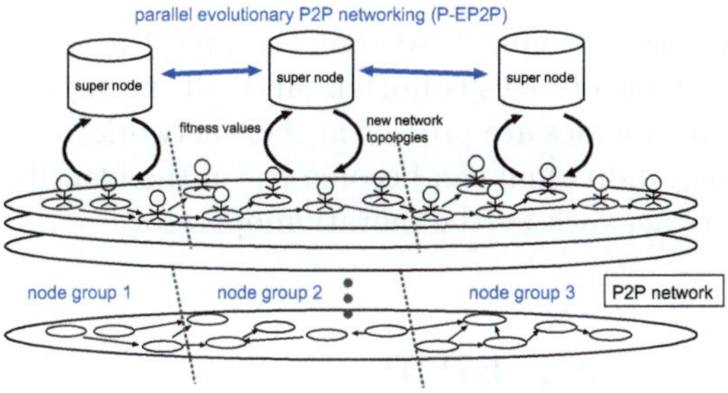

Fig. 9.8. Overview of the parallel evolutionary P2P networking technique [21].

9.4.2 Evolutionary Operators

In a super node of each node group, evolutionary operators are applied only to nodes of the group.

First, in each node group, the tournament selection is conducted using fitnesses assigned to individuals encoding network topologies. Since it can happen that the node groups provide different fitnesses for identical individuals, as mentioned above, selected individuals can be of different value in the node groups.

Second, NLX is applied to the selected individuals by the tournament selection in each node group. A range within which NLX is applied in each node group are the vector elements (the loci) that are correspondent to the nodes of that group. For example, suppose that a node a_1 in a node group 1 of focus makes a directed link to a node b_2 in another node group 2 and the node b_2 makes a directed link to a node c_3 in another node group 3. Then, we will consider a copy of the linkage among these nodes, $a_1 \rightarrow b_2 \rightarrow c_3$. In case that those nodes belongs to the same node group, the linkage of $a_1 \rightarrow b_2 \rightarrow c_3$ can be copied. However, in case that those nodes belong to different node groups, the linkage from the node a_1, which is in the node group 1 of focus, to the node b_2, which is outside the node group 1, is copied, but the linkage from the node b_2 to the node c_3, which is also outside the node group, cannot be copied.

In one attempt of NLX, when the number of times of recombination has not reached Q yet and a linkage from a node in a node group of focus to a node in another node group appears, a linkage from the node in that other group to some node can not be copied, as mentioned above. In this case, one new vector element is selected from all of the vector elements in the node group of focus to be copied. One attempt of NLX is finished when the total number of the vector elements copied becomes Q.

Finally, the mutation operator used herein changes each element value of the individual to one of the possible values at random, but whether the change occurs is determined with a given mutation rate. The element value of the individual represents an identifier of a node to which the node corresponding to the element position is linked, so that the mutation operator changes a node to which the node of focus is linked. Since all the super nodes exchange information on which nodes are present in their networks among them, every element value of the individual can become one of the identifiers for all of the nodes in the entire network by the mutation operator.

9.5 Simulations of P-EP2P

In this section, we examine the relationship between the number of node groups and adaptability of network topologies.

9.5.1 Simulation Model and Configurations

To focus only on the examination of the relationship between the number of node groups and the adaptability of network topologies as much as possible, we simplify other things. We consider that all of the nodes are always present in the network without leaving during the simulation period. In addition, it is assumed that all of the nodes excluding a node as a search object search the entire network for only one node to receive some service over the simulation period of time. The node as a search object does not conduct search. The above-mentioned assumptions are not practical, but the network in the simulation model is static and therefore changes in the network topologies affect the adaptability of the network topologies. A time unit is regarded as the period of time required for all of the nodes to complete one search.

The parameter values of the P-EP2P used in the simulations are listed in Table 9.3. The parameters whose values are changed in the simulations are just the number of nodes L, which takes 10^3, 10^4, 10^5, and the number of node groups N_G, which takes 1, 2, 5, and 10. Also, the number of nodes as a search object is just one and all of the nodes excluding the search object conduct only one search in a time unit, so that the search result for every node is always the same until the network topologies are changed by applying the evolutionary operators to all of the node groups at the same time.

9.5.2 Simulation Results

We observed the search failure rate of the present network topologies during a period of use, which is $T = 20$. Figure 9.9 shows the time-varying search failure rates for a variety of the number of nodes and the number of node groups. The results shown in Figure 9.9 are averages over 10 independent simulation runs.

Table 9.3. Parameters values of P-EP2P used in the simulations.

Parameter	Description	Value
L	number of nodes (genes)	10^3, 10^4, 10^5
N	number of P2P network topologies (individuals)	30
T	time period for which generated topologies are used	20
H_{max}	allowed number of hops for one search	5
D	number of directed links generated by a node in one topology	1
K	tournament size for the tournament selection	2
p_c	crossover rate	100%
p_e	probability with which recombination is conducted for a selected element in NLX	10%
Q	length of node linkage in NLX	5
p_m	mutation rate	0.5%
N_G	number of node groups	1, 2, 5, 10
G_k	number of nodes in the k-th node group	L/N_G

We can observe from Figure 9.9 that the degree of improvement in the search failure rate becomes smaller as the number of the node groups increases, for any number of nodes. One reason for this would be that NLX changes only a link between nodes in a node group of focus or a link from a node in a node group of focus to a node in other node group. In case that a node as a search object belongs to another node group and the evolutionary operators used cannot create a direct link from nodes in the node group of focus to the node as the search object, multiple links from nodes in the node group of focus to the node as the search object via several nodes in other groups need to be generated and maintained. However, such a linkage of nodes via several node groups cannot be created by the evolutionary operators conducted in one node group. Another reason would be that multiple node groups can assign different fitness functions to the identical network topology. If multiple node groups do so, it is hard for the evolutionary operators to create a linkage between a node as a search object and a node searching for the search object via several node groups.

Figure 9.10(a) shows an example of time-varying ratios of nodes linking to nodes in other node groups when the number of nodes is 10^3. We can observe from Figure 9.10(a) that the ratio is almost the same over the simulation period of time. This would be because novel links from nodes in a certain node group to other node groups are brought only by the mutation operator, which also changes such a link to a link between nodes in one node group. Figure 9.10(b) shows an example of time-varying fitnesses that are assigned to the identical network topology by all of node groups when the number of node groups is five and the number of nodes is 10^4. We can observe from Figure 9.10(b) that the five node groups gave different fitnesses to the same

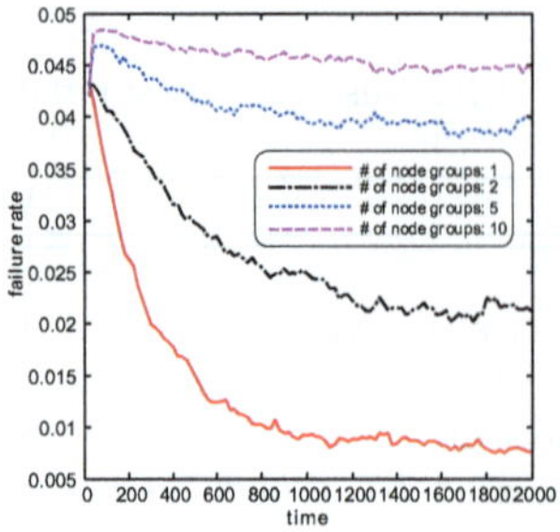

(a) $L = 1000$.

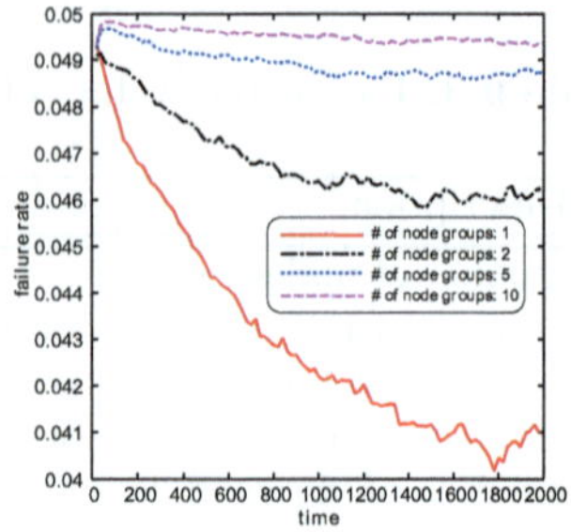

(b) $L = 10000$.

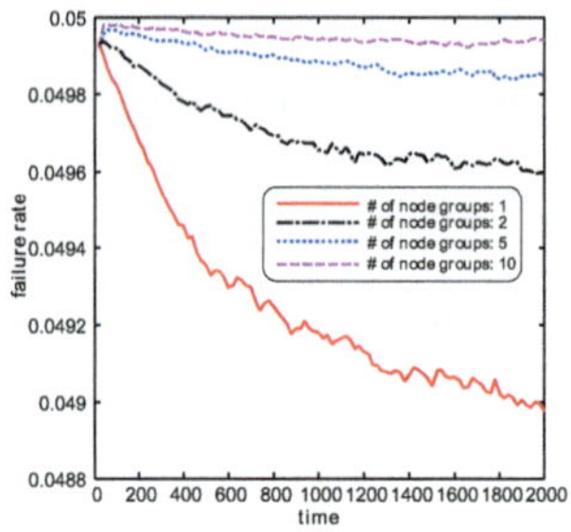

(c) $L = 100000$.

Fig. 9.9. The time-varying search failure rates for different number of node groups [21].

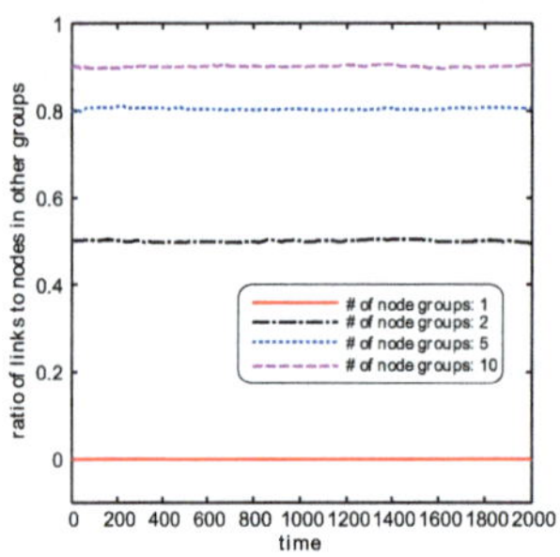

(a) Example of time-varying ratios of nodes linking to nodes in other node groups when the number of nodes is 1000.

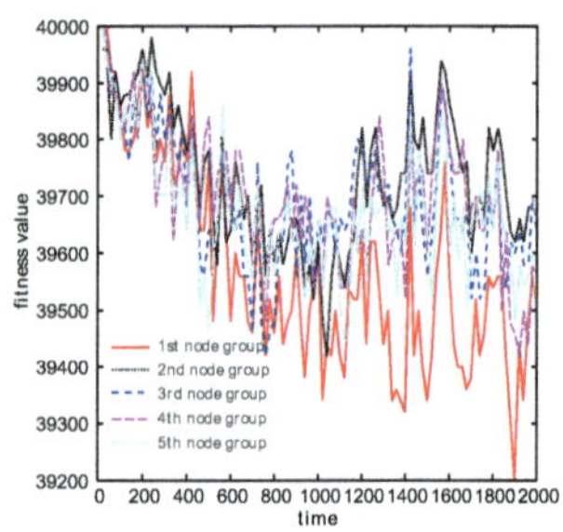

(b) Example of time-varying fitnesses that are assigned to the identical network topology by all of node groups when the number of node groups is five and the number of nodes is 10000.

Fig. 9.10. Examples for time variation of observed values [21].

network topology for most of the time and also that the difference between the fitnesses assigned by the five groups became larger with the time. In addition to this fact, probabilistic procedures included in the selection operator would make it hard to evolve the multiple network topologies consistently.

9.6 Conclusion

In the present chapter, we introduced an evolutionary P2P networking technique (EP2P) that dynamically and adaptively optimizes several P2P network topologies, in which all of the nodes are included at the same time, in an evolutionary manner. In addition, for two different evaluation scenarios, we examined through simulations whether the EP2P can provide reliable search capability in dynamic P2P environments, including participation and departure of nodes and time-variation of search objects. The simulation results suggested (1) that, with respect to reliable search capability, random topology reconstruction is better than topology reconstruction by EP2P when several types of search objects are present in the network at any moment and replicas of these search objects are created in the nodes, and (2) that topology reconstruction by the EP2P is better than random topology reconstruction when only a few types of search objects are present in the network at any moment and these search objects are not replicated. Moreover, for the scenario in which the EP2P was effective, we showed through simulations that when each node makes not one but several directed links to other nodes in a single network topology, the EP2P improves the reliable search capability. Finally, the number of directed links that yield more reliable search capability was found to depend on how often nodes leave and join the network.

The present chapter also introduced the parallel evolutionary P2P networking technique (P-EP2P) for bringing adaptability to large-scale networks. P-EP2P first divides an entire network into several smaller networks to avoid the overload of a super node and then applies EP2P to each of the small networks to make the entire network adaptive. The results of the simulations for evaluating P-EP2P suggested that there is a trade-off relationship between load balancing among super nodes and search reliability.

In order for EP2P to more quickly evolve network topologies that provide reliable search capability and to extend the circumstances in which the EP2P can induce evolutionary adaptation of network topologies, we will consider a method for holding diverse network topologies simultaneously that are specialized for different search objects. In addition, guidelines on the parameter values settings and an investigation of the relationship between the parameter values and the performance of EP2P are necessary. We also need to more clearly and practically define performance metrics for the load balancing among super nodes as well as for the search reliability for simulation studies of P-EP2P, and then seek methods especially for maintaining useful network topologies under these two performance metrics. Finally, although we showed the simulation results for evaluating the online approaches, EP2P and P-EP2P, in the present chapter, we need to evaluate EP2P and P-EP2P through online real experiments.

Acknowledgements. The present study was supported by the Japan Society for the Promotion of Science through a Grant-in-Aid for Young Scientists (B) (22700077).

References

1. LimeWire, `http://www.limewire.com/`
2. Napster, `http://www.napster.com/`
3. Phex, `http://www.phex.org/`
4. Skype, `http://www.skype.com/`
5. Bäck, T.: Evolutionary Algorithms in Theory and Practice: Evolution Strategies, Evolutionary Programming, Genetic Algorithms. Oxford University Press (1996)
6. Cheng, S.T.: Topological optimization of a reliable communication network. IEEE Transactions on Reliability 47(3), 225–233 (1998)
7. Clarke, I., Snadberg, O., Wiley, B., Hong, T.W.: Freenet: A distributed anonymous information storage and retrieval system. In: Proceedings of Workshop on Design Issue in Anonymity and Unobservability. International Computer Science Institute, Berkeley, CA, USA (2000)
8. Cohen, E., Shenker, S.: Replication strategies in unstructured peer-to-peer networks. In: Proceedings of the ACM SIGCOMM 2002 Conference on Applications, Technologies, Architectures, and Protocols for Computer Communication, Pittsburgh, PA, USA (2002)
9. Das, T., Nandi, S., Deutsch, A., Ganguly, N.: Bio-inspired Search and Distributed Memory Formation on Power-Law Networks. In: Rudolph, G., Jansen, T., Lucas, S., Poloni, C., Beume, N. (eds.) PPSN 2008. LNCS, vol. 5199, pp. 154–164. Springer, Heidelberg (2008)
10. Ganguly, N., Canright, G., Deutsch, A.: Design of an Efficient Search Algorithm for P2P Networks Using Concepts from Natural Immune Systems. In: Yao, X., Burke, E.K., Lozano, J.A., Smith, J., Merelo-Guervós, J.J., Bullinaria, J.A., Rowe, J.E., Tiňo, P., Kabán, A., Schwefel, H.-P. (eds.) PPSN 2004. LNCS, vol. 3242, pp. 491–500. Springer, Heidelberg (2004)
11. Imai, P., Tschudin, C.: Practical online network stack evolution. In: SASO 2010 Workshop on Self-Adaptive Networking (2010)
12. Knowles, J., Corne, D.: A new evolutionary approach to the degree-constrained minimum spanning tree problem. IEEE Transactions on Evolutionary Computation 4, 125–134 (2000)
13. Koo, S.G.M., Lee, C.S.G., Kannan, K.: A genetic-algorithm-based neighbor-selection strategy for hybrid peer-to-peer networks. In: Proc. of the 13th Intl. Conference on Computer Communications and Networks (ICCCN 2004), pp. 469–474 (2004)

14. Rong, L.: Multimedia resource replication strategy for a pervasive peer-to-peer environment. Journal of Computers 3(4), 9–15 (2008)
15. Lua, E.K., Crowcroft, J., Pias, M., Sharma, R., Lim, S.: A survey and comparison of peer-to-peer overlay network schemes. IEEE Communications Surveys & Tutorials 7(2), 72–93 (2005)
16. Lv, Q., Cao, P., Cohen, E., Li, K., Shenker, S.: Search and replication in unstructured peer-to-peer networks. In: Proceedings of the 16th International Conference on Supercomputing, New York, USA, pp. 84–95 (2002)
17. Merz, P., Wolf, S.: Evolutionary Local Search for Designing Peer-to-Peer Overlay Topologies Based on Minimum Routing Cost Spanning Trees. In: Runarsson, T.P., Beyer, H.-G., Burke, E.K., Merelo-Guervós, J.J., Whitley, L.D., Yao, X. (eds.) PPSN 2006. LNCS, vol. 4193, pp. 272–281. Springer, Heidelberg (2006)
18. Munetomo, M., Takai, Y., Sato, Y.: An adaptive network routing algorithm employing path genetic operators. In: Proceedings of the Seventh International Conference on Genetic Algorithms, pp. 643–649 (1997)
19. Neri, F., Kotilainen, N., Vapa, M.: A memetic-neural approach to discover resources in P2P networks. SCI, vol. 153, pp. 113–129 (2008)
20. Ohnishi, K., Oie, Y.: Evolutionary P2P networking that fuses evolutionary computation and P2P networking together. IEICE Transactions on Communications E93-B(2), 317–328 (2010)
21. Ohnishi, K., Oie, Y.: Parallel evolutionary P2P networking for realizing adaptive large-scale networks. In: The Second Workshop on Heuristic Methods for the Design, Deployment, and Reliability of Networks and Network Applications (HEUNET 2011) (SAINT 2011 Workshop) (2011)
22. Pournaras, E., Exarchakos, G., Antonopoulos, N.: Load-driven neighbourhood reconfiguration of gnutella overlay. Computer Communications 31(13), 3030–3039 (2008)
23. Srivatsa, M., Gedik, B., Liu, L.: Large scaling unstructured peer-to-peer networks with heterogeneity-aware topology and routing. IEEE Transactions on Parallel and Distributed Systems 17(11), 1277–1293 (2006)
24. Thampi, S.M., Chandra, S.K.: Review of replication schemes for unstructured P2P networks. In: Proceedings of IEEE International Advance Computing Conference IEEE (IACC 2009), Patiala, India, pp. 794–800 (2009)
25. Walkowiak, K., Przewoniczek, M.: Modeling and optimization of survivable P2P multicasting. Computer Communications 34(12), 1410–1424 (2011)
26. Zhou, G., Gen, M.: A note on genetic algorithm approach to the degree-constrained spanning tree problems. International Journal of Networks 30(2), 91–95 (1997)

Chapter 10
A Taxonomy of Evolutionary Inspired Solutions for Energy Management in Green Computing: Problems and Resolution Methods

Joanna Kołodziej, Samee U. Khan, and Albert Y. Zomaya

Abstract. Over the last years, the engineers, researchers, and vendors have teamed up to design and develop the intelligent models and algorithms that constrict the use of electrical energy in computing devices in the large-scale heterogeneous systems. This chapter realizes the need to present to the scientific community a current state of the art on research, current trends, and future work on evolutionary inspired solutions for green computing.

10.1 Introduction

In modern highly parametrized large-scale computing systems the quest for more powerful computational resources has enabled significant scientific discoveries. These systems are usually composed of thousands of various computing devices, data centers and numerous services with different routing and communication protocols and various local access policies. However, such advancement in the modernization and development of intelligent and scalable distributed computing systems has significantly strained the electrical energy

Joanna Kołodziej
Institute of Computer Science,
Cracow University of Technology ul. Warszawska 24, 31-155 Cracow, Poland
e-mail: jkolodziej@uck.pk.edu.pl

Samee U. Khan
Department of Electrical and Computer Engineering,
North Dakota State University, ND 58108, USA
e-mail: samee.khan@ndsu.edu

Albert Y. Zomaya
School of Information Technologies, University of Sydney,
Sydney, NSW 2006, Australia
e-mail: albert.zomaya@sydney.edu.au

J. Kołodziej et al. (Eds.): Advances in Intelligent Modelling and Simulation, SCI 422, pp. 215–233.
springerlink.com

resources, distribution, and protection systems. For all these reasons there is a great need to design, develop, and test devices, procedures, methodologies, and algorithms that constrict the use of electrical energy in computing devices.

This chapter approaches the surveying of the field from the perspective of evolutionary inspired solutions for energy management in energy-aware computing. The survey presented here will serve as a stepping stone for young researchers and also combines two research disciplines, namely evolutionary and "green" computing.

The remainder of the chapter is structured as follows. Sec. 10.2 presents a simple taxonomy of energy and resource management methods in large-scale heterogeneous computing systems. Evolutionary inspired optimization techniques in static and dynamic energy management are surveyed in Sec. 10.3 and Sec. 10.4. The chapter ends in Sec. 10.5 with summative analysis of reviewed evolutionary approaches, conclusions and discussion on possible further research directions.

10.2 Taxonomy of Energy Management in Future Generation Distributed Computing Systems

A lot of research projects have been done in the domain of energy aware resource management in today's large-scale computing system. Based on the taxonomy for cloud computing proposed in [11] the management methods in modern distributed computing systems can be classified into two main categories: static energy management (SEM) and dynamic energy management (DEM), as shown in Fig. 10.1.

At the hardware level of the class of the static management methods the system devices can be replaced by the low-power battery machines or nano-processors and the system workload can be effectively distributed. It allows to optimize the energy utilized for computing the applications, storage and data transfer by reducing the number of idle devices and idle periods of active processors. It is important to carefully consider the implementation of programs that are executed in the system in order to achieve a high and fast reduction in the energy usage. Even with perfectly designed hardware, poor software design can lead to significant power and energy losses. Therefore the process of compilation or code generation and the order of instructions in application source code may lead to the optimization of the energy management.

A wide class of the dynamic energy management methods is composed of the strategies for dynamic adaptation of the system performance to the current resource requirements and other parameters of the system's state. In this case the systems experience variable workloads in the working periods which allows the dynamic adjustment of power states according to current performance requirements. Similarly to static solutions, dynamic management methodologies can be distinguished by their application levels into hardware

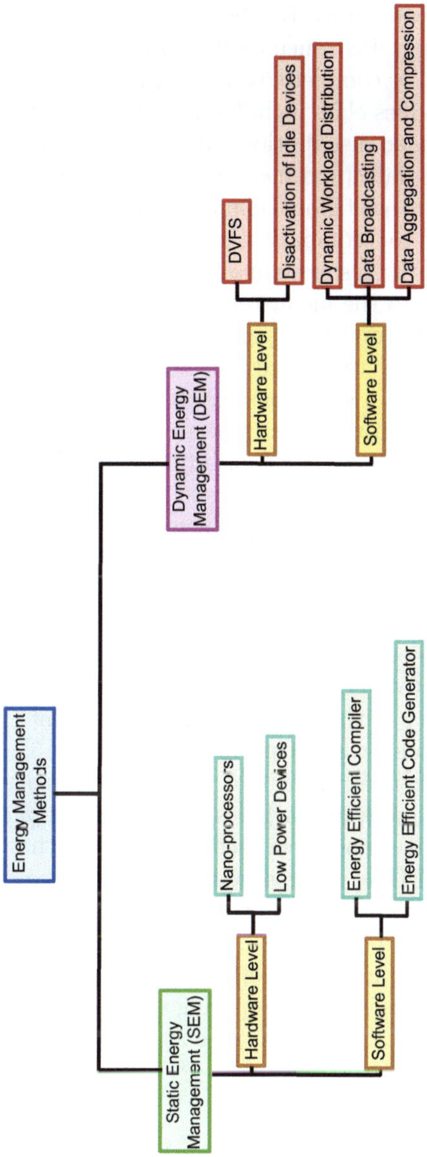

Fig. 10.1. Taxonomy of energy management in large-scale distributed computing systems

and software local categories. Hardware tools can be classified as Dynamic Performance Scaling (DPS), such as Dynamic Voltage and Frequency Scaling (DVFS); and partial or complete dynamic deactivation of inactive processors. The software techniques class includes all optimization techniques connected with dynamic workload distribution, efficient data broadcasting, data aggregation and dynamic data (and memory) compression.

The fast development of global communication technologies enables in fact unlimited access of the computing system users to the highly distributed resources. However, the resource management methods in such systems may be very difficult due to the incoherent different local policies at the system's and operational levels, and high parametrization and dynamics of the whole structure. Evolutionary techniques are very effective in finding optimal or near optimal solutions in such complex dynamic environments.

Still not so large family of energy-aware genetic-based optimization methods are presented in literature. Basically classical single population strategies are used as the energy optimizers. The adaptation of such methodologies for solving the optimization problems in large-scale dynamic environments requires an application of specialized genetic operators, such as partially matching or cycle crossovers and swap or rebalancing mutation mechanisms [30], [22]. The energy consumed by the system is usually just one component of a multi-objective fitness function. In such a case the Multi-objective Genetic Algorithm (MOGA) framework [10] seems to be a key solution to tackle the complexity of the optimization process. Ant Colony Optimization (ACO) [7] and Particle Swarm Optimization (PSO) [15] algorithms are useful in generating the optimal paths and tree structures in graph-based models of networks, multi-processor machines and parallel applications. Finally, just few approaches in grid and cloud scheduling show that island, hierarchical and cellular parallel GAs can essentially speed up the optimization process and improve the exploration and exploitation of the large search space.

This chapter surveys the recent and most promising evolutionary inspired solutions to the static and dynamic energy-aware resource management in modern large-scale distributed computing systems. The presented algorithms are characterized by the following three attributes:

- type of the algorithm;
- objective function;
- application area.

10.3 Static Energy Management: Code Optimizers in Embedded Systems

Static methods of energy management in distributed systems are usually employed at the hardware level. Energy absorbing devices can be replaced by the machines with low power batteries and processors. However, this is not so good area for the implementation of the evolutionary meta-heuristics. Some recent projects are focused on the extension of the system architecture by future generation nano-processors, which successfully reduce the energy consumption in the system. The most effective architecture optimization methodologies for such systems are driven by the evolutionary inspired meta-heuristics. The high effectiveness of those techniques is demonstrated mainly in the optimization of compilers and the generators of the source codes of applications at the software level.

Fig. 10.2 presents the basic genetic-based approaches to static energy management in embedded systems.

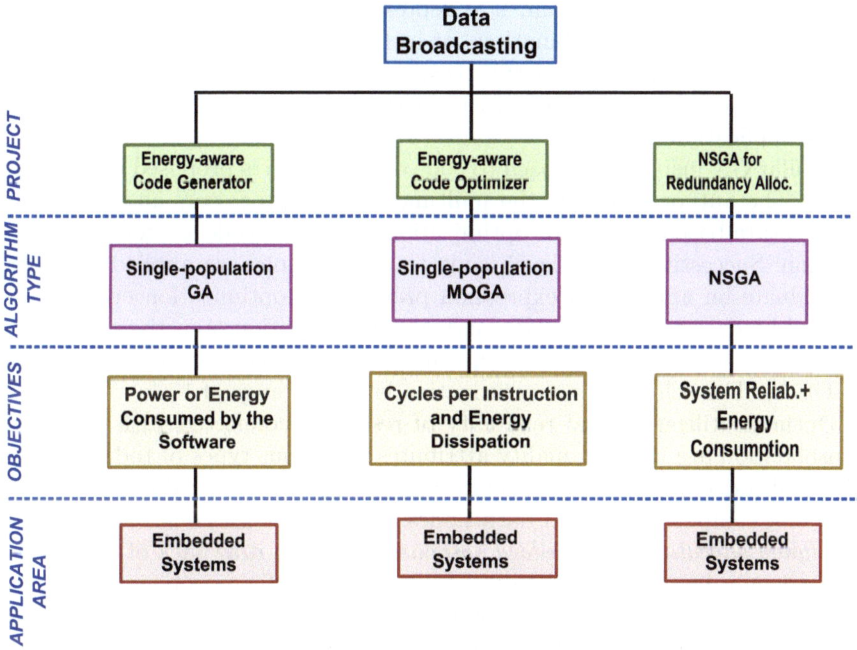

Fig. 10.2. Genetic-based methods for static resource management in embedded systems

Most of the embedded systems are composed of the digital signal processors (DSPs) that flexibly account for the modification of the system specification. However, many embedded applications are still prepared in assembly code. An implementations of such codes are time-consuming and inefficient in system energy utilization. For such reasons, there is a need for optimizing compilers and and application source codes to adapt them to the special architectures and thus to make them capable of exploiting the irregular architecture features of DSPs.

Lorenz et al. define in [24] an energy-aware code generator (GCG) based on single population genetic algorithm. This code generator reduces the energy consumption by suitable instruction selection and instruction scheduling. Energy aware compilation is done with respect to an instruction level energy cost model which is integrated into the code generator and simulator. The genetic algorithm module works on specialized chromosomes, that encode a set of basic blocks, which are created by using a simple decomposition procedure to the source program. Each such a block is represented by a node in data flow graph (DFG). Each gene of the chromosome represents an operation like a load or an addition. The values of a gene express an information about used registers, performed processor instruction, execution cycle (etc.), which are necessary to code generation. An objective function is defined as a consumed power or energy of a program. It is represented by values of average power dissipation of certain combinations of instructions. The authors used their method for SIMD instructions (SIMD=single instruction multiple data). The obtained results show the 30% of the energy reduction with 8% of reduction of the application code.

Similar GA-inspired approach to code optimization is proposed by Azzemi in [2]. The author considers the multimedia DSP processors and define an architecture-based parametric optimization of C source code for iterative compilation. Successive source-level, code transformations are applied in order to evaluate an application expression profile. The optimization criteria are defined by cycles per instruction and energy dissipation objectives. This bi-objective function is optimize by a simple multi-objective genetic algorithm (MOGA). The achieved energy reduction is in the range of 17%.

Optimal utilization and reliability of resources connected with the data consolidation are the key quality attributes in various types of today's complex embedded systems. The energy consumption may be reduced by replication of computational and data nodes. This problem is referred to as *redundancy allocation in embedded systems.* The main drawback of using this method may lie in additional energy requirements for supplying the replicated nodes. However, the key solution of the problem may be a trade-off analysis of resource reliability and energy consumption at an architectural level, and the employment of the redundancy allocation, which has a significant effect on both system quality attributes. In [27] Meedeniya et al. try to solve the redundancy allocation problem in the embedded systems by using the Markov Reward Model [14] for system representation. The authors optimized a bi-objective function with system reliability and energy consumption components by using the non-dominated sorting GA (NSGA). Each chromosome encodes a single redundancy allocation. Each allele in a chromosome represents a redundancy level for a system component. The achieved empirical results show that the proposed method can significantly reduce the energy consumption for a very small trade-off of reliability, which would definitely be an interesting information for the system designer.

10.4 Evolutionary Inspired Dynamic Data and Resource Management in Green Computing

Evolutionary-based solutions to dynamic energy management in large-scale distributed systems are primarily proposed as scalable and robust methodologies for scheduling and data processing in networking, cluster and grid computing. This section highlights the recent research in "evolutionary-driven" energy optimization in data management, namely data broadcasting and aggregation in wireless sensor networks, grid and cloud scheduling, where the voltage supply of the devices may be modulated in the system.

10.4.1 Energy Efficient Data Transmission

Data transmission (or data broadcasting) is, beyond the resource allocation and scheduling, a fundamental problem in large-scale data centers, intelligent networks and grid and cloud environments. The types of the implemented communication protocols have a great impact on the whole system performance. The problem of an efficient and energy-aware data broadcasting is especially essential in today's large-scale wireless networks such as ad-hoc and sensor networks, where the nodes, acting potentially both as routers and hosts, are equipped with antennas for sending and receiving information. Communication may be performed by one-to-one transmissions (single-hop) or using other nodes as relay stations (multi-hop). In both cases each sender node must adjust its emission power in order to reach the respective receiver node. Additionally, in the cases where energy is supplied by batteries, the network lifetime is limited by the batteries of the wireless devices. Therefore, energy saving is critical in all network operations.

Minimum Energy Broadcast (MEB) is defined as a problem of minimizing the energy during the data transfer. Formally, it can be formulated as the minimal spanning tree task ($T = (V, E_T)$) in the fully connected graph $G = (V, E)$ representing the system structure. The root of the tree is a source node for the data (signals) emission and the following energy emission function is minimized:

$$P(T) = \sum_{i \in V} \max_{(i,j) \in V_T} d(i, j)^{\alpha}, \tag{10.1}$$

where $d(i, j)$ is the Euclidean distance between the nodes i and j and α is a parameter that, depending on the environment, takes typically values between 2 and 4. It is assumed that the graph G for wireless networks is directed and $d(i, j)^{\alpha} < p_{max}$, where p_{max} is a maximal emission power in the system. If the antennas in the network nodes are directional, a beam width and a beam direction must be chosen for each node $i \in V$.

In classical cluster and grid systems the energy utilized for the data files transfer between two connected nodes is summarized (and then optimized) for all possible nodes pairs.

Figure 10.3 presents few recent evolutionary inspired approaches to energy aware data transfer in cluster system and wireless sensor networks.

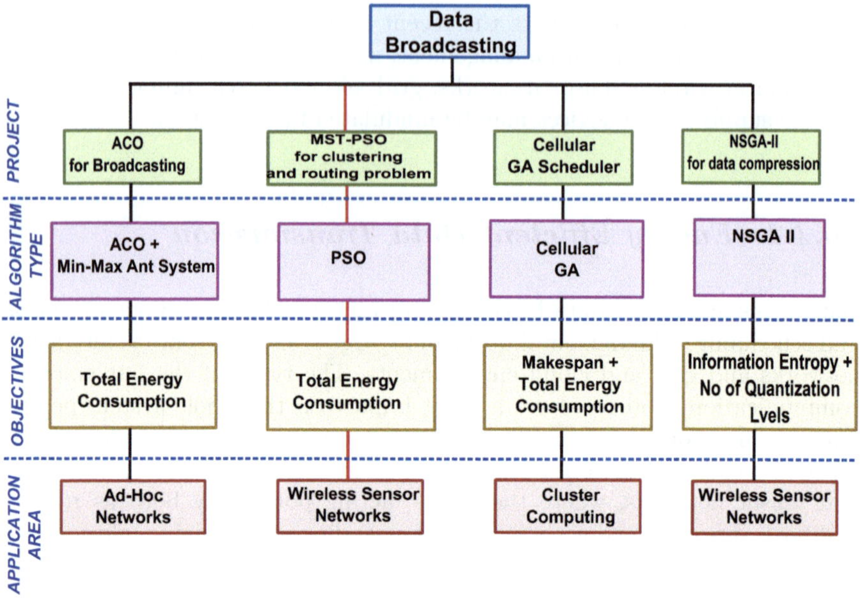

Fig. 10.3. Evolutionary solutions to energy-aware data broadcasting

In [13] Hernández, Blum and Francès address the problem of signal broadcasting in the ad-hoc networks. They consider the system with omni-directional and directional antennas. The emission energy defined in Eq. 10.1 is an objection function, which is globally optimized by using a specialized Ant Colony Optimization algorithm – Min-Max Ant System in the Hyper-Cube Framework [35]. At each iteration of the algorithm artificial ants construct a broadcasting tree rooted at the emission source node. Local search r-shrink algorithm is applied to each of these trees and the pheromone values may be updated by using also the best-so-far solutions. The power saving rates achieved in the experimental analysis is ab. 85 %, which makes the methodology spectacular solution for improving the network effectiveness in the reduction of the energy emission.

Cao et al. in [4] have considered a routing problem in wireless sensor networks (WSNs) and a case, in which a node and its cluster-head engage in a multi-hop communication. They used a particle swarm optimization (PSO) algorithm for the nodes clustering. A distance based minimum spanning tree of the weighted graph of the network is generated and the best connection between a node and its cluster-head is searched from all the optimal spanning trees on the criterion of energy consumption. Cluster-heads are elected based on the energy available to the nodes and the Euclidean distance to its neighbor node in the optimal tree. The results show that the PSO-based clustering methods ensure longer network life.

An interesting approach of cellular GA-based schedulers to cluster computing is presented by Guzek et al. in [12]. The authors consider a general scheduling problem of parallel application modeled by a directed acyclic graph (DAG) in a cluster of heterogeneous machines. The cellular algorithm is used primarily for sub-tasks clustering (the number of clusters copes with the number of processors in the machine) and scheduling. The privilege objection is makespan and the second criterion – total energy consumed during the inter-processor communication. This communication model is based on the classical delay model [32] and the energy utilized for a data transfer is measured for each CPUs connection in a parallel machine.

Dynamic data compression in the application codes seems to be a promising software tool for saving the energy used for the data propagation in wireless sensor networks. Compression methods exploit the data structure and reduce the data size. Marcelloni and Vecchio [26] perform a data compression on a network (single) node based on a differential pulse code modulation scheme with quantization of the differences between consecutive codes of the signal samples. The trade-off between a performance of compression algorithm and the amount of the lost information is determined by the set of quantization parameters. The authors employ the Non-dominated Sorting Genetic Algorithm II (NSGA–II) for optimizing the combinations of these parameters corresponding to different optimal trade-offs. The chromosomes in this approach encode quantizers defined by using the following parameters: (1) a width of the dead zone, (2) a width of the cell in the first granular subregion, (3) a number of cells in the first granular subregion, (4) a width of the cell in the second granular subregion, and (5) a number of cells in the second granular subregion.

The chromosomes are encoded as binary Gray strings. The granular regions are the regions with quantization levels. Information entropy and the number of distinct quantization levels (used in the quantizer are the optimization criteria. The evaluation analysis of the proposed method shows the 62% reduction of the energy consumed in data transmission.

10.4.2 Energy-Aware Data Aggregation in Grids, Clouds and Wireless Sensor Networks

Data aggregation is the combination of data from different datasets by using a specified aggregation function. This function can be defined as a duplicate data suppression, minima, maxima and average data indicator. A big amount of the energy in data centers is the idle power wasted when servers run at low utilization. Multiple data center applications may be hosted on a common set of servers. Also sensor nodes in wireless networks may generate significant redundant data. This allows for consolidation of application workloads on a smaller number of servers and aggregation of similar data packets from multiple network nodes that may increase the system utilization by save the energy.

In grid and cloud computing the problem of loading servers to a desired utilization level for each resource may be modeled as a multi-dimensional bin packing problem where servers are bins with each resource (CPU, disk, etc.) being one dimension of the bin. The bin size along each dimension is given by the energy optimal utilization level. Each hosted application with known resource utilizations can be treated as an object with a given size in each dimension. The ultimate goal of the consolidation algorithm is to pack all items to possible minimal number of bins. An objective function for such a problem can be defined as follows(see also [9]:

$$f = \sum_{v=0}^{n-1} y_v, \tag{10.2}$$

and is minimized subject to the following constraints:

$$\sum_{i=0}^{m-1} \bar{r}_{i,k} x_{i,v} \leq C_{v,k} y_v, \forall v \in \{0, \ldots, n-1\}, \forall k \in \mathbb{R} \tag{10.3}$$

$$\sum_{v=0}^{n-1} x_{i,v} = 1, \forall i \in \{0, \ldots, m-1\} \tag{10.4}$$

where

- n is the number of bins;
- m is the number of items;
- y_v is the bin variable which is 1 if the bin v is selected and 0 otherwise;
- $x_{i,v}$ is the allocation variable equals 1 if the item i is assigned to the bin v, and 0 otherwise;
- $C_{v,k}$ is the capacity of bin v of resource $k \in \mathbb{R}$;
- $\bar{r}_{i,k}$ is the i-th item maximum demand for resource $k \in \mathbb{R}$ over the last measurement period.

The condition 10.3 ensures that the capacity of each bin is not exceeded and constraint 10.4 guarantees that each item is assigned to at most one bin.

In wireless sensor networks signal processing methods may be used for data aggregation. In this case, it is referred to as data fusion where each node is capable of producing an accurate output signal by using some techniques such as beam forming to combine the incoming signals and reducing the noise in the output ones.

Selected genetic-based methods for data aggregation are reported in Fig. 10.4.

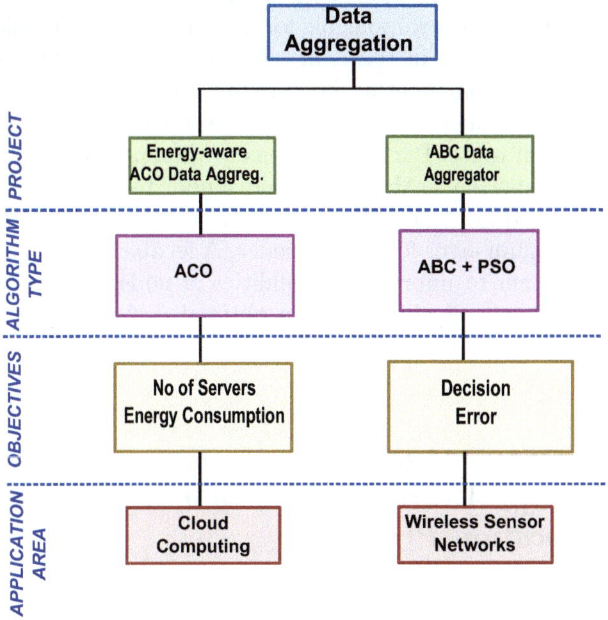

Fig. 10.4. Genetic-based methods for data aggregation in large-scale distributed systems

One of the most recent Ant Colony Optimization (ACO) approaches in data aggregation in cloud computing is presented by Feller et al. in [9]. The authors used the ACO algoriithm for the consolidation of virtual machines on the least number of physical nodes in the cloud system. The problem is interpreted as an instance of the multi-dimensional bin-packing (MDBP) problem. The fitness function is defined as a sum of boolean bin variables given by the Eq. 10.2. The authors follow the MAX-MIN Ant System (MMAS) Framework for updating the pheromone trials [35] for the ants. The power function is specified for each host in the system for estimating the energy consumed by a workload placement. This power function is defined as a linear function $P(u)$ of the host utilization u, that is to say:

$$P(u) = (P_{max}, \ldots, PPidle) \times u + P_{idle}, \tag{10.5}$$

where P_{idle} and P_{max} stand for the average power values when the system is idle and fully utilized, respectively. Computational results show the 4.1% of energy conservation on average 4.7% of hosts.

An efficient data-aggregation in wireless sensor networks can be achieved by the determination of optimal local thresholds in the decisions made by the networks fusion center for detecting the events. Each sensor node in the network collects local observations corrupted by noise and sends a summary to a fusion center, which is responsible for making the final decision. Thresholding may lead to a gain in terms of bandwidth and energy consumed by the system. Veeramachaneni et al. [37] present a hybrid of ant-based control and PSO (ABC-PSO) method for the local threshold management to achieve an optimal decision route. Partial solutions to the optimization problem are constructed by artificial ants that move from a node to another and define the paths of network nodes. Then PSO algorithm identifies the thresholds and achieves the minimum error for the sequence. A feedback on this is presented to ants to help them to improve the qualities of node sequences to achieve optimal thresholds on all nodes and an optimal decision route (hierarchy) that assure minimum energy expenditure.

10.4.3 Dynamic Voltage and Frequency Scaling in Energy-Aware Scheduling and Resource Allocation Problems

Scheduling in traditional distributed systems has been mainly studied for system performance parameters without data transmission requirements. With the emergence of data grids (DGs) and data centers, data-aware scheduling has become a major research issue. Today's data centers arise quite naturally to support needs of scientific communities to share, access, process, and manage large data collections geographically distributed.

Computing devices (CPUs) are the major energy "consumers" in a data center. The energy of the system is utilized for the tasks execution, data storage at the data hosts, data transmission, decoupling of data from processing and data replication.

Power and total energy consumption can be reduced by lowering the supply voltage of CPUs using Dynamic Voltage Scaling (DVS) or Dynamic Voltage/Frequency Scaling (DVFS) methods [25]. It is assumed that each machine in the system (it can be a data or/and computing node) is equipped with a DVS module, which allows to modulate its supply voltage and operating frequency. Instead of complete deactivation of the processor, its clock frequency along with adjustments of the supply voltage can be gradually reduced or increased in cases when the resource is not fully utilized.

The energy consumption model in a data center is usually based on the power consumption model in complementary metal-oxide semiconductor (CMOS) logic circuits. The power consumption of a CMOS-based microprocessor is defined as a sum of the capacitive, short-circuit and leakage power. The most significant factor is the capacitive power, which can be interpreted as the dynamic power consumption P_d of a CPU i in a data center and can be calculated in the following way:

$$P^i_d = A_i \cdot C_i \cdot v_i^2 \cdot f_i, \tag{10.6}$$

where A_i is the number of switches per clock cycle, C_i is an effective switched capacitance of the circuits, v_i is the supply voltage and f_i is the clock frequency. For constant values of parameters A_i and C_i for a given CPU it can be assumed that power P^i_d is proportionate to v_i^2. Since the DVS mechanism has been considered to scale up and down CPU frequency, the execution time of a task on a given machine will significantly vary according to the CPU frequency. The decrease in the execution time usually is a result of approximately proportional increase in CPU frequency. The energy consumption of the processor i can be expressed as processor power multiplied by execution time of the operation (application). In the general case the energy can be defined as follows[1]:

$$E_i = \int_0^{completion[i]} P^i_d(t)dt, \tag{10.7}$$

where *completion*[*i*] is a completion time of the processor *i*.

It can be observed from the above equations that the decrease in voltage supply can reduce the energy per operation in a quadratic manner, and unfortunately, it may significantly slow down the completion time of the operation.

Table 10.1 shows the parameters for 16 typical DVFS levels and three main "energetic" categories for CPUs considered in the most of the publications on DVFS approaches.

Table 10.1. DVFS levels for three CPU classes

	Class I		**Class II**		**Class III**	
Level	**Volt.**	**Rel.Freq.**	**Volt.**	**Rel.Freq.**	**Volt.**	**Rel.Freq.**
0	1.5	1.0	2.2	1.0	1.75	1.0
1	1.4	0.9	1.9	0.85	1.4	0.8
2	1.3	0.8	1.6	0.65	1.2	0.6
3	1.2	0.7	1.3	0.50	1.9	0.4
4	1.1	0.6	1.0	0.35		
5	1.0	0.5				
6	0.9	0.4				

The detailed energy model description in grid and cloud data centers can be found i.e. in [17] and [11].

[1] The integration is replaced by the summation in the case of the discrete time process.

The DVFS technique is classified as an effective hardware dynamic energy optimizer in resource allocation and scheduling problems in large-scale distributed systems. The energy-aware scheduling is usually considered as a multi-objective global optimization problem with makespan and total energy consumption as the main criteria. In most of the DVFS approached the scheduling has been defined as classical or dynamic load balancing problem. In such cases linear, dynamic and goal programming are the major optimization techniques (see i.e. [23], [38], [19], [17], [20]). Recent evolutionary-based approaches that apply DVFS to reduce energy consumption are presented in Fig. 10.5. The total energy utilization in the system is a component of the fitness functions.

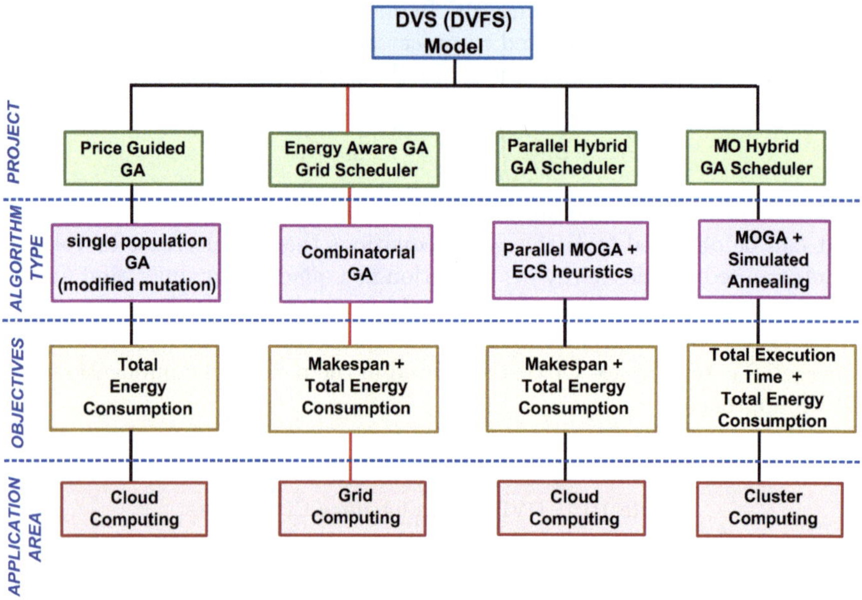

Fig. 10.5. Evo-based meta-heuristic for energy aware scheduling with modular voltage supply

In [33] and [34] Shen et al. present a *shadow price* technique for improving the genetic operations in standard GA used as a scheduler in computational cloud. The "shadow price" for a pair task-machine is defined as an average energy consumption per instruction for the processor that can operate at different voltage levels. Then the classical move and swap mutation operations are used for an optimal mapping of tasks to machines. The fitness function for such GA scheduler is expressed as a total energy consumption.

Total energy consumed by a computational grid is the key criterion in independent batch scheduling problem adressed by Kołodziej, Khan and Xhafa in [21]. The expected times of the execution of tasks on the machines in the system are estimated by using the *Expected Time to Compute* matrix model [1]. There are two GA-based schedulers developed for makespan and energy consumption optimization. The authors consider two scenarios, where all machines works at the highest voltage level and are switched to the sleep mode in idle periods, and the case of operating at different voltage levels under optimal makespan constraint. The schedulers were experimentally evaluated in static and dynamic grid environment. In both cases the modulation of the voltage supply of the machines reduced the energy consumption by 25–30 % in average.

Kessaci et al. in [16] present two versions of multi-objective parallel Genetic Algorithm (MOPGA) hybridized with energy-conscious scheduling heuristics (ECS). The GA engine is based on the concepts of island GA and multi-start GA models. The authors consider parallel applications represented by a directed acyclic graph (DAG), which are mapped onto multi-processors machines. The voltage and frequencies of the processors are scaled up at 16 discrete levels and genes in GA chromosomes are defined by the task-processor labels and processor voltage. The objective function is composed of two criteria: privileged makespan and total energy consumption in the system. The reduction of the energy utilization achieved in the experimental analysis is about 47.4%.

The solution presented in [16] is dedicated to general computing and embedded systems. An application of such methodology in computational cloud is demonstrated by Mezmaz et al. in [28]. The energy conservation rate in cloud system is very similar to the results achieved in the general large-scale cluster models.

Another hybrid GA approach is presented by Miao et al. in [29]. The authors propose a multi-objective genetic algorithm which is hybridized with simulated annealing for the improvement of the local solutions for the scheduling problem in cluster computing.

10.5 Conclusions

This chapter surveyed the recent research results related to the implementations of evolutionary inspired methodologies supporting the energy and power management in modern large-scale distributed computing systems, such as wireless networks, grid, cloud and cluster computing systems. Although, the genetic meta-heuristics are still not the most popular solutions to key green computing problems, the results of the empirical analysis presented by a wide community of authors confirmed high effectiveness of these methods in fast and significant reduction (in the range of 6 % to 85%) of the energy consumed in the system.

Fig. 10.6 shows a simple summative analysis of the evolutionary approaches to energy management in today's most popular distributed computing systems.

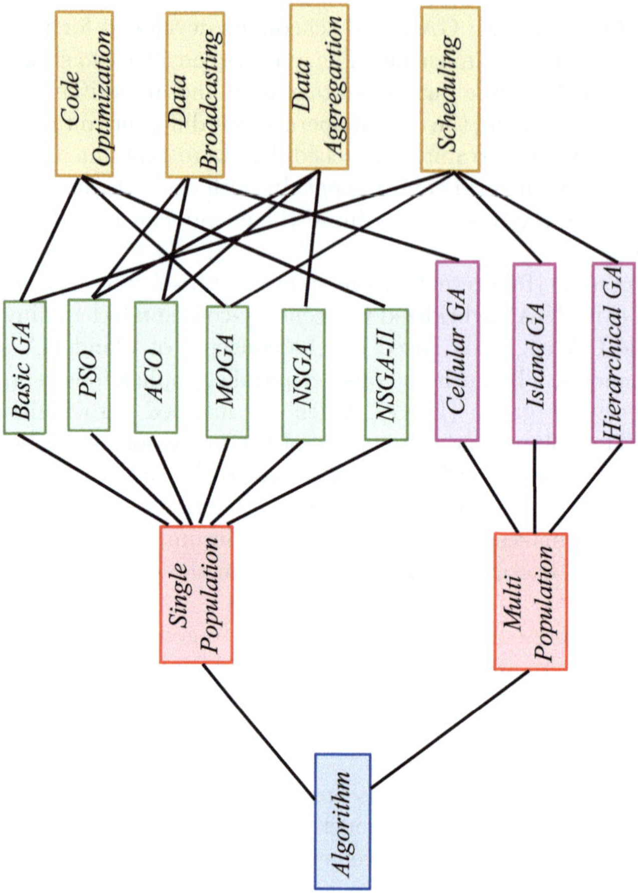

Fig. 10.6. Summative analysis of evolutionary-based approaches to energy management in distributed computing systems

The algorithms are classified into single- and multi-population methods. The first group contain all popular single population genetic techniques for global optimization. They may be used for solving all problems addressed in this chapter and they work in all types of dynamic environments we considered in this study. The second class of multi-population algorithms is very small, that confirms an early stage of the research on such approaches.

An emergence of new generation IT systems, grow environmental concerns and imply new challenges in efficient management of huge packages of highly parameterized data. An incorporation of new additional criteria into the energy-aware data and resource management in future generation distributed systems may expose limitations in effectiveness of existing solutions at both hardware and software levels. It certainly implies a need of development of new models and meta-heuristic optimization techniques which can tackle the higher complexity of the system components, new access polices and conditions and users' preferences and requirements. A promising research direction, which can make a significant progress in green computing, may be an utilization of game-theoretical models and evolutionary-based resolution methods for supporting the decisions of the system users and resource providers. Simple cooperative games with Nash-bargained solutions have been already developed and successfully applied in energy-aware scheduling in grids and data centers (see [18], [36]), which can be a strong background for an expansion of evolutionary inspired solvers of such models.

References

1. Ali, S., Siegel, H.J., Maheswaran, M., Hensgen, D.: Task execution time modeling for heterogeneous computing systems. In: Proceedings of Heterogeneous Computing Workshop, pp. 185–199 (2000)
2. Azzemi, N.Z.: A Multiobjective Evolutionary Approach for Constrained Joint Source Code Optimization. In: Proc. of ISCA 19th International Conference on Computer Application in Industry (CAINE 2006), Las Vegas, Nevada, USA, pp. 175–180 (2006)
3. Beloglazov, A., Buyya, R., Lee, Y.C., Zomaya, A.Y.: A Taxonomy and Survey of Energy-Efficient Data Centers and Cloud Computing Systems. Advances in Computers 82, 47–111 (2011)
4. Cao, X., Zhang, H., Shi, J., Cui, G.: Cluster heads election analysis for multi-hop wireless sensor networks based on weighted graph and particle swarm optimization. In: Proc. of the 4th International Conference on Natural Computation (ICNC), vol. 7, pp. 599–603 (2008)
5. Chandran, J.J.G., Victor, S.P.: Optimized Energy Efficient Localization Technique in Mobile Sensor Networks. IACSIT International Journal of Engineering and Technology 2(2), 149–156 (2010)
6. Diaz, C.O., Guzek, M., Pecero, J.E., Danoy, G., Bouvry, P., Khan, S.U.: Energy-aware Fast Scheduling Heuristics in Heterogeneous Computing Systems. In: Proc. of ACM/IEEE/IFIP International Conference on High Performance Computing and Simulation (HPCS), Istanbul, Turkey, July 2011 (2001)
7. Dorigo, M., Stützle, T.: Ant Colony Optimization. MIT Press, Cambridge (2004)
8. Fard, G.H.E., Monsefi, R.: A Fast Multi-objective Genetic Algorithm based Approach for Energy Efficient QoS-Routing in Two-tiered Wireless Multimedia Sensor Networks. Modern Applied Science 4(6), 101–112 (2010)
9. Feller, E., Rilling, L., Morin, C.: Energy-Aware Ant Colony Based Workload Placement in Clouds. INRIA Report RR-7622, Rennes, France (2011)
10. Fonseca, C.M., Fleming, P.J.: An overview of evolutionary algorithms in multiobjective optimization. Evolutionary Computation 3(1), 1–16 (1995)
11. Garg, S.K., Yeo, C.S., Anandasivam, A., Buyya, R.: Energy-Efficient Scheduling of HPC Applications in Cloud Computing Environments. CoRR abs/0909.1146 (2009)

12. Guzek, K., Pecero, J.E., Dorrosoro, B., Bouvry, P., Khan, S.U.: A Cellular Genetic Algorithm for Scheduling Applications and Energy-aware Communication Optimization. In: ACM/IEEE/IFIP International Conference on High Performance Computing and Simulation (HPCS), Caen, France, pp. 241–248 (2010)
13. Hernández, H., Blum, C., Francès, G.: Ant Colony Optimization for Energy-Efficient Broadcasting in Ad-Hoc Networks. In: Dorigo, M., Birattari, M., Blum, C., Clerc, M., Stützle, T., Winfield, A.F.T. (eds.) ANTS 2008. LNCS, vol. 5217, pp. 25–36. Springer, Heidelberg (2008)
14. Katoen, J.P., Khattri, M., Zapreev, I.S.: A Markov reward model checker. In: Proc. of the QEST: International Conference on the Quantitative Evaluation of Systems, pp. 243–244. IEEE Computer Society (2005)
15. Kennedy, J., Eberhart, R.: Particle swarm optimization. In: Proc. of the IEEE International Conference on Neural Networks, November 27-December 1, vol. 4, pp. 1942–1948 (1995)
16. Kessaci, Y., Mezmaz, M., Melab, N., Talbi, E.-G., Tuyttens, D.: Parallel Evolutionary Algorithms for Energy Aware Scheduling. In: Bouvry, P., González-Vélez, H., Kołodziej, J. (eds.) Intelligent Decision Systems in Large-Scale Distributed Environments. SCI, vol. 362, pp. 75–100. Springer, Heidelberg (2011)
17. Khan, S.U.: A Goal Programming Approach for the Joint Optimization of Energy Consumption and Response Time in Computational Grids. In: Proc. of the 28th IEEE International Performance Computing and Communications Conference (IPCCC), Phoenix, AZ, USA, pp. 410–417 (2009)
18. Khan, S.U., Ahmad, I.: A Cooperative Game Theoretical Technique for Joint Optimization of Energy Consumption and Response Time in Computational Grids. IEEE Transactions on Parallel and Distributed Systems 20(3), 346–360 (2009)
19. Khan, S.U.: A Self-adaptive Weighted Sum Technique for the Joint Optimization of Performance and Power Consumption in Data Centers. In: Proc. of the 22nd International Conference on Parallel and Distributed Computing and Communication Systems (PDCCS), Louisville, KY, USA, pp. 13–18 (September 2009)
20. Kliazovich, D., Bouvry, P., Khan, S.U.: DENS: Data Center Energy-Efficient Network-Aware Scheduling. In: Proc. of ACM/IEEE International Conference on Green Computing and Communications (GreenCom), Hangzhou, China, pp. 69–75 (December 2010)
21. Kołodziej, J., Khan, S.U., Xhafa, F.: Genetic Algorithms for Energy-aware Scheduling in Computational Grids. In: Proc. of the 6th International Conference on P2P, Parallel, Grid, Cloud and Internet Computing (3PGCIC 2011), Barcelona, Spain, October 26-28 (2011) (article in perss)
22. Kołodziej, J., Xhafa, F.: Enhancing the genetic-based scheduling in computational Grids by a structured hierarchical population. Future Generation Computer Systems 27, 1035–1046 (2011), doi:10.1016/j.future.2011.04.011
23. Lee, Y.C., Zomaya, A.Y.: Minimizing Energy Consumption for Precedence-Constrained Applications Using Dynamic Voltage Scaling. In: Proc. of the 9th IEEE/ACM International Symposium on Cluster Computing and the Grid (CCGrid), Shanghai, China, pp. 92–99 (2009)

24. Lorenz, M., Wehmeyer, L., Dräger, T.: Energy aware Compilation for DSPs with SIMD instructions. In: Proc. of Languages, Compilers and Tools for Embedded Systems: Software and Compilers for Embedded Systems LCTES/SCOPES 2002, pp. 94–101 (2002)
25. Lorch, J.R., Smith, A.J.: Improving dynamic voltage scaling algorithms with pace. In: 2001 ACM SIGMETRICS International Conference on Measurement and Modeling of Computer Systems, pp. 50–61 (2001)
26. Marcelloni, F., Vecchio, M.: Enabling energy-efficient and lossy-aware data compression in wireless sensor networks by multi-objective evolutionary optimization. Information Sciences 180, 1924–1941 (2010)
27. Meedeniya, I., Buhnova, B., Aleti, A., Grunske, L.: Architecture-Driven Reliability and Energy Optimization for Complex Embedded Systems. In: Heineman, G.T., Kofron, J., Plasil, F. (eds.) QoSA 2010. LNCS, vol. 6093, pp. 52–67. Springer, Heidelberg (2010)
28. Mezmaz, M., Melab, N., Kessaci, Y., Lee, Y.C., Talbi, E.-G., Zomaya, A.Y., Tuyttens, D.: A parallel bi-objective hybrid metaheuristic for energy-aware scheduling for cloud computing systems. J. Parallel Distrib. Comput. (2011) (in press), doi:10.1016/j.jpdc.2011.04.007
29. Miao, L., Qi, Y., Hou, D., Dai, Y.H., Shi, Y.: A multi-objective hybrid genetic algorithm for energy saving task scheduling in CMP system. In: Proc. of IEEE Intl. Conf. on Systems, Man and Cybernetics (ICSMC 2008), pp. 197–201 (2008), doi:10.1109/ICSMC.2008.4811274
30. Michalewicz, Z.: Genetic Algorithms + Data Structures = Evolution Programs. Springer (1992)
31. Min, R., Furrer, T., Chandrakasan, A.: Dynamic voltage scaling techniques for distributed microsensor networks. In: Proc. IEEE Workshop on VLSI, pp. 43–46 (2000)
32. Rayward-Smith, V.J.: UET scheduling with unit interprocessor communication delays. Discrete Applied Mathematics 18(1), 55–71 (1987)
33. Shen, G., Zhang, Y.Q.: A New Evolutionary Algorithm Using Shadow Price Guided Operators. Applied Soft Computing 11(2), 1983–1992 (2011)
34. Shen, G., Zhang, Y.-Q.: A Shadow Price Guided Genetic Algorithm for Energy Aware Task Scheduling on Cloud Computers. In: Tan, Y., Shi, Y., Chai, Y., Wang, G. (eds.) ICSI 2011, Part I. LNCS, vol. 6728, pp. 522–529. Springer, Heidelberg (2011)
35. Stützle, T., Hoos, H.: Improvements on ant-system: Introducing max-min ant system. In: Proc. of the Artificial Neural Networks and Genetic Algorithms Conference, pp. 245–249. Springer, Wien (1996)
36. Subrata, R., Zomaya, A.Y., Landfeldt, B.: Cooperative power-aware scheduling in grid computing environments. J. Parallel Distrib. Comput. 70, 84–91 (2010)
37. Veeramachaneni, K., Osadciw, L.A.: Swarm intelligence based optimization and control of decentralized serial sensor networks. In: Proc. of the IEEE Swarm Intelligence Symposium, pp. 1–8 (2008)
38. Zomaya, A.Y.: Energy-Aware Scheduling and Resource Allocation for Large-Scale Distributed Systems. In: Proc. of the 11th IEEE International Conference on High Performance Computing and Communications (HPCC), Seoul, Korea (2009)

Chapter 11
A Simulation Model for Mechanisms, Heuristics and Rules for P2P Systems

Ciprian Dobre, Florin Pop*, and Valentin Cristea

Abstract. Large-scale P2P distributed resources that aggregate and share resources over wide-area networks present major new challenges for scientists. This chapter focuses on the challenges to enable scalable, high-level simulation of applications, middleware, resources and networks to support scientific and systematic study of large scale distributed applications and environments, specifically designed for P2P systems. Its contributions are on the intelligent control of distributed P2P systems, where simulation is used in decision making as a way to predict future performance under some control law in question.

11.1 Introduction

Over the Internet today, computing and communications environments are more complex and chaotic than classical distributed systems, lacking any centralized organization or hierarchical control. Peer-to-Peer systems provide a good substrate for creating, for example, large-scale data sharing, content distribution and application-level multicast applications [23]. In the last years, extensive research was conducted towards overcoming scaling problems with unstructured P2P net-works, such as Gnutella, where data placement and overlay network construction are essentially random. Still, evaluation of such solutions is hard to achieve because of the size of the systems and the non-deterministic nature of computing and data resources involved [4].

Ciprian Dobre · Florin Pop · Valentin Cristea
University POLITEHNICA of Bucharest, Romania
e-mail: ciprian.dobre@cs.pub.ro, florin.pop@cs.pub.ro,
valentin.cristea@cs.pub.ro

* Corresponding author.

J. Kołodziej et al. (Eds.): Advances in Intelligent Modelling and Simulation, SCI 422, pp. 235–267.
springerlink.com

The field of modeling and simulation was long-time seen as a viable alternative to develop new algorithms and technologies and to enable the development of large-scale distributed systems (LSDS), where analytical validations are prohibited by the nature of the encountered problems. The use of discrete-event simulators in the design and development of LSDS is appealing due to their efficiency and scalability. Their core abstractions of process and event map neatly to the components and interactions of modern-day distributed systems and allow the design of realistic scenarios. Compared with the alternative of implementing the new technology directly in real-world to demonstrate its viability, the simulation of distributed systems is a far better alternative because it achieves faster validation results, minimizing the costs involved by the deployment process [9].

The focus of this chapter is on the modeling and simulation of large scale P2P systems as a support for *intelligent control of distributed P2P systems.* We describe alternatives to designing and implementing simulation instruments to be used in the validation of distributed P2P technologies. The chapter is based on the experience accumulated by the authors in developing a generic model for the simulation of distributed system technologies, integrating components and mechanisms to create realistic simulation experiments of large scale systems. The original model proposed in the MONARC simulator incorporates all the necessary components and characteristics that allow the complete and accurate design of realistic simulation experiments of complex distributed architectures, consisting of many resources and various technologies, ranging from data transferring to scheduling and data replication, with resources working together to provide a common set of characteristics [10].

The chapter presents the design characteristics of a simulation model that extends the original model of MONARC to allow evaluating P2P mechanisms and technologies. We present the characteristics of P2P systems that influenced the design process of the proposed model. We demonstrate that it includes the necessary components to describe various actual distributed P2P technologies, and provides the mechanisms to describe from network overlays, to the evaluation of different strategies in job scheduling procedures [2] and intelligent algorithms. As a case study, we demonstrate that the proposed simulation model contains the necessary components and characteristics to allow complete and accurate design of specific P2P simulation experiments. We present experiments designed to evaluate scheduling procedure. We also present a simulation study designed to evaluate the properties of several network overlays.

We also present a critical comparison study of the most important simulation projects involved in the modeling of P2P systems, based on existing research [29]. The analysis highlights their specific characteristics, from the types of simulation models or internal simulation design to the implementations of the respective instruments. We demonstrate that, although the use of a particular simulation instrument depends very much on the scope of the simulation being conducted and the skills of the user, they all cover important aspects of distributed systems, allowing exploration of different areas of parameter space. Among these projects, we demonstrate that the model proposed in MONARC is the most generic one, being capable to handle a wider range of simulation scenarios. It also offers the highest number of capabilities, being able to consider many parameters, capabilities and components that any other existing simulation instrument for P2P systems.

The chapter is organized as follows. Section 11.2 presents open issues for large scale P2P systems and Section 11.3 the mechanism for performance optimization in large scale P2P systems. Then Section 11.4 presents the simulation models and proposed tools for LSDS. Section 11.5 describes the general framework of MONARC and in Section 11.6 the MONARC extension for P2P is highlighted. Some experimental evaluations are presented in Section 11.7. The conclusions are presented in the last section of this chapter.

11.2 P2P Issues and Their Influence on the Simulation Model

This section presents specific characteristics of Large Scale P2P systems, and their successful integration into the proposed simulation model. Such particular aspects were previously identified by authors [13] as having a major impact on the effectiveness and deployment of P2P systems and applications:

- *Decentralization.*
- *Scalability.*
- *Anonymity.*
- *Self-organization.*
- *Cost of ownership.*
- *Ad hoc connectivity.*
- *Performance.*
- *Security.*
- *Transparency.*
- *Usability.*
- *Fault-resilience.*
- *Interoperability.*

These characteristics of the P2P systems positively influenced the development process of the simulation model. Various studies also identified specific elements of the simulation model that enables the correct modeling of an LSDS environment. Authors of the study presented in [35] present the features that must be implemented by a simulation model in order to allow the correct modeling of a Grid environment. The identified set of features consists of: *multi-tasking IT resources*, *job decomposition*, *task parallelization*, *heterogeneous resources*, *resource scheduling*, and *resource provisioning*. We present the identified feature of a generic P2P environment simulator model.

Multi-tasking IT resources: Processors, database servers, network links, and data storage devices are all pre-emptive multi-tasking resources. Tasks submitted to such a resource go immediately to the processing queue containing all other tasks that are being processed by the resource. Each of these tasks get processed by the resource for a pre-defined time-slice and then put back to the processing queue (if not waiting for other resources). As a result of multi-tasking, the submission of a new task can change the completion time of existing tasks being processed by the resource. The simulation model incorporates processing units, database servers, network links, and data storage devices. The processing unit has a queue of all the tasks being concurrently processed by the resource. An interrupt mechanism ensures modeling of tasks concurrency. The time needed to complete a task takes into account the number of other competing tasks. In this way the multi-tasking requirement is one of the characteristics implemented in the simulation model. A detailed simulation of the task management within each resource is far too time-consuming when considering grid environments with hundreds of resources simulated for the duration of weeks. Instead, when a new task is submitted to a resource, the simulation framework performs a good approximation of the multi-tasking behavior by re-estimating the completion time of all tasks being processed by that resource.

Job decomposition: Each job from a workload may be composed of multiple resource requirements. The terminology of "task" is generally used to define a single resource requirement within a job. Each resource requirement may be for a specific configuration of a server or a database. For example, a job may be composed of three tasks: get data from a customer database, perform data mining on a compute server, and add the results to a sales database. Job decomposition is important to model because grid designs typically focus on providing a certain type of resource (e.g., compute servers) on a grid and overlook the impact on other resources (e.g., databases, network bandwidth) needed by the job. The simulated job is programmed to handle specific actions. A job can, for example, be programmed by the user to request data from a database server, to perform some computation using the obtained records and then send the results for further processing to another job. The tasks performed by a job can be correlated with the tasks performed by another one. The dependencies between the jobs can be specified in the form of DAG structures in the simulation model.

Task parallelization: Each task in a job (decomposed as described above) may be parallelizable. This is often the case with grid workloads. Parallelization could be of two types: "Embarrassingly parallel" tasks can simply be split up into as many chunks as available resources. The second category consists of parallel tasks that are more constrained and consist of a specific number of parallel paths, regardless of the number of available resources.

A simulated job can start new simulated jobs, each one performing specific tasks. This, correlated with the job decomposition characteristic implemented using the DAG structures, can be used to handle both described cases of task parallelization.

Heterogeneous resources: Grid resources can be of various vendor platforms, models, and operating systems. Therefore, the processing time of a task on a resource is subject to its performance benchmarks. A resource may be associated with multiple performance benchmarks that are relevant for different types of computing tasks. The computing unit specified in computing power in the form of generic SI95 units, accommodating for a wide-range of performance benchmarks. The computation, data and network models all take into account resources with various amounts of resources to model the heterogeneity of resources in the simulation experiments.

Resource scheduling: The grid simulator must be able to model scheduling policies as used by various resource brokers to determine which resource should process an arriving task. For that, the simulation model provides a meta-scheduling functionality, allowing the execution the simulated jobs in a distributed manner. The meta-scheduler can incorporate a wide-range of user-defined scheduling algorithms. Local scheduling algorithms can also be added in any user-defined experiments.

Resource provisioning: The ability to provision resources for processing particular types of tasks is another feature to be simulated. These provisioning policies could be either calendar based (e.g., a department needs a server to be an email server during 9AM to 5PM every workday and can release it to be a data-mining processor on the grid at other times) or based on a more dynamic policy that monitors workload arrivals and resource usage and reacts accordingly. Simulation of such provisioning policies in conjunction with the resource scheduling policies would allow grid designers to determine whether the respective policies are aligned and consistent with each other. Also, it would provide the desired grid performance in terms of resource availability, workload throughput and processing times. This is why the simulation model allows the insertion of background jobs to handle the execution of specific resource provisioning tasks. Such jobs can be used in dependency with other simulated jobs. In addition, the database server can perform programmed actions, being modeled as a special task in the simulation model. It can simulate special operations such as data archiving on a calendar based designed policy. The automation of resource provision is particularly important to the simulation model because it can be used to experiment with various data replication algorithms and provide flexibility to the simulation scenarios [3]. In traditional client-server models the information is concentrated in centrally located servers and distributed to client computers and, for example, access rights and security are more easily managed. The centralized systems topology yields inefficiencies, bottlenecks, and wasted resources and even if hardware performance and cost have improved, centralized repositories are expensive to set up and hard to maintain.

One of the more important ideas of *decentralization* is the emphasis on the usersównership and control of data and resources. In a fully decentralized system, every peer is an equal participant making the implementation of the P2P models difficult because there is no centralized server with a global view (see Fig. 11.1). This is why many P2P file systems are built as hybrid approaches. This is the of Napster, where there is a centralized directory of the files but the nodes download files directly from their peers [11].

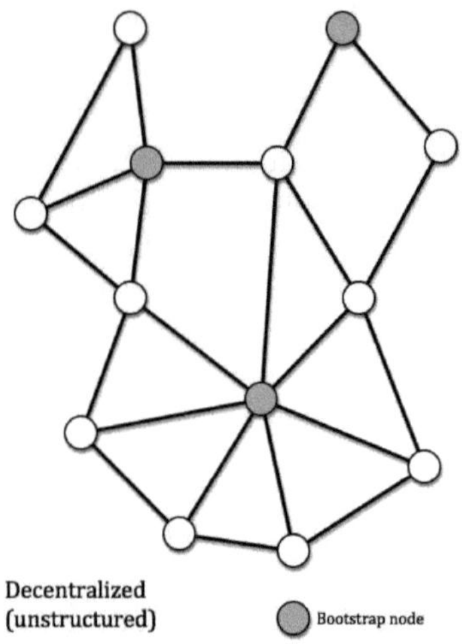

Fig. 11.1. Unstructured overlay: an overlay is represented as a graph that describes how the nodes are connected with each other.

In fully decentralized file systems (Freenet, Gnutella), finding the network becomes difficult. In Gnutella, new nodes must know the address of another node or use a host list with known IP addresses of other peers. The node joins the network of peers by establishing a connection with at least one peer currently in the network. Then, it can begin discovering other peers and cache their IP addresses locally. One method to categorize the autonomy of a P2P system is through the "pure P2P" versus "hybrid P2P" distinction [28].

An immediate benefit of decentralization is improved *scalability*. Scalability is limited by factors such as the amount of performed centralized operations (e.g, *synchronization* and *coordination*), the amount of maintained state, the inherent parallelism and the programming model that is used to represent the computation. Achieving good scalability should not be at the expense of other features, such as determinism and performance guarantees. For instance, hybrid P2P systems, such as Napster, keep some amount of the operations and files centralized.

Searching may fail even when an object exists, making the behavior of the system nondeterministic. Systems such as CAN, Chord, Oceanstore, and PAST, dictate a consistent mapping between an object key and hosting node [33]. Thus, an object can always be retrieved as long as the hosting nodes can be reached. Nodes in these systems compose an overlay network, where each node maintains information about a small number of other nodes. This approach limits the amount of state that needs to be maintained. The logical topology of the overlay provides some guarantees on the lookup cost.

A goal is to allow people to use systems without concern for legal or other ramifications and to guarantee that censorship of digital content is not possible. There are three kinds of *anonymity* involved in communicating pair: sender anonymity, which hides the senderś identity; receiver anonymity, which hides a receiverś identity; and mutual anonymity, which hides the identities of the sender and receiver, are hidden from each other and other peers [18].

It is important to understand the degree of anonymity a certain technique can achieve. There is a spectrum of anonymity degrees that cover absolute privacy, beyond suspicion, probable innocence and provably exposed. Beyond suspicion means that even though an attacker can see evidence of a sent message, the sender appears no more likely to be the originator of that message than any other potential sender in the system. There are six techniques for enforcing different kinds of ano-nymity with different kinds of constraints:

- *Multicasting* (or *broadcasting*) can be used to enforce receiver anonymity. An entity that is interested in obtaining a document subscribes to the multicast group and its identity is hidden for the sender and other members of the group. The party that possesses the document sends the document to the group. This technique can take advantage of the underlying network that supports multicast (e.g., Ethernet or token ring) [21].
- *Spoofing* the senderś address. For connectionless protocols such as UDP, the senderś anonymity can be enforced by spoofing the senderś IP address but this requires the protocol changing and this is not always feasible, because most ISPs now filer packets originating from invalid IP addresses.
- *Identity Spoofing.* Anonymity can be ensured by changing the identity of a communicating party. For example, in Freenet, a peer passing a file to a requestor can claim to be the owner of the content. The responder is possibly innocent, from an attackerś point view, because there is a nontrivial probability that the real responder is someone else ([34]).
- *Covert paths.* Two parties communicate through some middle nodes. A party that wishes to hide its identity prepares a covert path with the other party as the end of the path. The covert paths can use store/forward or persistent connection. By varying the length of the covert paths and changing the selected paths with different frequency, different degrees of anonymity can be achieved.
- *Intractable aliases.* The client can open an account and be recognized upon returning to the opened account, while hiding the true identity of the client from the server. Techniques of this kind ensure sender anonymity

and rely on a trusted proxy server. The degree of anonymity that can be achieved falls in between absolute privacy and beyond suspicion.

- *Non-voluntary placement.* A publisher forces a document onto a hosting node using, for example, consistent hashing and because the placement is non-voluntary, the host cannot be held accountable for owning the document.

In P2P systems, *self-organization* is needed because of scalability, fault resilience, intermittent connection of resources, and the ownership cost. Those systems scale unpredictably in terms of the number of systems or users, and the load, causing an increased probability of failures, requiring self-maintenance and self-repair (see Fig. 11.2 for different topologies properties) [25].

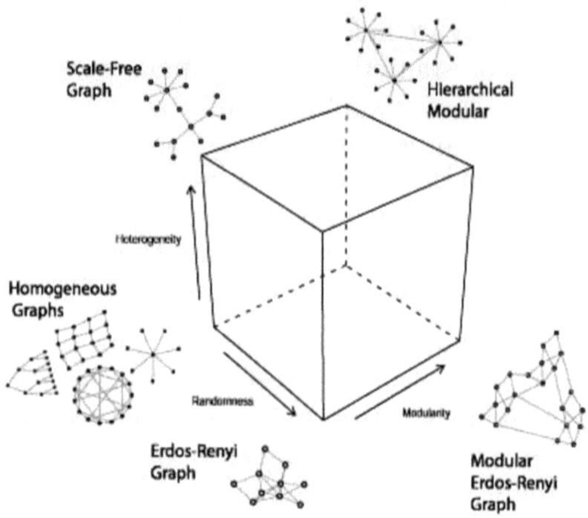

Fig. 11.2. The space of network topologies properties. (Inspired by [25]).

Similar reasoning applies to intermittent disconnection because it is hard for any predefined configuration to remain intact over a long period of time. Adaptation is required to handle the changes caused by peers connecting and disconnecting from the P2P systems. It is costly to have dedicated equipment for managing such a fluctuating environment, so the management is distributed among the peers.

Shared ownership reduces the cost of owning the systems and the content, and the maintaining cost. This is applicable to all classes of P2P systems but it is most obvious in distributed computing. For example, SETI@home is faster than the fastest supercomputer in the world, yet at only a fraction of its cost [12].

The ad-hoc nature of connectivity has a great impact and P2P systems and applications need to be able to handle systems joining and withdrawing from the pool of available systems. For example, in distributed computing, the parallelized applications cannot be executed on all systems all of the time because some of the systems will be available all of the time, some will be available part of the time, and some will be not be available at all.

In systems with higher guarantees, such as service-level agreements, the ad-hoc nature is reduced by redundant service providers. In collaborative P2P systems and applications, the ad-hoc nature of connectivity is evident because collaborative users are increasingly expected to use mobile devices, making them more connected to Internet and available for collaboration. To handle this situation, collaborative systems support transparent delay of communication to disconnected systems, having proxies delegated on networks to receive messages, or having other sorts of relays on the sending system or somewhere in the network that will temporarily hold communication for an unavailable system. P2P systems and applications need to be designed to tolerate sudden disconnection and ad-hoc additions to groups of peers. The P2P architectures taxonomy is presented in Fig. 11.3.

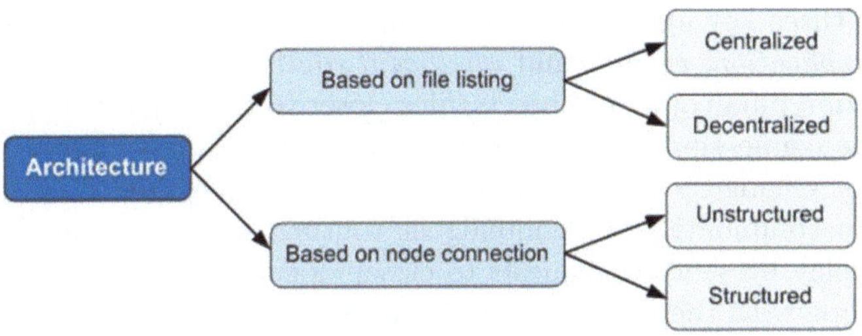

Fig. 11.3. P2P Architectures Taxonomy.

P2P systems aim to improve *performance* by aggregating distributed storage capacity and computing cycles of devices spread across a network. Performance is influenced by three types of resources: processing, storage, and networking. Networking delays can be significant in wide area networks. Bandwidth is a major factor when a large number of messages are propagated in the network and large amounts of files are being transferred among many peers. This limits the scalability of the system.

In centrally coordinated systems (e.g., Napster, Seti@Home) coordination between peers is controlled by a central server, although the peers also may later contact each other directly. These systems are vulnerable to the problems facing when using centralized servers. To overcome such limitations, different hybrid P2P architectures have been proposed to distribute the functionality of the coordinator in multiple indexing servers that cooperate with each other to satisfy user requests. DNS is an example of a hierarchical P2P system that improves performance by defining a tree of coordinators, with each coordinator responsible for a peer group. Communication between peers in different groups is achieved through a higher level coordinator.

In decentralized coordinated systems, such as Gnutella and Freenet, there is no central coordinator and communication is handled individually by each peer. They use message forwarding mechanisms search for information and data, but they end up sending a large number of messages over many hops from one peer to another. Each hop contributes to an increase in the bandwidth on the communication links and to the time required to get results for the queries. The bandwidth for a search query is proportional to the number of messages sent, which in turn is proportional to the number of peers that must process the request before finding the data.

11.3 Performance Optimization in Large Scale P2P Systems

There are three key approaches to optimize performance of large scale P2P systems: *replication*, *caching* and *intelligent routing* [17].

Replication puts copies of objects/files closer to the requesting peers, minimizing the connection distance between the peers requesting and providing the objects. Changes to data objects have to be propagated to all the object replicas. In combination with intelligent routing, replication minimizes the distance delay by sending requests to closely located peers. Replication also helps to cope with the disappearance of peers.

Caching reduces the path length required to fetch a file/object and therefore the number of messages exchanged between the peers. Reducing such transmissions is important because the communication latency between the peers is a serious performance bottleneck facing P2P systems. Efficient caching strategies can be used to cache large amounts of data infrequently. The goal of caching is to minimize peer access latencies, to maximize query throughput and to balance the workload in the system. The object replicas can be used for load balancing and latency reduction.

Intelligent routing and *network organization.* To fully realize the potential of P2P networks, it is important to understand and explore the social interactions between the peers. Adamic ([1]) explored the power-law distribution of the P2P networks, and introduced local search strategies that use high-degree nodes and have costs that scale sub-linearly with the size of the network. [22] determine "good" peers based on interests, and dynamically manipulate the connections between peers to guarantee that peers with a high degree of similar interests are connected closely. Establishing a good set of peers reduces the number of messages broadcast in the network and the number of peers that process a request before a result is found.

P2P systems came up with different naming and discovery schemes, but there is also a requirement for administration transparency. Users are typically non-experts and, thus, the P2P software should not require any significant set up or configuration. P2P systems should be network and device transparent. They should work on the Internet, intranets, and private networks, using high-speed or dial-up links. They should also be device transparent, which means they should work on a variety of devices, such as PDAs, desktops, and cell phones. Another form of transparency is related to security and mobility. Automatic and transparent authentication of users and delegation to user proxies can significantly simplify users' actions. Supporting mobile users and disconnection in particular, can enable users to work independently of whether and how they are connected to the Internet or intranets.

Some current P2P systems (e.g., Groove) provide special nodes, called relays, that store any updates or communication temporarily until the destination reappears on the network. Others (e.g., Magi) queue messages at the source, until the presence of the destination peer is detected. Another problem is non-availability of resources. This may occur either because the resource is unreachable because of a network failure or because the peer hosting the resource has crashed/gone offline. While the former may be resolved by routing around the failure and is already supported by the Internet, the latter requires more careful consideration. Replication of crucial resources helps alle-viate the problem. A challenging aspect of P2P systems is that the system maintenance responsibility is completely distributed and needs to be addressed by each peer to ensure availability, different from client-server systems, where availability is a server-side responsibility.

In the past, there were different ways to approach interoperability, such as standards IEEE, common specifications, and common source code, open-source and de facto standards. In the P2P space, some efforts have been made to improve interoperability, but interoperability is still not supported. The P2P Working Group is an attempt to gather the community of P2P developers together and establish common ground by writing reports and white papers that would enable common understanding among P2P developers.

The JXTA effort approaches interoperability as an open-source effort, by attempting to impose a de facto standard. A number of developers are invited to contribute to the common source tree with different pieces of functionality. Only a minimal underlying architecture is supported as a base, enabling other systems to contribute parts that may be compatible with their own implementations. A number of existing P2P systems have already been ported to the JXTA base [14].

11.4 Simulation Models and Tools for LSDS

This section analyzes existing work in constructing methodologies for analyzing and comparing various simulation tools targeting LSDS.

The various properties of any simulator designed to handle LSDS related technologies fall in two main categories: the *simulation taxonomy* and the *design taxonomy*. The simulation taxonomy analysis the simulation tools according to the adopted simulation models, while the design taxonomy categorizes the simulation tools according to their design and implementation.

The *simulation taxonomy* comprises five properties, as presented in the Fig. 11.4. Motivation indicates the major target of a simulation tool. According to this taxonomy a modeling instrument can be used to study various scheduling algorithms, to study various replication and data movement optimizations, or to study a particular model of an LSDS system. This category considers only the upper most motivation. If, for example, we consider the case of a simulation tool designed specifically to study various scheduling algorithms, we can observe that for this case the simulator must also provide additional support, such as simulated underlying networks or processing nodes. If the underlying LSDS components are also simulated, then a scientist could also evaluate various other file replication algorithms (assuming the possibility to also simulate data warehouses for example). But generally such modifications require great amount of work, and, except for the cases when the original developers of a simulator redesigned it at some point to comprehend some different classes of Grid related problems, such developments are virtually non-existent. The majority of the Grid simulation projects were developed in the context of the validation of the LHC experiments and their proposed running conditions. For this reason we often see as possible motivation the categories identified by the authors of the taxonomy described in [36]: Data Transport, Data Replication or Scheduling issues.

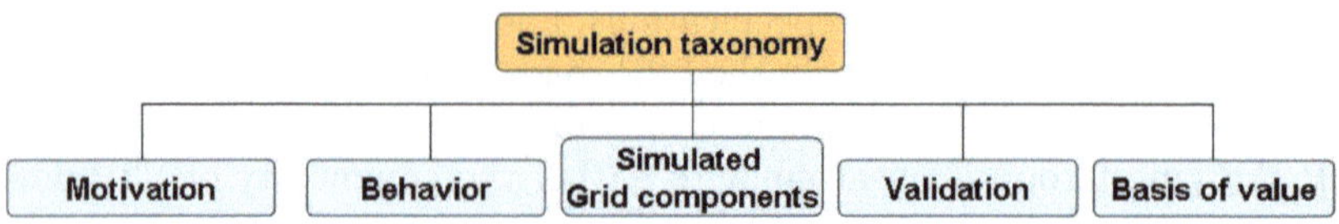

Fig. 11.4. The simulation taxonomy.

The *behavior taxonomy* classifies the modeling instruments based on how the simulation proceeds. A deterministic simulation has no random events occurring, so repeating the same simulation will always return the same simulation results. In contrast, a probabilistic simulation has random events occurring, so repeating the same simulation often returns different simulation results [15, 8, 9].

The simulators for LSDS must be designed according with criteria grouped under the following headings:

- Simulator Architecture
- Usability
- Scalability
- Statistics
- Underlying Network Simulation
- System Limitations

Using the categories of the proposed presented taxonomy, in this section we present an analysis of the properties of most representative modelling instruments for LSDS. One goal of this study is to present the most relevant related work in the field of modelling and simulation of distributed systems. One other objective of the analysis is to evaluate on a real case scenario the capabilities of the presented taxonomy to correctly investigate the capabilities of various simulation instruments for LSDS systems.

DHTSim is a discrete event simulator for structured overlays, specifically DHTs. It is intended as a basis for teaching the implementation of DHT protocols, and as such it does not include much functionality for extracting statistics. It is implemented as discrete event based message passing within the JVM [26]. *P2PSim* is a discrete event packet level simulator that can simulate structured overlays only. It contains implementations of six candidate protocols: Chord, Accordion, Koorde, Kelips, Tapestry and Kademlia. *Overlay Weaver* is intended to be a toolkit for easy development and testing of P2P protocols. It provides functionality for simulating structured overlays only and does not provide any simulation of the underlying network [20]. *PlanetSim* is a discrete-event overlay network simulator written in Java. It supports both structured and unstructured overlays, and is packaged with Chord-SIGCOMM and Symphony implementations. *SimGrid* [16] is a simulation toolkit that provides core functionalities for the evaluation of scheduling algorithms in distributed applications in a heterogeneous, computational Grid environments mainly. It aims at providing the right model and level of abstraction for studying Grid-based scheduling algorithms and generates correct and accurate simulation results. *GridSim* [24] is a grid simulation toolkit developed to investigate effective resource allocation techniques based on computational economy. *OptorSim* [32] is a Data Grid simulator designed for evaluating optimization in data access technologies for Grid environments. It adopts a Grid structure based on a simplification of the architecture proposed by the EU DataGrid project. *ChicagoSim* [19] is a simulator designed to investigate scheduling strategies in conjunction with data location. It is designed to investigate scheduling strategies in conjunction with data location.

A summary report of presented simulation tools and others are provided in Table 11.1 (web references are also available for each simulator).

None of these simulators present general solutions to modeling dependability technologies for LSDS. They focus on providing evaluation methods for the traditional research in this domain, which up until recently targeted the development of functional infrastructures. However, lately, the importance of dependable distributed systems was widely recognized and this is demonstrated by the large number of research projects initiated in this domain. Our proposed simulation model aims to provide the means to evaluate a wide-range of solutions for dependability in LSDS [31].

Security in particular has never been properly handled by any of these projects be-fore. The only currently existing simulator that offers the possibility to evaluate security solutions designed for distributed systems is G3S

Table 11.1. Analysis of simulation tools.

Simulator	Support	Scalability (max nodes)
Peersim	Structured and unstructured overlays	10^6
P2PSim	Structured overlay	3000
Neurogrid	Structured and unstructured overlays	≤ 20
Narses	Unstructured overlays	≥ 3
Querycycle	Unstructured overlays	–
GPS	Structured and unstructured overlays	–
DHTSim	Structured overlay	–
Overlay Weaver	Structured overlay	4000
Gnutellasim	Unstructured overlays	600
OMNeT++	Structured and unstructured overlays	1000
Planetsim	Structured overlays	100000
OverSim	Structured and unstructured overlays	100000
SimGrid	Structured overlays	≥ 3
GridSim	Structured overlays	≥ 3
OptorSim	Structured overlays	≥ 3

(Grid Security Services Simulator) [27]. It aims to support various authentication mechanisms such X.509 certificates and Kerberos tickets. It also includes mechanisms for disseminating security threats, for evaluating various access control policies, etc. The simulator uses the simulation model proposed in GridSim. Similar to this model, we too support all the mechanisms found in G3S and several others. In addition we offer the possibility of evaluating security in a more general context, considering the entire context of distributed systems, with its specific characteristics.

In addition, our proposed simulation model targets the generic evaluation of de-pendable distributed systems. The model is able to simulate, for example, various faults occurring in such systems, which allows researchers experiment with abnormal behavior of any of the systemś components. This can be coupled with various security enforcement or fault recovery solutions. We argue that a correct evaluation of dependability in distributed systems should provide a complete state of the entire distributed system. Because of the complexity of the LSDS, involving many re-sources and many jobs being concurrently executed in heterogeneous environments, there are not many simulation tools to address the general problem of LSDS computing. The simulation instruments tend to narrow the range of simulation scenarios to specific subjects, such as scheduling or data replication. The simulation model provided by MONARC is more generic that others, as demonstrated in [10, 8]. It is able to describe distributed system technologies, and provides the mechanisms to describe concurrent network traffic, to evaluate different strategies in data replication, and to analyze job scheduling procedures [5].

11.5 MONARC Simulation Engine

MONARC (MOdels of Networked Analysis at Regional Centers) is a simulator for realistic evaluation of large distributed computing systems [7]. The MONARC engine is structured on several layers. The basic one provides all the components of the system and their interactions. The largest one is the regional center, described by its location on a map (latitude and longitude), and which contains a farm of processing nodes (central processing units), database servers, mass storage units and local and wide area networks. Each network is described by speed and by a wide area network or router that connects it. Each CPU is described by power, memory capacity, and address on network, name of the network where it is connected and the maximum speed of communication in the network. Another component that models the behavior of the application is the "Activity" object. It is used to generate jobs based on different scenarios. A job is another basic component, scheduled for execution on a CPU unit by a "Job Scheduler" objects. Over this basic layer, it is possible to build a wide range of models, with different level of complexity (one or multiple regional centers, each with different hardware configuration, depending on the system that we want to simulate).

With this structure it is possible to build a wide range of models, from the very centralized to the distributed system models, with an almost arbitrary level of complexity (multiple regional centres, each having different hardware configuration and possibly different sets of replicated data). The analysis of the characteristics of various LSDS architectures was essential in the design process of the simulation model. It influenced the decision on the type of components and interactions required to completely and correctly model various Grid related experiments. Table 11.2 explains the meanings of some of the components used by the simulation framework.

For constructing the P2P overlay, we decided to use a single regional center with many CPU, on each of them running an application composed of two jobs. This application represents one peer in the system. We had to define an activity to generate these jobs. The activity reads the graph of the peers from a configuration file and then, based on this graph, generates the jobs, each of them knowing the topology or at least a part of it (its neighbors). In such a simulation all CPU units belong to the same local area network, and normally, if node x wants to communicate with node y, even if in the graph no edge connects them, they transfer data directly. By constructing the overlay, two nodes communicate directly, if an edge exists between them, or indirectly, using the path between them, if no edge connects them.

The most important classes of MONARC, grouped based on their appurtenance to a particular simulation model, are presented in Figure 11.5. In the figure the classes are grouped on several packages. The classes comprising the simulation engine (in light grey) were presented in the previous section. The classes comprising the job model (in red) together with the scheduling model

Table 11.2. Simulation concepts for P2P Systems.

COMPONENT	SHORT DESCRIPTION
Data Container	Contains Objects of a single type in defined range.
Data Base	Contains sets of Containers. It is associated with an AMS Server.
Data Base Catalogue	Provide the mechanism to locate any Object and the AMS server which can retrieve this object.
Mass Storage Unit	High capacity unit, but slow, connected in LAN. Performs data access (W/R) in an OODB Model. Handles data storage on local disks or tape.
CPU Node	Typical processing Node, having a defined processing power, memory, and I/O channel. Allows concurrent execution of multiple jobs.
I/O Link (Link Port)	Describes the quality of I/O connection of each component on LAN. Allows simultaneously multiple transfers. It is the basic entity where any simulated transfer can start off.
LAN	Specifies how individual components are connected in LAN.
WAN	Specifies the connectivity for the wide area network. An internet like naming scheme is used.
Router	Specifies a router used to connect two or more WANs together in the simulation.
Protocol	Describe the modality in which the transfer is simulated.
Farm	A set of CPU Nodes, AMS servers, Mass Storage Unit and a Job Scheduler.
Job	Specifies typical tasks used in the simulation.
Active Job	Used by simulation system to perform a user defined job. It is dynamically allocated to a CPU Node when load constrains are satisfied.
Activity	A generic object used as a loadable module which defines a set of jobs and how they are submitted for execution.

(in light green), the ones included in the data model (in green), and the ones belonging to the network model (in orange) are all presented in this section.

In the diagram, the activity represents the users of the system being modeled. They generate data processing jobs based on different scenarios. Each regional center can have one or more (or none) such activities submitting jobs to it. The farm entity handles the CPU units (processing nodes) of the regional center. The farm is an active object by itself. Its role is to wait for job submissions coming from the activity objects and to call the appropriate job scheduler to handle the execution. The CPUUnit describes the behavior of a computing station belonging to the modeled system. It is characterized by the amount of available CPU power (measured using the SPEC benchmark) and the amount of memory. The memory occupation can make use of a paging mechanism, if required. This ensures the correct modeling of the type of behavior specific to several existing operating systems. The amount of CPU power to be used is allocated to the processing jobs according to an indicated priority. The user can specify which jobs must be processed with a higher priority than others. This translates into more amount of power being allocated for some prioritized processing jobs.

An extension to the CPU unit is represented by the CPUCluster. This object models a cluster of CPU units. Besides CPU power and memory, it is also characterized by the available disk space. This class can even simulate the failure of CPUs. The CPUCluster is well suited for modeling large LSDS environments, while the original CPUUnit object is more appropriate to model low scale distributed systems, consisting of only several nodes.

In MONARC, with every simulation step, the simulator executes the following operations in a loop described as follow:

```
1. Look at each simulation task and:
    a. If the task is in the created state, assign
    it to a worker thread from the pool and change
    the task's state to ready
    b. If the task is in the ready state, restart
    its execution by making a V() on the semaphore
    c. If the task is in the finished state,
    remove it
2. Wait until all the tasks that were running
   block again or finish their execution
```

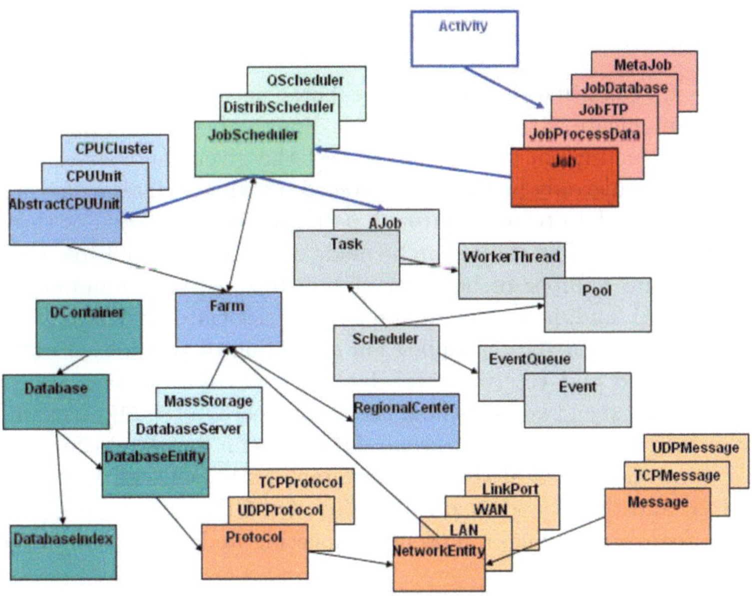

Fig. 11.5. The components of the MONARC simulation framework.

```
3. Process the events:
    a. Take from the future queue the event(s)
    with the minimum time stamp. The simulation
    time advances, becoming equal to that
    time stamp.
    b. For each event taken from the queue,
    look for the destination task.
    If it is waiting for an event (i.e., it is
    in the waiting state), deliver the event
    to the task.
    Else, put the event into the deferred queue.
```

The distribution of jobs to the appropriate processing resources is being handled by a specialized JobScheduler. MONARC 2 provides two already defined algorithms for the scheduling of jobs, a centralized and a distributed one. We present these entities, as well as the details regarding the implementation of the data, network and task processing simulation models, in the next sections.

For instance, for a P2P overlay, in the next figure we illustrate a way of connecting five peers in a system:

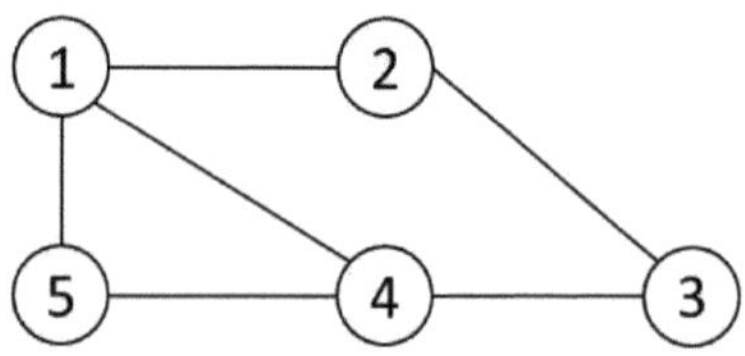

Fig. 11.6 Example of P2P overlay topology.

All nodes are connected in the same local area network. Normally, if *peer1* wants to communicate with *peer3*, it does it directly, because a link connects them. By constructing the overlay, we respect the given topology and make the information move only on the links associated with the edges in the graph. So, *peer1* transfers data to *peer3* through *peer2*, or through *peer4*, or through *peer5* and then *peer4*. The choice can be made depending on multiple factors, such as minimum number of hops, maximum link speed, minimum traffic (balanced traffic), and, generally, minimum cost associated with the link.

Each peer is defined by an application consisting of two jobs. First jobś purpose is to send and receive useful data, and also find a path based on minimum cost between two peers (e.g. peer1 and peer3 in the above example) and the second job's purpose is to send data between first kind of jobs (e.g. *peer2* for *peer1* and *peer3* in the above example). We will call first kind of jobs JobP2P and the second kind **JobRelay**. On CPU unit 1 runs, besides JobP2P 1, *JobRelay 1* which represents connecting node for other two peers (e.g. 2 and 5).

In the next section we discuss about two algorithms implemented for constructing this overlay. In the first algorithm, each JobP2P knows the entire topology of the system (the graph associated). It calculates locally the path to a remote peer where it wants to transfer data, based on a chosen criterion and then, sends the data through that path to the destination. *JobRelay* does not hold any information; its aim is to listen for messages from peers or from other relays and then send them to the next hop in the path. In the second algorithm each *JobPeer* knows only its neighbors. For finding a path to a destination, a peer broadcasts a message to the neighbors. If a neighbor does not know the route to the destination it broadcasts the request to all its neighbors. When a node holds information about the destination desired, it replies it unicast to the source that made the request. When the answer reaches the source peer, it can send the data to the remote peer selected. The second algorithm is appropriate to the idea of gossiping protocol as peers does not know the topology of all the system. Yet, both algorithms bring advantages and disadvantages and in the next section both topics will be discussed. We will choose the best algorithm in terms of performance, scalability and flexibility, over which the gossiping protocol will be implemented.

11.6 MONARC Extensions with an Overlay for P2P Systems

The discovery of the resources is simulated in MONARC by the resource catalogue, a hardware configuration file, which includes global parameters, such as the number of simultaneous threads and the name of the regional center, and detailed description of the center which includes its location, all the central processing units, the local, wide area network and the routers that it manages.

The descriptions of the resources are loaded at the beginning of the simulation. The configuration file also specifies the name of the Activity object used to generate data processing jobs based on our scenario. The Activity object loads the topology of the P2P system from a configuration file and overrides the *pushJobs* method according to the loaded simulation scenario. The file that specifies the graph contains the number of the nodes in the system and each of them is described by index, flag and color. Then, it follows edges definition, with the index of vertices connected by it, a color and a cost. The loader of topology has a method that returns the graph described in the file. The Activity object, in the overridden method *pushJobs*, generates, for each node in the graph, a *JobP2P* and a *JobRelay* objects. These objects receive, in the constructor, the node associated, and besides that, each *JobP2P* receives the topology. A P2P and a relay job associated with the same index of the node are running on the same CPU, each of them sending and receiving messages on different ports. Therefore, we will open four ports on a CPU, two of them for each different job running on it, one for sending and one for receiving. The *Activity* object also generates a *Job* responsible for closing the relays, as those jobs are waiting for messages and continue the transfer by forwarding them, in an infinite loop. In conclusion, when a P2P job finishes its processing and transfer, it sends a message to *JobCloseRelays* (which is running on a different CPU). After this job receives a number of messages equal to the number of nodes in the system, it sends a message with null data to each *JobRelay*, so they will take the decision to stop.

A *JobP2P* class has five methods: *computePath*, *sendData*, *receiveData*, *stopRelays* and *run*. It extends the basic *Job* class. The *computePath* method is used to find the path between the current node and a remote one, based on our topology. It receives as parameters the index of the destination and the name of the algorithm used to generate an optimal path. It returns the path according to the algorithm desired. The *sendData* method is used to simulate the sending of data to the next node in the path. It calls the *computePath* method to find the way to the destination. Then, a package is made, with the path and the data to transfer and is sent to the next hop. The relays unpack the data received, remove the current hop in the path and forward the message to the next node, until the message reaches its destination. The *receiveData* method is used to simulate the receive of data from a relay or a P2P neighbor. The *run* method overrides that one defined in the super class, and its responsibility is to send and receive data from remote P2P job. This method is overritten by the gossiping algorithm implemented in a class that extends *JobP2P*. Before the end of the *run* method, a message to the job responsible for closing the relay is send.

A *JobRelay* object receives a message, unpacks the transferred data, removes the first node from the containing path, and finally sends the new message to the next relay or P2P job (if the next hop in the way to the destination is the destination itself). It runs in a loop and stops only when a null data is received.

A *JobCloseRelay* waits for a number of messages equal to the number of nodes in the graph and then it sends a null data message to all relays in the system to stop running.

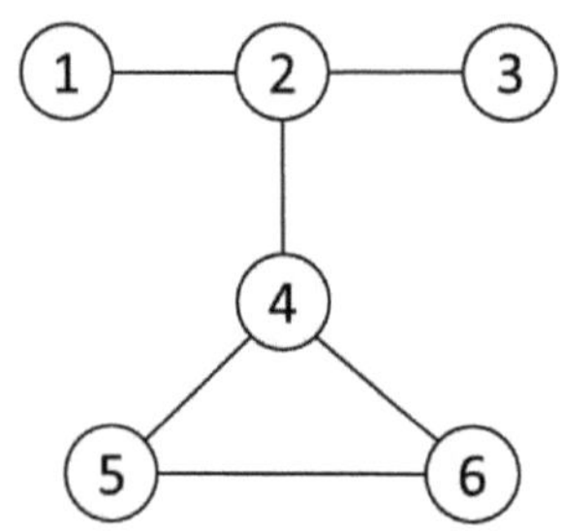

Fig. 11.7 Graph for the basic algorithm for overlay construction.

In Figure 11.7 we defined a topology for a P2P system. The rule of sending messages is node x to node $x+1$ modulo the number of vertices. We will take for example two nodes which communicate directly, by being neighbors in graph, and two which communicate indirectly, with the help of relays met in the path to the destination. *JobP2P 1* (which runs on *CPU 1*, having the address *10.0.0.2*) sends the data directly to *JobP2P 2* (which runs on *CPU 2*, having the address *10.0.0.3*). The final sending makes use of the port calculated using the rule:

```
Port of P2P job, for sending = (index of job) + 3
```

The receiving port for a P2P job is composed as:

```
Port of P2P job, for receiving = (index of job) + 1
```

In this case *JobP2P 1* sends the data starting from *10.0.0.2*, port *4*, to *JobP2P 2*, on *10.0.0.3*, port *3*. *JobP2P 6* should send information to *JobP2P 1*, following the previously described rule. So, it locally calculates the optimal path to the destination, and considering the fact that each link costs the same value, the resulting way will be *(4, 2, 1)*. *JobP2P 6* will send a package consisting of the list of nodes to the destination, from which we removed the next hop *(4)* and the specific data. Therefore the package is *(2, 1) / data*. It will be send from *JobP2P 6* to *JobRelay 4*.

JobRelay listens for information on port:

```
Port of Relay job, for receiving = (index of job) + 2
```

Therefore, *JobP2P 6* sends the data from *10.0.0.7*, port *9*, to *JobRelay 4*, on *10.0.0.5*, port *6*. *JobRelay 4* unpacks the information received in the message, sees that the next hop to the destination is *2*, removes it from the path, makes a new package with the remaining list and the useful data *((1) + data)* and transfers it to the *JobRelay 2*.

```
Port of Relay job, for sending = (index of job) + 4
```

JobRelay 4 sends from *10.0.0.5*, port *8*, to *JobRelay 2*, on *10.0.0.3*, port *4*. *JobRelay 2* removes the remaining node from the list, checks it for emptiness and sends a new message with the data received to *JobP2P 1*, connected directly to node *2*.

The disadvantage of this approach is the fact that each *JobP2P* object should keep in memory all the topology of the system, so the fact that the algorithm works or not depends strictly on the local memory capacity of each peer. Also, another drawback is the difference with how the gossiping protocol should be designed. In such protocols, a node cannot control and have knowledge about all the system from the beginning; the information is achieved in time. So, it is suitable for a node to keep in memory only the list of neighbors. These two disadvantages are removed with the algorithm described in the next section.

AODV Algorithm for Overlay Construction. The second algorithm is Ad hoc On-Demand Distance Vector (AODV) Routing, a routing protocol for mobile ad hoc networks (MANETs) and other wireless ad-hoc networks. It is a reactive routing protocol, meaning that it establishes a route to a destination only on demand. In contrast, the most common routing protocols of the Internet are proactive, meaning they find routing paths independently of the usage of the paths. AODV is, as the name indicates, a distance-vector routing protocol.

In AODV, the network is silent until a connection is needed. At that point the network node that needs a connection broadcasts a request for connection. Other AODV nodes forward this message, and record the node that they heard it from, creating an explosion of temporary routes back to the needy node. When a node receives such a message and already has a route to the desired node, it sends a message backwards through a temporary route to the requesting node. The needy node then begins using the route that came first to it.

Following the model described in the previous section, we also had to define an *Activity* for generate the jobs. In this case the relays have to receive information about direct neighbors, as they need to manipulate the routing table. When a *JobP2P* wants to transfer data to a remote node with which it is not connected, it makes a request to the *JobRelay* associated and asks for a route to the destination desired. If the relay does not have the answer, a request is broadcast to all the neighbors and the AODV algorithm is applied.

A *JobP2P* has the following methods: *sendRequest*, *sendData*, *receive*, *stopRelays*, and *run*. The *run* method overrides the parent method of the superclass *Job*, and describes the behavior of the peers. The *sendRequest's* purpose is to send a request to the relay running on the same CPU unit for a specific destination. The *sendData* method sends data to a relay, which represent a temporary node in the way to the desired gossip partner, if no edge connects them, or otherwise directly to the peer. The *receive* method has two purposes: one is to receive a route requested from the associated relay and the second is to receive data from a partner. After the peers finish their job, they send a message to the task responsible with stopping the relays.

A *JobRelay* has to manage the routing table of each application running on a CPU unit, and also has to keep in the memory the requests that were made by a source node and could not yet been accomplished (pending requests looking for a route to a specific destination). This is necessary as, when an answer is find, it should be replied backwards, unicast to the source. So, for each request that we broadcast to all neighbors, we keep in memory the source and destination, to identify it, and also, the next temporary step to the initial requesting source. The relays also have to monitor the sequence number of each request to a specific target, to avoid loops. So, they maintain a vector with a number of positions equal to the number of vertices of the graph and on position x we keep the sequence number of the request to destination x. The class has three methods: *receive*, *sendData* and *run*. The behavior of the object is specified by the *run* method and it always should first receive a request and then send some data according to the situation. There are five requirements that are able to be managed and all are listed below:

- Listen for request from the complementary P2P job which asks for a route. If the relay knows a next hop to the destination required, it sends back the reply, otherwise it broadcasts the request to all neighbors. The *JobRelay* forwards the route to the P2P job, when it becomes known.
- Manage requests from other relays looking for a route. The situation is similar to the one above: if the destination can be reached from that point, a reply is send backwards; otherwise a request is broadcast to other temporary nodes directly connected.
- Forward useful data received to a specific destination. If the destination is the CPU unit where this relay runs, the data is sent to the partner *JobP2P*, and the transfer is accomplished.
- Transfer backwards a route reply to a specific requesting source. It is possible that many replies will be received, for the same target, but we know that the first one is the fastest, so we keep it in the routing table and remove it from the pending requests. Therefore, all the other answers corresponding to the same target will be dropped.
- Stop running when a null data is received. This means that all peers finished and sent a message to the *JobCloseRelays* (the job responsible for closing the relays).

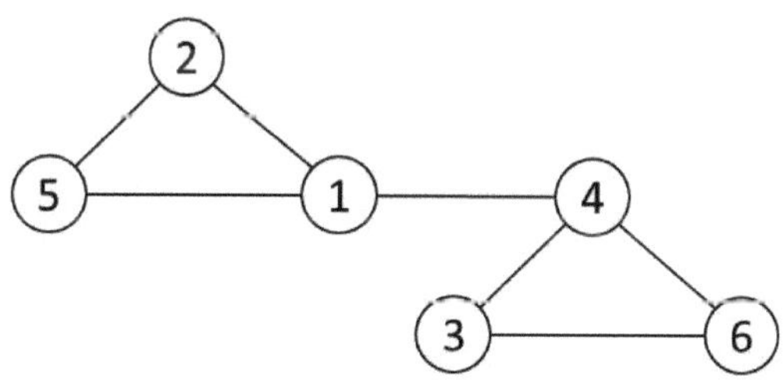

Fig. 11.8 Graph for the AODV algorithm for overlay construction.

The next section provides an example and it is explained how the algorithm works on a specific topology. We will explain how *peer6* can transfer data to *peer1* and also, *peer2* to *peer3*. *Peer6* delegates its associated relay with the responsibility to find the next hop in the way to *1*. *Relay 6* checks its routing table and as no entry resides in it for the destination required, the request is forwarded to all neighbors. *Relay 4* has the route to *1*, as an edge directly connects them, so an answer is given back to *relay 6* in order to set the entrance in the routing table and then to *peer6*. After receiving the next hop for the target destination requested, *JobP2P 6* sends the data to the relay part of application *4*, which forwards the message to *peer1*.

Peer2 delegates the complementary relay with finding the route to *3*, the same way described above. Therefore, this one communicates the request to

5 and *1*. None of them has an entry in the routing table for the final remote index, so both broadcast the request to all neighbors except the source of the message. Having this said, *relay 1* forwards the call to relays *5* and *4*. First of them ignores it because it has just processed on its CPU the same request identified by a sequence number associated with the destination. Similarly, *relay 1* drops the requirement solicited by *5*. Before forwarding the message, *relay 1* has to keep in memory some fields for identifying the associated reply containing the route desired, for sending it unicast backwards. So, we will create a pending object with the requesting source and its destination and also, the next hop in the way to the source (source: *peer2*, destination: *peer3*, next hop to *peer 2*: *relay 2*). *Relay 4* knows how to reach the destination (as *3* is connected with *4* and some entrances in the routing table are set at the beginning of the job for all the neighbors), so it sends back to *relay 1* the reply with the route. After receiving it, *node 1* removes the pending object set for this request, fix in the routing table the next hop to the final target solicited, which is *4*, and forward the reply to the next temporary node in the way to the initial source. As this node is right the relay associated to the requesting peer (*node 2*), it sets properly the entrance in his routing table (next hop: *1*, destination: *3*) and sends the reply to its peer. Then *peer2* transfers data to *relay 1*, which forwards it to *relay 4*, and this one to *peer3*.

As in the first algorithm described, two peers connected by an edge in the graph, change information directly, avoiding relays.

This algorithm brings some advantages over the first one, as peers keep in memory only their neighbors. But, in spite of the simplicity brought by the first *JobRelay* designed, the second corresponding job does not only keep memory busy with different information related to the routing process, it also has to manage much more requests. Another drawback of the algorithm implemented is network flooding with messages. Besides that, the number of context switches increases, as for reaching a peer that is not connected to a certain node; we have to make a request to the associated relay. Moreover, when a node broadcasts data, for every message sent, a context switch appears. In the first algorithm the routing decisions were strictly made by the sending peer, so it is avoided any overhead brought by communication between peers and relays corresponding to the same application, and by multiple sending of route request.

11.7 Simulation Experiments

In the simulation experiments, the regional center (P2P) is first described by latitude and longitude, parameters which do not present interest in our simulation as we only have a single regional center with multiple jobs. Then, the name of the local area network is specified, its speed, the name of a CPU and also of the activity object, both described below.

In the CPU section are specified 55 CPU, each with the same features: power, di-mension of memory in megabytes, the address in the local area network (first CPU has the address 10.0.0.1, the second 10.0.0.2, etc) and link maximum speed. We use the first unit (0) for hosting the job responsible for closing the relays and the others for the peers and relays specified by the topology.

In the activity section is specified the name of the activity class that generates the jobs. We varied several parameters, such as maximum simultaneous threads, CPU power, memory dimension and maximum link speed, to analyze which of the two algorithms works better in terms of scalability and performance.

For the first series we chose a honeycomb topology with three chains and we calculated each peer's destination according to the formula:

```
ID (Destination) = (ID (Source) + 1) modulo (topology
dimension)
```

A configuration file for a simulation test is presented below:

```
queue_type = vector
max_simultaneous_threads = 1000
regional0 = P2P
network_fairness = true

[P2P]
latitude  = 45.0
longitude = 27.0
initial_pool_size = 0
lan0 = LAN
lan0_max_speed = 1000.0       \# Mbps
cpu_unit0 = cpu0
activity0 = activity

[cpu0]
from = 0
to = 54
cpu_power = 1.0                   \# SI95
memory = 20.0                   \# MB
page_size = -1.0                  \# doesn't matter
link_node = 10.0.0.1            \# the address
link_node_max_speed = 10.0  \# Mbps
link_node_connect = LAN

[activity]
class_name = P2PAODVActivity
```

Table 11.3. Results (all times are in ms).

				Basic algorithm		AODV algorithm	
max_threads	cpu_pow	mem	link_node_max_speed	ExecTime	SimTime	ExecTime	SimTime
1000	1.0	20.0	10.0	2404	88.83	36917	247.78
1	1.0	20.0	10.0	2747	88.83	43437	242.74
1	4.0	20.0	10.0	2567	88.83	43114	243.50
1000	4.0	20.0	10.0	2470	88.83	40674	243.37
1000	4.0	10.0	10.0	2844	88.83	43073	242.81
1000	4.0	100.0	10.0	2330	88.83	40225	245.99
1000	4.0	100.0	100.0	4297	9.09	113475	38.01
1000	4.0	100.0	1000.0	3327	2.25	484706	33.78

As seen in Table 11.3, the simulation time remains constant, when the maximum link speed of a node does not vary. In the first six experiments, we varied the maximum simultaneous threads, the power of the CPU and the memory dimension and the only output parameter that has changed visibly is the execution time, which varied in an expected way (for example, when the maximum simultaneous number of threads decreases, the execution time increases, when the CPU power increases, the execution time decreases, when memory is smaller, the time is greater). We observe an interesting fluctuation of the simulation time when we vary the fourth parameter, as it decreases sharply, with a high percentage on every unit added to the input parameter.

Time variations appear the same way on both algorithms, but we can clearly ob-serve that the real time on the basic algorithm is with at least one unit smaller than on the AODV algorithm, and the simulation time with three times less on the average in the first, than in the second algorithm. We continue to analyze the algorithms and provide a series of tests performed on the same topology (honeycomb with 54 nodes), but this time the destination ID is calculated using the formula:

```
ID (Destination) = (ID (Source) + 8) modulo (topology
dimension)
```

We expect both time parameters analyzed to grow, as in the first series of tests, the way of generation the destination id (*source id + 1*) caused 15 direct communications (for example *peer 1* with *peer 2* and all the peers of the first chain, beside *peer 6*, which transfers data to *peer 7*, on the second chain, or peers (14, 15), (39, 40), etc. All these pairs transfer data directly, so they finish working - the sending part - immediately after they are scheduled for processing), 28 communications separated by one hop (pairs (12, 13), (27, 28), etc), 10 by two hops (pairs (7, 8), (32, 33), etc) and one by four hops (pair (54, 1)). Thus, reaching the destination desired is much easier and faster this way. In the second series of tests, we have two direct communications ((8, 16), (9, 17)), two communications separated by one hop (10, 18), five by two hops (1, 9), five by three hops (11, 19), two by four hops (50, 4), six by five hops (14, 22), ten by six hops (18, 26), four by seven hops (26, 34), three by eight hops (30, 38), nine by nine hops (33, 41) and six by ten hops (45, 53). We can easily observe that the communication will be slower in this case, as in the basic algorithm, for reaching a destination, a package has to pass through multiple relays compared to the first case, and in the AODV algorithm, in addition to this reason, the reply to a route request will come later, as a node who knows the path to a specific remote index is far from the source.

The conclusions of this set of experiments are similar to the ones above, as before changing the maximum speed of a node link, the simulation time remains at constant values, on both algorithms, and, even though the difference is not significant, it is smaller on the basic algorithm. Moreover, when the link speed increases, the simulation time decreases. We can also state the major difference between the real times measured, in the AODV algorithm being with at least one unit greater.

The next section provides some graphics for a better view of how the jobs evolve, of the time when peers start ending and when the job responsible for closing the relays starts executing the commands, of how a CPU hosting a peer and a relay works with I/O and of how these jobs are completed. The tests are made on the first case of the second set of tests (*max_simultaneous_threads* = *1000*, *cpu_power* = *1.0*, *memory* = *20.0*, *link_node_max_speed* = *10.0* and *destination_id* = *source_id* + *8*).

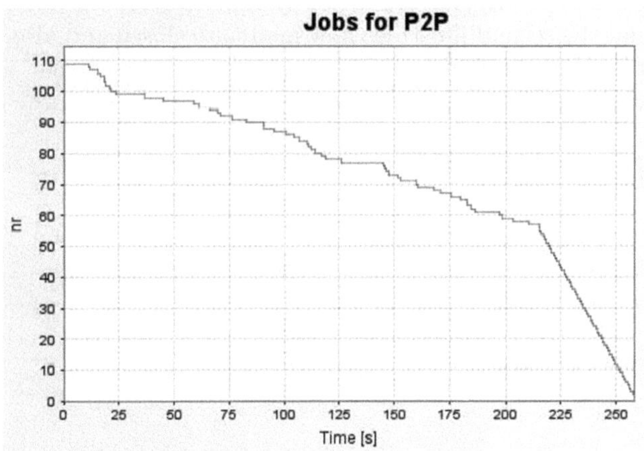

Fig. 11.9. Graph of running jobs for the basic algorithm for overlay construction, 54 nodes.

The first pair of figures (Fig. 11.9 and Fig. 11.10) represents the graphics of the jobs simulated. Initially, in the system 109 jobs are generated (54 peers, 54 relays and one job responsible for relays ending). Given the fact that the numbers of pairs which communicate through a certain number of hopes are proportional distributed, on the basic algorithm (Fig 11.9) we observe an almost linear decrease of the jobs (peers) until a straight line appears, which represents relays closure. All the peers calculate the path to the destination locally, so the ending of a peer depends on how fast its information reach the destination (and this depends on how many hops it has to cross) and when it receive data from a source peer. For example, *peer 16* receives information from *peer 8*, which is directly connected, and sends data to *peer 24* which is five nodes away. Normally, this peer finishes among the first, this fact depending also on the way the jobs are scheduled for processing and on network characteristics. In contradiction, we take, for example, *peer 40* which communicates with *peer 32* (eight hops away) and *peer 48* (nine hops away). This job stops processing later. After the simulated time 220 sec, the relays are closed by the job responsible, each of them being commanded at a fixed time variation (the time between a sending made by the *JobCloseRelays* and a receiving made by the destination relay). This explains the straight line after that time. The difference brought by the AODV algorithm is the fact that jobs start finishing later, as for finding the path to a certain destination, a request should be broadcasted in the system and a reply should be sent back. This happens twice for each *job X*, one search being made for a remote peer desired by *X* and one by a source node seeks the path to *X*. This explains straight horizontal line at the beginning of the simulation and the late start of peers ending.

Figures 11.11 and 11.12 are complementary to the graphics above, as the first two show how many jobs are in the system at every second of the simulated time and the second illustrate how many are closed and also, submitted in the system. The straight red line emphasizes the fact that all jobs are generated when simulation starts.

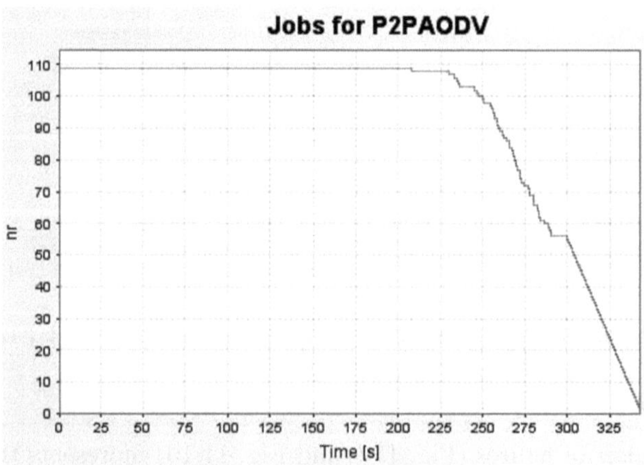

Fig. 11.10. Graph of running jobs for the AODV algorithm for overlay construction, 54 nodes.

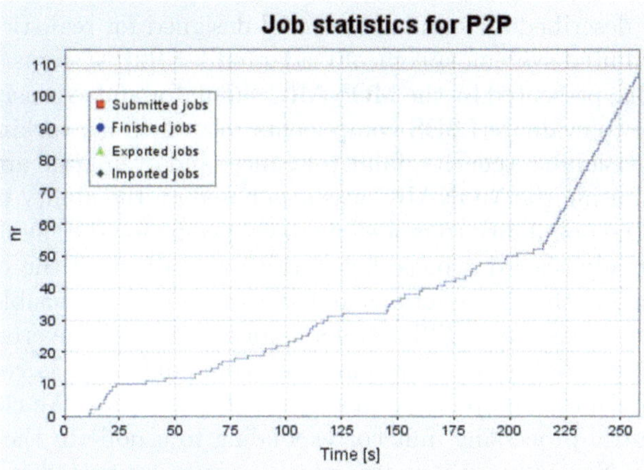

Fig. 11.11. Job statistics for the basic the basic algorithm for overlay construction, 54 nodes.

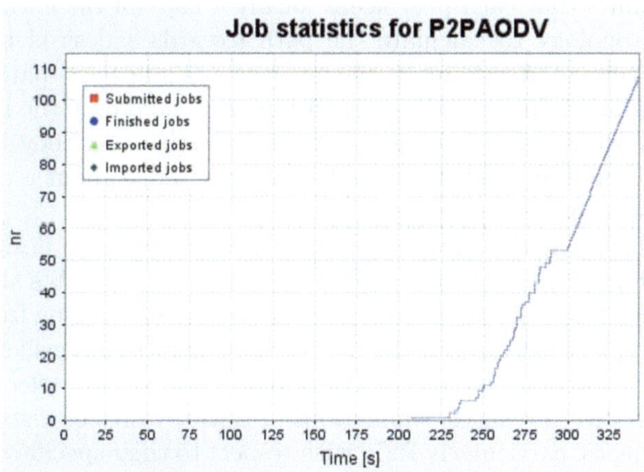

Fig. 11.12. Job statistics for the basic the AODV algorithm for overlay construction, 54 nodes.

11.8 Conclusions

The purpose of chapter was to present the general characteristic for P2P simulation, a short analysis of simulation model and existing simulation tools, and to present the design of an algorithm to monitor the resources in a P2P system. Distributed systems have evolved exponentially last decade, so a centralized system to manage them would be difficult and expensive to maintain. Thus, we needed a fully decentralized mechanism.

In this we described an simulation model designed for realistic evaluation of large distributed system, specifically adapted for P2P systems. The model extends the one presented by the MONARC simulator and extensively used in the literature to evaluate LSDS components related to networking, scheduling, data provisioning, security, fault tolerance, and even large architectures. The choice of using the MONARC was also based on the ability to easily extend it with the possibility to model overlays specific for P2P systems, as the simulator already offered a basic layer which contained all the components of the system and their interactions and which was easy extensible according to our specific needs. For a P2P network, constructing the overlay means to respect a given topology of a system when communication is realized. The topology is given by a graph of connections between nodes. We chose to allocate each central processing unit corresponding to a node to the same local area network. Normally, two nodes, which are not connected in the graph, exchange data directly, being in the same LAN. By respecting the topology, they communicate through the path that connects them.

Two algorithms were designed in order to extend this simulation tool. The first one assumes that each peer keeps locally a copy of the entire topology. It uses this topology to calculate the path towards a desired target when sending the package. Accordingly, the message will follow the path computed from the beginning. In the second algorithm peers hold only the list of direct connected neighbors. For finding a path to a destination, they broadcast a message in the system, and the peers having knowledge about a path to that destination send backwards corresponding replies.

The simulation model adopted by MONARC follows the design of many actual distributed system technologies. The model incorporates all the necessary components to conduct various simulation studies, ranging from scheduling to data replication or data transfer technologies. In extensive evaluation experiments we demonstrated that the simulation model is perfectly adapted to be used to design realistic simulations of a wide-range of distributed systems technologies, particularly P2P, with respect to their specific components and characteristics.

Acknowledgements. The research presented in this paper is supported by national project "DEPSYS - Models and Techniques for ensuring reliability, safety, availability and security of Large Scale Distributes Systems", Project CNCSIS-IDEI ID: 1710. The work has been co-funded by the Sectoral Operational Programme Human Resources Development 2007-2013 of the Romanian Ministry of Labour, Family and Social Protection through the Financial Agreement POSDRU/89/1.5/S/62557.

References

1. Adamic, L.A., Lukose, R.M., Puniyani, A.R., Huberman, B.A.: Search in power-law net-works. Physical Review E. v046135 i64 (2004)
2. Olteanu, A., Pop, F., Dobre, F., Cristea, F.: An adaptive scheduling approach in distributed systems. In: 2010 IEEE International Conference on Intelligent Computer Communication and Processing (ICCP 2010), Cluj-Napoca, Romania, pp. 435–442 (2010)
3. Eremia, B., Dobre, C., Pop, F., Costan, A., Cristea, V.: Simulation model and instrument to evaluate replication techniques. In: International Conference on, P2P, Paralel, Grid, Cloud and Internet Computing (3PGCIC 2010), Fukuoka, Japan, pp. 541–547 (2010)
4. Xie, C., Chen, G., Vandenberg, A., Pan, Y.: Analysis of hybrid P2P overlay network topology. Comput. Commun. 31(2), 190–200 (2008)
5. Dobre, C.: A General Framework for the Modeling and Simulation of Grid and P2P Systems. In: Antonoupoulos, N., Exarchakos, G., Li, M., Liotta, A. (eds.) Handbook of Research on P2P and Grid Systems for Service-Oriented Computing: Models, Methodologies, and Applications, pp. 657–686. Information Science Reference, IGI Global (February 2010) ISBN: 978-1615206865
6. Dobre, C., Stratan, C.: MONARC 2 - distributed systems simulation. In: 14th International Conference on Control Systems and Computer Science, pp. 145–149. Ed. Politehnica Press, Bucharest (2003)
7. Dobre, C., Pop, F., Cristea, V.: A fault-tolerant approach to storing objects in distributed systems. In: International Conference on, P2P, Paralel, Grid, Cloud and Internet Computing (3PGCIC 2010), Fukuoka, Japan, pp. 1–8 (2010)
8. Dobre, F., Pop, V., Cristea, V.: New Trends in Large Scale Distributed Systems Simulation. In: Proc. of the 2009 International Conference on Parallel Processing Workshops (ICPPW 2009), pp. 182–189. IEEE Computer Society, Washington, DC (2009)
9. Dobre, C., Pop, V., Cristea, V.: Simulation Framework for the Evaluation of Dependable Distributed Systems. Scalable Computing: Practice and Experience, Scientific International Journal for Parallel and Distributed Computing (SCPE) 10(1), 13–23 (2009)
10. Dobre, C., Cristea, V.: A Simulation Model for Large Scale Distributed Systems. In: Proc. of the 4th International Conference on Innovations in Information Technology (Innovations 2007), Dubai, United Arab Emirates (November 2007)
11. Stutzbach, D., Rejaie, R., Sen, S.: Characterizing unstructured overlay topologies in modern P2P file-sharing systems. IEEE/ACM Trans. Netw. 16(2), 267–280 (2008)
12. Anderson, D.P., Cobb, J., Korpela, E., Lebofsky, M., Werthimer, D.: SETI@home: an experiment in public-resource computing. Commun. ACM 45(11), 56–61 (2002)
13. Milojicic, D.S., Kalogeraki, V., Lukose, R., Nagaraja, K., Pruyne, J., Richard, B., Rollins, S., Xu, Z.: Peer-to-peer computing. Technical report (2002)
14. Antoniu, G., Bougé, L., Jan, M., Monnet, S.: Large-Scale Deployment in P2P Experiments Using the JXTA Distributed Framework. In: Danelutto, M., Vanneschi, M., Laforenza, D. (eds.) Euro-Par 2004. LNCS, vol. 3149, pp. 1038–1047. Springer, Heidelberg (2004)

15. Pop, F., Dobre, C., Godza, G., Cristea, V.: A Simulation Model for Grid Scheduling Analysis and Optimization. In: Proc. of PARELEC Conference, Bialzstok, Poland, pp. 133–138 (September 2006)
16. Casanova, H., Legrand, A., Quinson, M.: SimGrid: a Generic Framework for Large-Scale Distributed Experimentations. In: Proc. of the 10th IEEE International Conference on Computer Modelling and Simulation (UKSIM/EUROSIM 2008), Cambridge, UK, pp. 126–131 (2008)
17. Wang, H., Takizawa, H., Kobayashi, H.: A dependable Peer-to-Peer computing platform. Future Gener. Comput. Syst. 23(8), 939–955 (2007)
18. Zhang, J., Duan, H., Liu, W., Wu, J.: Anonymity analysis of P2P anonymous communication systems. Comput. Commun. 34(3), 358–366 (2011)
19. Ranganathan, K., Foster, I.: Decoupling Computation and Data Scheduling in Distributed Data-Intensive Applications. In: Int. Symposium of High Performance Distributed Computing, Edinburgh, Scotland (2002)
20. Shudo, K., Tanaka, Y., Sekiguchi, S.: Overlay Weaver: An overlay construction toolkit. Comput. Commun. 31(2), 402–412 (2008)
21. Walkowiak, K., Przewoniczek, M.: Modeling and optimization of survivable P2P multicasting. Comput. Commun. 34(12), 1410–1424 (2011)
22. Ramanathan, N., Kohler, E., Estrin, D.: Towards a debugging system for sensor networks. Int. J. Netw. Manag. 15(4), 223–234 (2005)
23. Rodriguez, P., Tan, S.-M., Gkantsidis, C.: On the feasibility of commercial, legal P2P content distribution. SIGCOMM Comput. Commun. Rev. 36(1), 75–78 (2006)
24. Buyya, R., Murshed, M.: GridSim: A Toolkit for the Modeling and Simulation of Distributed Resource Management and Scheduling for Grid Computing. Journal of Concurrency and Computation: Practice and Experience (CCPE) 14 (2002)
25. Sole, R.V., Ferrer-Cancho, R., Montoya, J.M., Valverde, S.: Selection, tinkering, and emergence in complex networks. Complex 8(1), 20–33 (2002)
26. Naicken, S., Livingston, B., Basu, A., Rodhetbhai, S., Wakeman, I., Chalmers, D.: The state of peer-to-peer simulators and simulations. SIGCOMM Comput. Commun. Rev. 37(2), 95–98 (2007)
27. Naqvi, S., Riguidel, M.: Grid Security Services Simulator G3S) - A Simulation Tool for the Design and Analysis of Grid Security Solutions. In: Proc. of the First International Conference on e-Science and Grid Computing, Melbourne, Australia (2005)
28. Lui, S.M., Kwok, S.H.: Interoperability of peer-to-peer file sharing protocols. SIGecom Exch. 3(3), 25–33 (2002)
29. Dinh, T.T.A., Theodoropoulos, G., Minson, R.: Evaluating Large Scale Distributed Simulation of P2P Networks. In: Proc. of the 2008 12th IEEE/ACM International Symposium on Distributed Simulation and Real-Time Applications (DS-RT 2008), pp. 51–58. IEEE Computer Society, Washington, DC (2008)
30. Cristea, V., Dobre, C., Stratan, C., Pop, F., Costan, A.: Large-scale Distributed Computing and Applications: Models and Trends, 390 p. Ed. Information Science Publishing (2010) ISBN : 978-1615207039
31. Cristea, V., Dobre, C., Pop, F., Stratan, C., Costan, A., Leordeanu, C.: Models and Techniques for Ensuring Reliability, Safety, Availability and Security of Large Scale Distributed Systems. In: 3rd International Workshop on High Performance Grid Middleware, 17th Intern. Conference on Control Systems and Computer Science, Bucharest, Romania, pp. 401–406 (May 2009)

32. Venters, W., et al.: Studying the usability of Grids, ethongraphic research of the UK particle physics community. In: UK e-Science All Hands Conference, Nottingham (2007)
33. Joung, Y.-J., Wang, J.-C.: Chord2: A two-layer Chord for reducing maintenance overhead via heterogeneity. Comput. Netw. 51(3), 712–731 (2007)
34. Takano, Y., Isozaki, N., Shinoda, Y.: Multipath Key Exchange on P2P Networks. In: Proc. of the First International Conference on Availability, Reliability and Security (ARES 2006), pp. 748–755. IEEE Computer Society, Washington, DC (2006)
35. Bagchi, S.: Simulation of grid computing infrastructure: challenges and solutions. In: Proc. of the 37th Conference on Winter Simulation (WSC 2005), Winter Simulation Conference, pp. 1773–1780 (2005)
36. Venugopal, S., Buyya, R., Ramamohanarao, K.: A taxonomy of Data Grids for distributed data sharing, management, and processing. ACM Comput. Surv. 38(1), Article 3 (June 2006)

Part IV
Economic and Biological Approaches

Chapter 12
An Economics-Inspired Noise Model in Spatial Games with Reputation

Dhruv Gairola and Siang Yew Chong

Abstract. Games are useful mathematical constructs to model real-world problems involving strategic interactions in various contexts such as politics, economics, and biology. Understanding specific conditions that lead to cooperation between self-interested individuals is an important issue in the study of real-world interactions. Although noisy and spatial interactions have been incorporated into models of complex interactions to better reflect those found in the real-world, most past studies consider simple extensions whereby interactions between all individuals are equally noisy. Here, we study a novel economics-inspired noise model based on the notion of *psychic distance* that reflects real-world interactions. The *psychic noise* that affects interactions between individuals depends on their psychic distance (e.g., cultural difference). Results from extensive computer simulations using a multi-agent system framework to investigate the impact of various constructions of noisy interactions indicate that noise typically has a negative impact on cooperation. However, a certain condition produces results reminiscent of the *psychic distance paradox*, where an increase in the noise level leads to an increase in the level of cooperation.

12.1 Introduction

In the field of behavioral sciences, researchers have long been intrigued by specific conditions that promote cooperation among selfish individuals. Findings indicate that cooperation tends to remain transient and abruptly breaks down to selfishness, sometimes with no apparent cause [13]. Various studies

Dhruv Gairola · Siang Yew Chong
School of Computer Science, University of Nottingham, Malaysia Campus, Jalan Broga, 43500 Semenyih, Malaysia
e-mail: dhruv.gairola.89@gmail.com, Siang-Yew.Chong@nottingham.edu.my

J. Kołodziej et al. (Eds.): Advances in Intelligent Modelling and Simulation, SCI 422, pp. 271–293.
springerlink.com

have been made to investigate how and why these behavioral interactions occur. They drive our understanding of behavior not only between highly complex social beings such as humans (e.g., Lumsden [33] analyzed the communal conflict in Cyprus between the Greek and Turkish Cypriots), but also between plants and other animals (e.g., Axelrod and Hamilton's [7] observation on the behavior of fig wasps). In particular, games are used as frameworks to model precisely this sort of behavior.

Games are useful mathematical constructs to abstract and model real-world problems involving strategic decision-making in a variety of contexts ranging from politics and economics to biology among others [4]. In particular, they capture the intrinsic properties of these problems through the specification of rules that constrain strategies to certain behaviors (legal moves they can make as responses to stimuli such as historical plays), goals for strategies to meet (to win the game), and rewards under finite resources (payoffs) [13].

Although noisy and spatial interactions have been incorporated into models of complex interactions to better reflect those found in the real-world, most past studies consider simple extensions whereby interactions between all individuals are equally noisy. For example, the noise level is fixed for every interaction [46, 19]. Here, we propose a more realistic and flexible noise model where the level of noise that affects interactions depends on how *different* individuals are. Our economics-inspired noise model is based on the notion of psychic distance [9, 27]. For example, individuals are more likely to misinterpret one another (higher psychic noise level) if they come from more different cultural backgrounds (larger psychic distance) compared to those who come from a similar cultural background.

We conduct a systematic empirical study to investigate various constructions of noisy interactions based on psychic distance in the learning of behaviors. Our computer simulations make use of multi-agent systems to abstract the complex interactions as a collection (population) of autonomous agents that interact with one another. Co-evolutionary learning provides the framework for agents to learn behaviors through an adaptation process on strategy representations based solely on strategic interactions with other agents in the population that are modeled as game-plays.

We incorporate psychic noise into the donation game that models interactions having short encounters. In our empirical studies, we consider functions that reflect different relationships between psychic distance and the probability of receiving an unintended signal (noise). Our results indicate that noise is mostly detrimental to cooperation. Nearly all the noise functions are identical in this respect, even when mutation is introduced when evolving the population. Our results are not affected by varying the population size. Certain functions, however, show increased cooperation as noise increases. This observation is reminiscent of the psychic distance paradox. Our empirical studies suggest that the noise model based on psychic distance better reflects real-world interactions.

The rest of this chapter is organized as follows. Section 12.2 starts with a general introduction to standard game models used to study cooperation and then describes the extensions investigated in our study. The next section discusses the psychic distance and the origin of the psychic noise. After that, our empirical results are presented and discussed. Finally, the conclusion is presented along with various suggestions for future studies.

12.2 Complex Interactions

There are various models of interactions that describe cooperation between inherently selfish individuals. One popular model is that of the Prisoner's Dilemma, which describes a game between two apprehended men (who have jointly violated the law) and are separately given two alternatives to choose from: either to confess to the crime C, or to not-confess F. Any of the four possible scenarios ($C-C$, $F-F$, $C-F$, $F-C$) confers a specific payoff value to either player. In the Iterated Pisoner's Dilemma (IPD) game (popularized by Axelrod's famous open tournament [2]), players play successive rounds against each other so that the results of prior interactions influence their respective strategies. From his studies, Axelrod showed that cooperative strategies could be successful in certain scenarios (e.g, the *tit-for-tat* strategy has proven viable in many instances [3]).

Extensions are often incorporated into models of complex interactions that are usually inspired from aspects of interactions found in the real-world. In our study, we are motivated to investigate the notion of psychic distance and its effect in noisy interactions. As a natural consequence of incorporating a distance measure (i.e., psychic distance), a spatial two dimensional toroidal grid is constructed so that various neighbourhood structures could be experimented with (larger neighbourhoods include players who are a further psychic distance away).

For highly abstract model of interactions, one can turn to evolutionary game theory for solutions [34, 43]. This approach considers a setting that involves an infinitely large population of players that are allowed to use a set of *predefined inheritable* strategies and compete for payoffs that decide their average reproductive success. Despite the use of highly abstract models of real-world interactions, they are often analytically intractable. In contrast, we consider the alternative to simulate and model the learning of complex behaviors that are often found in real-world interactions. We use multi-agent systems to abstract complex interactions as a population of autonomous agents (players) that interact with one another. Co-evolutionary learning provides the framework for players to learn and adapt their behaviors through strategic interactions (game-plays). Many past studies [13, 26, 15, 17, 14] have used this approach to study the behavioral outcomes involving complex interactions that better reflect real-world interactions rather than to use highly abstract models and attempt to solve them analytically.

12.2.1 Noisy Behavior

Noise, defined as the *faulty transmission of strategy choices* [5], plays an important part in the modelling of real-world interactions. For example, among animals, there exist species whose sense of individual recognition is impaired, which suggests that they are incapable of determining the history of interaction with other players [5]. Such animals are prone to making mistakes in judgement due to this impairment (which is a manifestation of noise). Even if this recognition is not impaired, noise is used to model the inherent imperfection that is characteristic of all animals. Even between nations, noise can be an important issue [12, 6]. One example is that of the South Korean airliner, which was shot down by the Soviets after it mistakenly flew into Soviet territory [24]. This incident sharply strained relations between the Americans and Soviets, manifested by the escalation of cold war tensions [24].

Axelrod and Dion [5] observed the existence of two types of noise (misimplementation and misperception) that can affect a player's response to his opponents play. With misimplementation, the player who made the mistake is aware of his error, but the opponent is unaware, and hence, unable to discriminate between the mistake and an intentional cooperation or defection. In contrast, misperception is when the opponent believes that another choice was made by the player. A spinoff of misperception, *noisy channels* is a situation where both players are aware that there is a possibility of mistake, but neither of them knows when this mistake will occur.

Julstrom [28] experimented with noise in a two-choice IPD environment, and found that noise can diminish the level of cooperation in the population. Chong and Yao [15] studied noise with multiple levels of cooperation and found that noise at low levels can be more conducive to cooperation than without noise. Their results are consistent with Axelrod and Dion's [5] predictions that strategies tend to be more forgiving when interactions have low level noise. In contrast, above a certain high level, increasing the noise was detrimental to cooperation. However, they also confirmed that if the system has sufficient behavioral diversity, cooperation can still be evolved with relative ease.

12.2.2 Spatiality

In most real-world interactions, an individual will not have the chance to interact with everybody in the population. Furthermore, an individual does not have an equal chance of interacting with everybody. In this case, the notion of spatiality can be employed, where individuals are allowed only to interact within a defined neighbourhood.

One common method of implementing spatiality is via a two-dimensional grid environment, where each cell in the grid represents an individual of the population. Additionally, this environment exists as a toroidal space, where the corners and edges of the grid wrap around, connecting individuals that reside in the corners and the edges. Each cell is restricted to interactions with individuals that reside within a fixed neighbourhood, where this restriction introduces the element of spatiality. This model is also referred to as a cellular automaton [45] and is used in this study.

Such parochial interactions have shown to have a positive impact on the evolution of cooperation for the NIPD game [42, 31]. This might be due to locally occurring reproduction, which allows successful strategies to produce clusters of similar individuals. Clustering sometimes allows strategies to propagate in an environment where these strategies might not have survived without clustering [11]. For example, the *tit-for-tat strategy* described earlier can invade a population of strategies that always defect [4, 35].

12.2.3 Indirect Reciprocity

In today's highly mobile and globalized world, interactions occur between people who may not have previously met each other. Indeed, humans are not merely limited to assessing their opponents on the mere basis of direct interactions [37]. The concept of indirect reciprocity introduces reputation as a key measure of previous cooperative behavior (for each player) [37, 30] in IPD games. Even if players have not interacted with each other before, through each other's reputation, they are able to judge the nature of their previous interactions. Cooperation between current partners depends on this prior behavior with others [17].

Direct reciprocity refers to the scenario when players interact with each other in the immediate vicinity. If one player cooperates with another, the other returns the favour directly to the cooperator. Here, cooperation is related to repeated interactions between the same players. In contrast, indirect reciprocity involves no such direct interactions. Strictly, there are two versions of indirect reciprocity- the upstream version (based on a "recent positive experience") and the more common downstream version (based on reputation) [39].

In the upstream version, when one player helps another, this other player feels encouraged and helps an altogether different player in return. In the case of the downstream version, when a player cooperates with a third party, the player's reputation increases. A higher reputation encourages cooperative behavior by the opponent due to this previous altruistic act.

Formal investigations have shown that reputation encourages cooperative behavior [39]. However, upscale reciprocity is a more nuanced concept [1, 10]. In both cases, cooperation by the player is an attempt at returning the favour. In the upstream case, opponents are rewarded for what somebody else did while in the downstream case, opponents are rewarded by a player who did not profit by what they did. When studies mention indirect reciprocity, they usually refer to the downstream version, i.e., based on reputation.

Measuring reputation can be achieved by two methods: image scores [37] and standing [30]. In both methods, if a player cooperates, its reputation increases. However, when a player does not cooperate, their methods differ. In image scoring, the player's reputation decreases [38]. In standing, the player's reputation only decreases if its action is unwarranted (e.g., the opponent has a favourable reputation); otherwise, it remains the same [30].

12.2.4 The Donation Game

In this study, we are concerned with modelling indirect reciprocity only, while excluding the effect of direct reciprocity. An ideal approach would thus be to use the donation game to model the mechanism of indirect reciprocity through image scoring [30]. In this mechanism, two players are selected for an interaction, where one is a donor, while the other is a receiver. The donor is selected randomly from the population while the receiver is selected randomly from the donor's neighbourhood. The donor then assesses the receiver and based on the receiver's reputation, decides whether to cooperate or defect. If the reputation (image score) of the recipient is greater than or equal to the strategy value of the donor, the donor cooperates, paying a cost c while the recipient obtains a benefit b (with $b > c$ [38]). The key is to restrict the number of interactions so that the chance of a donor ever meeting the same receiver again is negligible, negating the effects of direct reciprocity.

Following [38], the donor's image score is increased by one unit on cooperation while the recipient's image score remains the same. If the donor refuses to cooperate, no payoff is awarded or deducted from either the donor or the recipient. However, the donor's image score is deducted by one unit (the recipient's image score remains unchanged). It is possible that some players may never get to be chosen for game playing. The benefit value b is set to 1.0 while the cost value c is set to 0.1 (to avoid negative payoffs, we add 0.1 [38]).

12.2.5 Implementation

For our studies, we consider the simple bit string representation for the player strategies in comparison with more complicated strategy representations (e.g., Finite State Machines [22], neural networks [18], etc). As a consequence, the mutation operator performs bit-flips on the strategies to match the allowed strategies defined in our experiment. A fitness proportionate selection operator (roulette wheel selection operator [8]) is also required to evolve the population to the next generation. Note that the selection operation is spatial, i.e., selections occur within the defined neighbourhood, not within the whole population.

Following [38], agent image scores (integers) range between −5 and +5 (inclusive) while strategy values (integers) range between −5 and +6 (inclusive). A strategy of −5 represents unconditional cooperators, while +6 represents defectors. Any strategy below or equal to a strategy value of 0 is deemed cooperative, since any strategy that lies in this range will cooperate with any player who has not played a game yet. Any strategy above 0 is non-cooperative, since any strategy in this range will refuse to cooperate with any player who has not played a game yet.

In Nowak and Sigmund's model, the chance of one player meeting the same opponent again is negligible. However, in this case, spatiality is introduced in the model, which means that the chance of repeated interactions is higher since the opponent is chosen from a restricted structure. This is important because indirect reciprocity considers modelling the situation where two players interact at most once with each other [39]. If players interact with each other more than once, then a player can hold the opponent responsible for past actions, i.e., direct reciprocity comes into play also. However, the implementation here considers only 125 games, which is small enough to discount the influence of direct reciprocity. Thus, the only mechanism that can have an impact on the behavioral outcome is that of indirect reciprocity.

12.3 Psychic Noise: A Novel Approach

Noise is an integral part of any interaction or communication, hence, it is important that we model noise convincingly. Most studies represent noise as occuring with fixed probabilities in interactions. However, our study argues that this probability is not the same for all interactions. Psychic distance is a well documented economic concept. By incorporating psychic distance into our experiments, we have designed a novel form of noise, called psychic noise, which is not only grounded on real-world observations and research, but also presents a far more flexible model compared to conventional noise models.

12.3.1 Psychic Distance

A term invented in 1956, psychic distance was used to describe certain aspects of complexity faced by multi-national companies when they exported goods and services to other countries [25]. In 1956, Beckerman [9], while studying the various factors that affected trade, noted psychic distance as a possible barrier to trade, a view supported by Linnemann too [32].

Psychic distance, according to Lee [29], results from perceptions about "cultural and business differences, where business differences are said to be attributed to differences in language, education, business practices, political and legal systems, economic environment, religious, and industry structure".

Even though the term was introduced by Beckerman, it was popularized by Johanson and Vahlne [27]. They claimed that these differences affect business activity because they restrict the flows of information between companies and the foreign market, thereby disturbing the firm's understanding of the foreign market [36]. Since then, various authors have used cultural distance as a synonym for psychic distance [40].

For the purpose of this study, psychic distance alludes to some fundamental behavioral difference between individuals, rather than a purely cultural difference. This difference gives rise to noise in interactions for our experiments. Psychic distance is represented by the shortest physical straight line distance between two individuals who reside in the cellular automata grid as individual cells. Since individuals are point sources, this distance can be measured using the centre point of each of the cells. When measuring the shortest distance between two centre points, one has to take note of the toroidal environment of the cellular automata too, i.e., the shortest route may result from going *around* this toroid as opposed to going across it.

12.3.2 Definition of Psychic Noise

Psychic noise is the noise due to the psychic distance between interacting players. Psychic noise is an important consideration in an increasingly globalized world, where interactions between diverse people is becoming a common occurrence. With more such interactions, there is a strong need to model such interactions that are more reflective of the real-world that cannot be accomplished using conventional noise models.

We note that psychic noise does not imply anything about cooperation between neighbours. That is to say, if two neighbours are a large psychic distance apart, this does not mean that the chance of cooperation is less. This merely means that the extent of noise present in the communication channels between them will be larger.

12.3.3 Modelling Psychic Noise

Using the image scoring mechanism, donors assess adversaries based on their own strategy and their opponent's image score. In their first experiment with the image scoring mechanism, Nowak and Sigmund [38] assumed that every individual's reputation was known to everybody else within the neighbourhood. This naturally implies the use of some form of communication between individuals before the game is played. However, where there is communication, one can expect noise, and it is not reasonable to assume that everybody's signal is communicated perfectly. In later experiments, Nowak and Sigmund indeed acknowledged this and decided that each individual would have a perception of their opponent's image score, which may not be an updated score. While this is a perfectly valid interpretation, in this study, however, the communication signal itself is examined for modelling noise.

Communication requires not just a signal but also a medium. Transmission through a medium can be thought of as transmission through a noisy channel that encapsulates the psychic distance between two individuals. In essence, an individual uses the signals received from the opponent to decide whether to cooperate or defect. If the opponent's signal is affected by noise, then the individual will make a mistake in his choice. As such, modelling the effect of psychic noise in this communication should detail a relationship between the psychic distance d_p and the probability of receiving an unintended signal P_{s_u}.

We already know that the relationship between d_p and P_{s_u} is a strictly increasing one (as psychic distance increases, psychic noise increases). This relationship holds over the domain $0 \le d_p \le d_{\max}$ where $d_{\max}$ is the largest psychic distance possible between two neighbours in the largest neighbourhood possible.

The following *threshold values* is considered: 0.001, 0.00325, 0.00550, 0.00775, 0.01, 0.0325, 0.0550, 0.0775, 0.1, 0.325, 0.550, 0.775 and 1.0. These threshold values represent the highest level of noise that is possible in a particular experiment for a particular noise function. This threshold is hit at d_{max}. A threshold value of 1.0 implies that at $d_{\max}$, $P_{s_u} = 1$. The chosen threshold values allow us to experiment with a large variety of noise levels, so that a wide range of behavior can be ascertained.

The lowest noise value in every game is 0, which occurs when an individual is not communicating with anybody, i.e., $d_p = 0$. Hence, in every game, there are two fixed noise values: one is 0 (which occurs when $d_p = 0$) and the other is the threshold value (which occurs at $d_{\max}$). As a result, every psychic noise function has to pass through the two coordinates (0, 0) and ($d_{\max}$, t) where t is the threshold value.

For our case studies, we investigate seven different noise functions having different characteristics. Figure 12.1 shows individual graphs for these noise functions, where noise functions (1)-(5) show the probability of error (noise level) increasing with psychic distance but at different rates of increase.

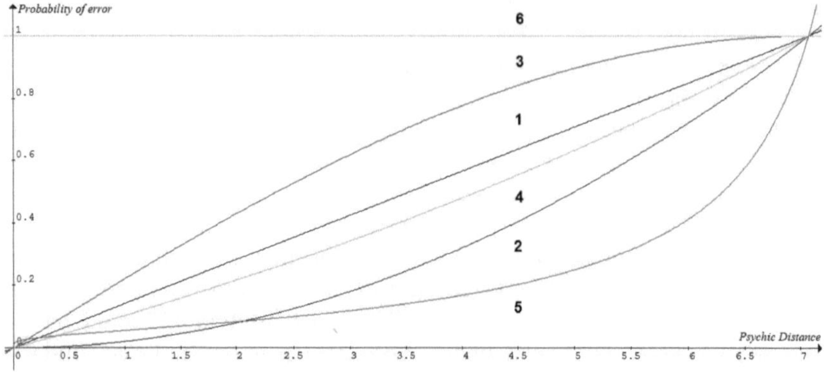

Fig. 12.1. Plots of the probability of error (noise level) against psychic distance for various noise functions with the threshold level set to 1.0. (1) linear, (2) polynomial, (3) sine, (4) exponential, (5) sound (inspired by the use of sound waves for communication), (6) conventional (a constant function), and (7) noiseless.

The following gives the equations for the noise functions used:

$$P_{s_u}(d_p) = \frac{t}{d_{\max}} \times d_p \tag{12.1}$$

$$P_{s_u}(d_p) = \frac{t \times d_p^2}{50} \tag{12.2}$$

$$P_{s_u}(d_p) = t \times \sin \frac{\pi \times d_p}{2 \times d_{\max}} \tag{12.3}$$

$$P_{s_u}(d_p) = 10^{\log \frac{(t+1)\times d_p}{d_{\max}}} - 1 \tag{12.4}$$

$$P_{s_u}(d_p) = \frac{1}{60 - 20 \times \log \frac{10^{\frac{1-60\times t}{-20\times t}}}{d_{\max}} \times d_p} \tag{12.5}$$

$$P_{s_u}(d_p) = t \tag{12.6}$$

$$P_{s_u}(d_p) = 0 \tag{12.7}$$

where t is the threshold value and $d_{\max}$ is the largest psychic distance possible between two neighbours in the largest neighbourhood possible.

12.4 Case Studies

12.4.1 Noise-Free Spatial Interactions

Past studies [38] have investigated how cooperative behaviors can be evolved through the mechanism of indirect reciprocity for interactions having short encounters. However, these studies did not consider localized interactions. The motivation for our first case study is to investigate the impact of spatial interactions having short encounters (e.g., single-round interactions) on the evolution of cooperation. In particular, we incorporated spatial interactions within the multi-agent systems framework on image scoring strategies, which have been studied previously [38].

We have conducted experiments for six different neighbourhood sizes: 5, 9, 25, 49, 81, and 100. The neighbourhood size of 5 is a Von Neumann neighbourhood while the others take the form of the Moore neighbourhood structure. In all experiments, the population size is set to 100. As such, for a population of 100 placed on the grid with a neighbourhood of size 100, the spatial structure has no effect. The population is evolved for 150 generations, which is sufficiently long to observe evolutionary results, e.g., persistent periods of cooperation. In each generation, a total of 125 donation games are played. Note that the total number of games played remains the same regardless of neighbourhood sizes. This is to ensure that only the neighbourhood size have an impact on the evolutionary outcome. The probability a random pair is selected to play the game remains unchanged regardless of the neighbourhood size. Our experiment settings are similar to [38], with the only difference being the incorporation of spatial interactions. All experiments are repeated for 250 independent runs.

Figure 12.2 shows the results of the average over 250 runs for the population average payoff for each experiment setting with different neighbourhood sizes. The population average payoff values indicate the degree of cooperation in the evolved population. The maximum average payoff value of 1.1 is obtained in a fully cooperative population (since in each game, the donor obtains a payoff of 0.1 and the recipient receives a payoff of 1.0). As such, Figure 12.2 indicates that the evolved population are highly cooperative given the high population average payoff values. Furthermore, the near zero standard error indicates that in all the runs, the population has evolved to cooperation.

Our result for the experiment with a neighbourhood size of 100 is similar to earlier study [38] using the same settings. The incorporation of spatial interactions in our experiments (neighbourhood sizes of 5, 9, 25, 49, and 81) has no significant impact on the level of cooperation. Past studies [26] have reported that spatial interactions promote cooperation. Consequently, introducing spatial interactions on the setting that is already conducive for the evolution of cooperation has no further impact, which is confirmed by our results.

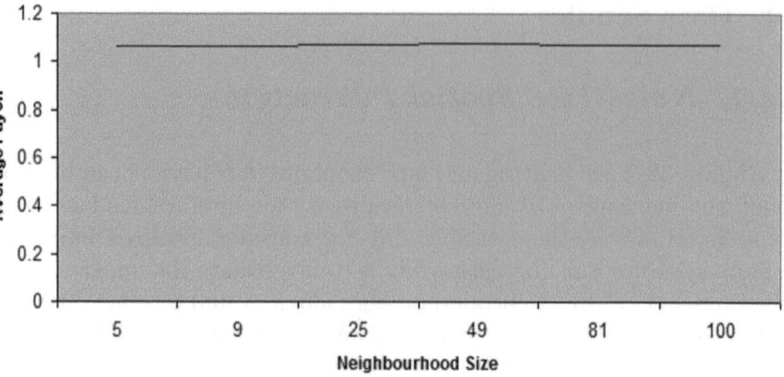

Fig. 12.2. Average payoff against neighbourhood size.

12.4.2 Noisy Spatial Interactions

However, complex real-world interactions are typically noisy. Furthermore, noise has a negative impact on the evolution of cooperation [15]. As such, our case study here will consider noise in spatial interactions having short encounters. In particular, we will investigate our economics-inspired psychic noise model and investigate the impact on the evolution of cooperation. We note in this case study, spatial neighbourhoods take on a new meaning. Since our noise model is dependent on the psychic distance between two neighbours, larger neighbourhoods introduce neighbours that are a further psychic distance away. The probability that an interaction is affected by noise increases with a larger neighbourhood size, especially for those that are located further away towards the edge of the neighbourhood.

As such, there are three factors that affect psychic noise in our model: 1) increasing the noise threshold, 2) increasing the neighbourhood size, and 3) changing the characteristic of the underlying noise function, particularly on the rate at which the noise level (probability that a noisy interaction occurs) increases with an increase in the psychic distance. The following summarizes our experimental results whereby six neighbourhood sizes, six noise functions (Figure 12.1), and all of the 13 threshold values were varied.

In general, increasing the noise level above a threshold level of 0.1 has a negative impact on the evolution of cooperation (i.e., lower population average payoff is observed in Figure 12.3). As an example, results for the experiment set involving the use of the linear noise function show that for nearly all the neighbourhood sizes, an increase in the noise threshold is followed by a decrease in the population average payoff (top-left, Figure 12.3). At lower threshold levels of 0.1 and below, our results indicate little variation in the population average payoff. Closer inspection on the evolved strategy

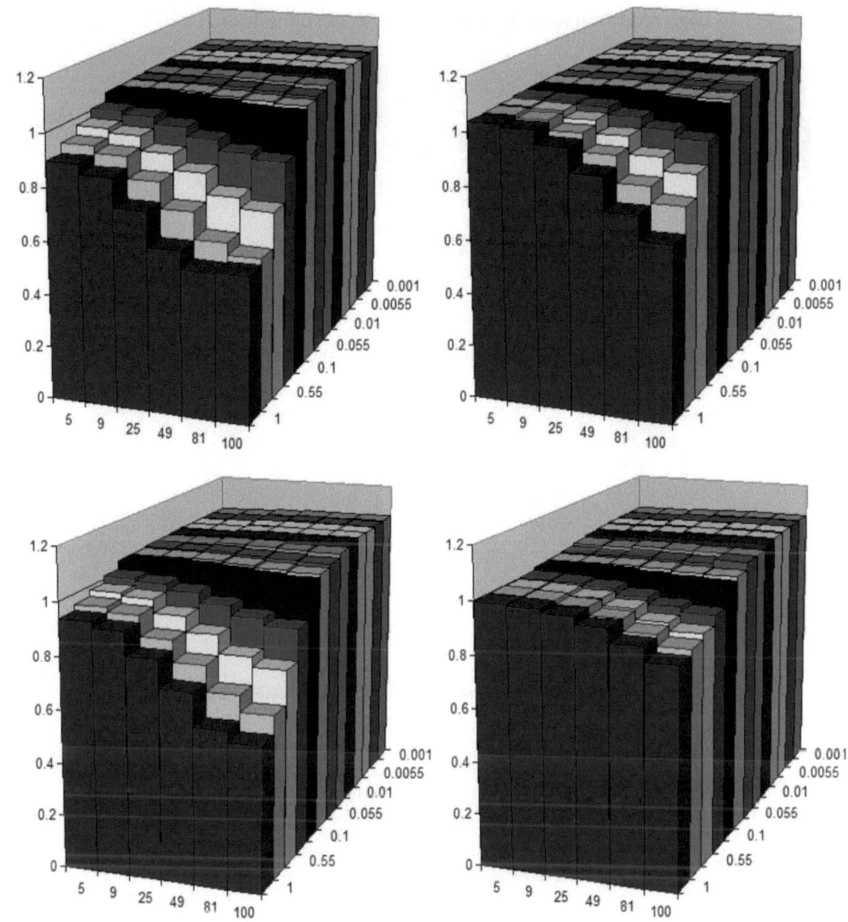

Fig. 12.3. Neighbourhood Size (x-axis) versus Threshold (y-axis) versus Average Payoff (z-axis) for the following functions (left to right starting from the top): linear, polynomial, exponential and sound.

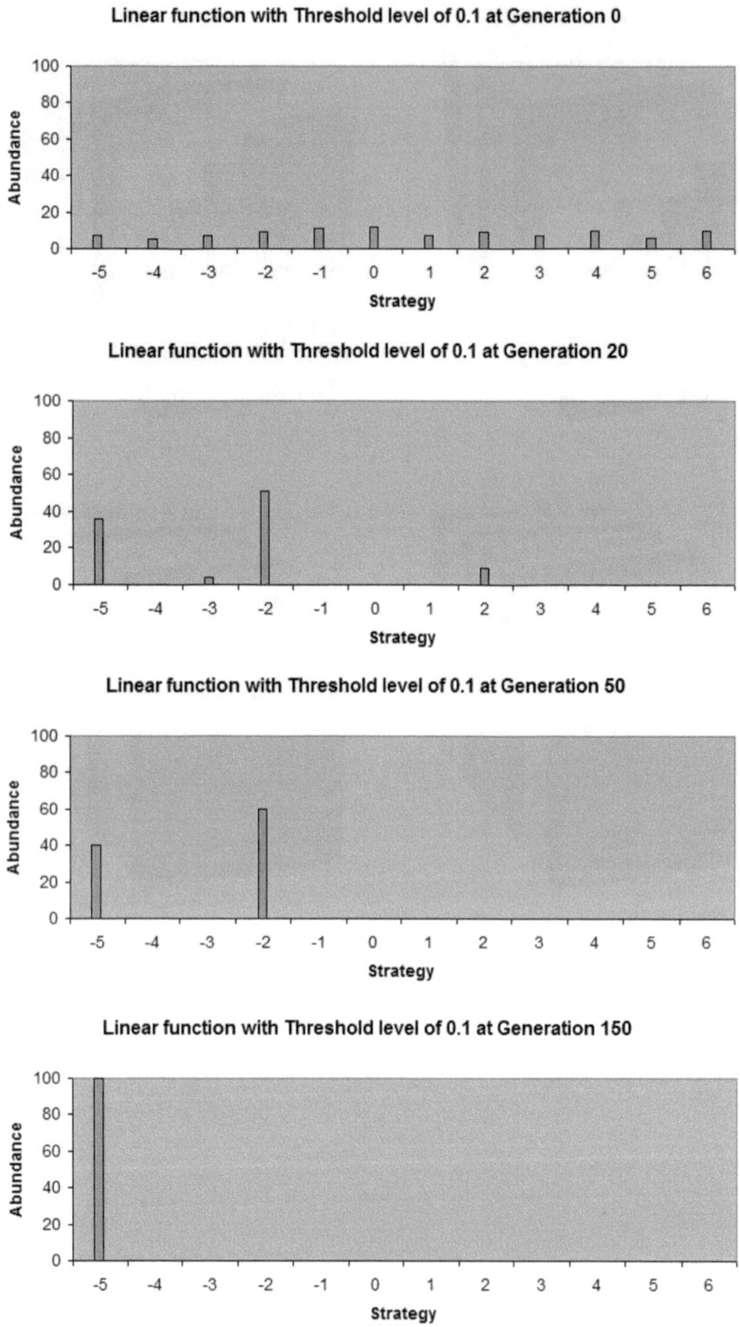

Fig. 12.4. Changes in the proportion of strategies for an evolution cycle with the linear function set to a threshold level of 0.1.

behavior reveals the population evolving to naive cooperators (Figure 12.4). This is consistent with other past studies that studied different models of cooperation (IPD games) where low level of noise has no significant effect on the evolution of cooperation but higher level of noise has a negative impact on the evolution of cooperation [15].

Our results also indicate that decreasing neighbourhood size has a positive impact on the evolution of cooperation. In particular, Figure 12.3 shows that as the neighbourhood size decreases, the population average payoff increases regardless of the noise level in the interactions. This suggests that localized interactions provided by spatial structures facilitate the evolution of cooperation for interactions having short encounters. Our results are also consistent with past studies that investigate other models of cooperation (IPD games) [26].

Finally, our results indicate that noise functions can have different impact on the evolution of cooperation. In particular, noise functions that are characterized by a higher rate of increase in noise levels when psychic distance increases (e.g., sine against linear) have a more negative impact on the evolution of cooperation (Figure 12.5). This is consistent with our expectation that for noise functions that introduce higher noise level for any psychic distance, which is given by the higher rate of increase in noise levels, the higher level of noise will lead to lower cooperative play in the evolving population.

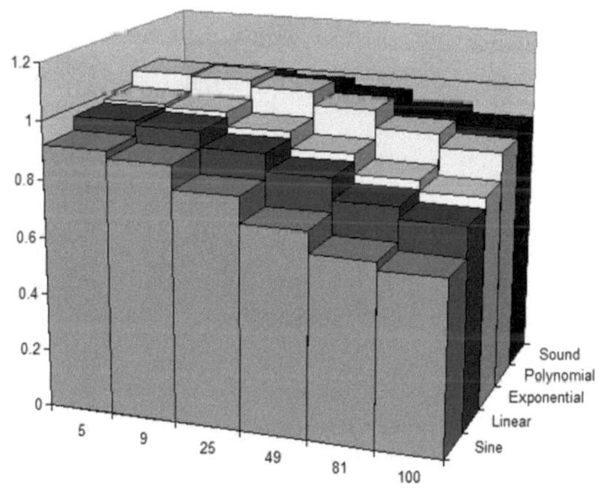

Fig. 12.5 Neighbourhood Size (x-axis) versus Functions (y-axis) versus Average Payoff (z-axis) for the threshold level of 0.55.

12.4.2.1 Mutation

We have not considered the effect of mutation in the evolutionary process in the previous experiments. Here, we introduce various levels of mutation in the experiments at different levels of 0.1%, 1%, 5% and 10%. Mutation results with a single parent generating a single offspring having a different image scoring strategy. A higher mutation level gives a higher probability that the offspring differs from the parent. In general, results for experiments with low mutation levels are qualitatively similar to the results of experiments without mutation. However, as the mutation level is increased, the evolution of cooperation increases as indicated by the increase in population average payoff (Figure 12.6). We note that our use of a simple bit string representation of strategies allow mutation to directly affect behaviors. Subsequently, our results are consistent with past studies [15] whereby increasing the level of behavioral diversity in the population has a positive impact on the evolution of cooperation.

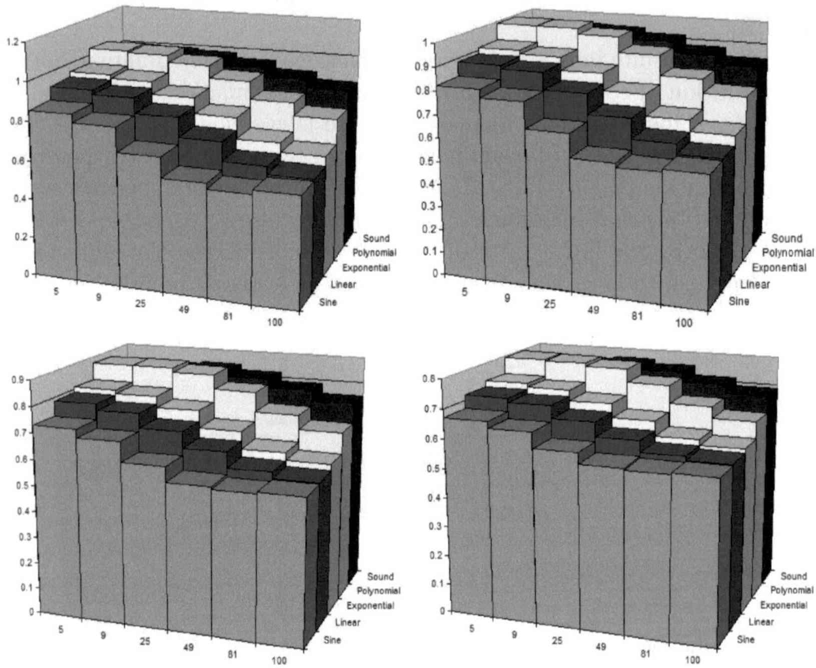

Fig. 12.6. Neighbourhood Size (x-axis) versus Functions (y-axis) versus Average Payoff (z-axis) at a threshold level of 0.775 for the following mutation factors (left to right starting from the top): 0.1%, 1%, 5% and 10%.

12.4.2.2 Various Population Sizes

We also investigate different population sizes in our experiments to determine whether population size can have an impact on the evolution of cooperation. We experiment with two other population sizes: 400 and 900. We have ensured that settings such as neighbourhood size are adjusted so that the probability a random pair is selected to play the game remains the same regardless of the change in population size. Our results show qualitatively little difference between experiments that use different population sizes (Figure 12.7).

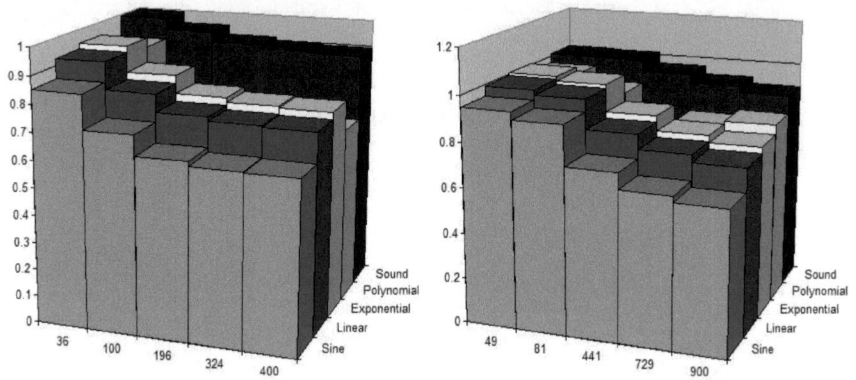

Fig. 12.7. Neighbourhood Size (x-axis) versus Average Payoff (y-axis) versus Functions (z-axis) at a threshold level of 0.55 for the following population sizes (left to right): 400 and 900.

12.4.2.3 Psychic Distance Paradox

Results for the experiment set involving the use of the sine noise function exhibit a certain peculiarity at the threshold level of 1.0 (which is the highest threshold level). At threshold levels below 1.0, results involving the use of sine noise function show the same trend as with other noise functions: the population average payoff decreases with increasing noise level (e.g., when the threshold level of 0.55 is used as shown in Figure 12.7). At the threshold level of 1.0, we observe consistently the population average payoff increasing at very high noise level that is introduced as a result of large neighbourhood interactions after an initial decrease across all the experiments. Such a concave shape can be seen in results for all three population sizes setting (Figure 12.8) and for all the four mutation factors setting (Figure 12.9).

At smaller neighbourhood sizes, the average payoff decreases at first, but increases as the neighbourhood sizes get larger. This is unlike the behavior observed for all the other functions, where the average payoff decreases with an increase in the noise level. This particular observation can be explained in the context of the psychic distance paradox, where less cooperative interactions are observed with neighbours who are psychically closer as compared to neighbours who are psychically further away. For instance, in Figure 12.8, with the population size of 100, it is clearly visible that neighbourhood sizes

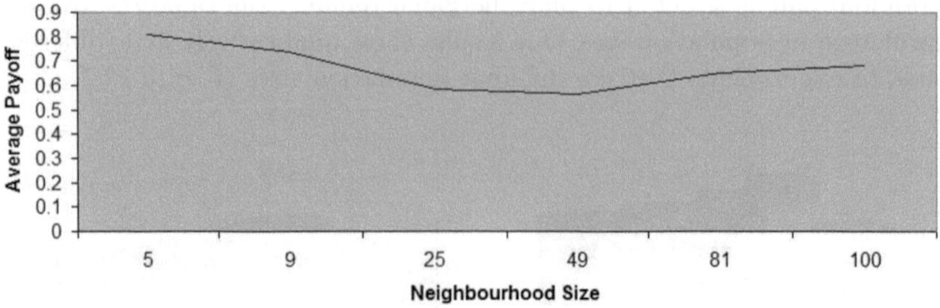

(a) 100

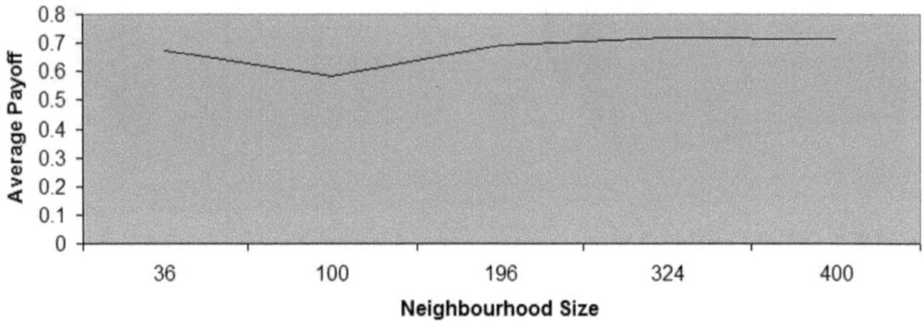

(b) 400

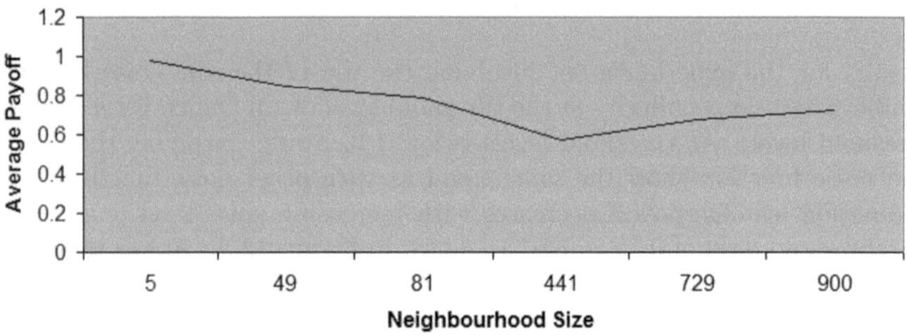

(c) 900

Fig. 12.8. Neighbourhood Size versus Average Payoff for the sine function for the following population sizes (starting from the top): 100, 400 and 900.

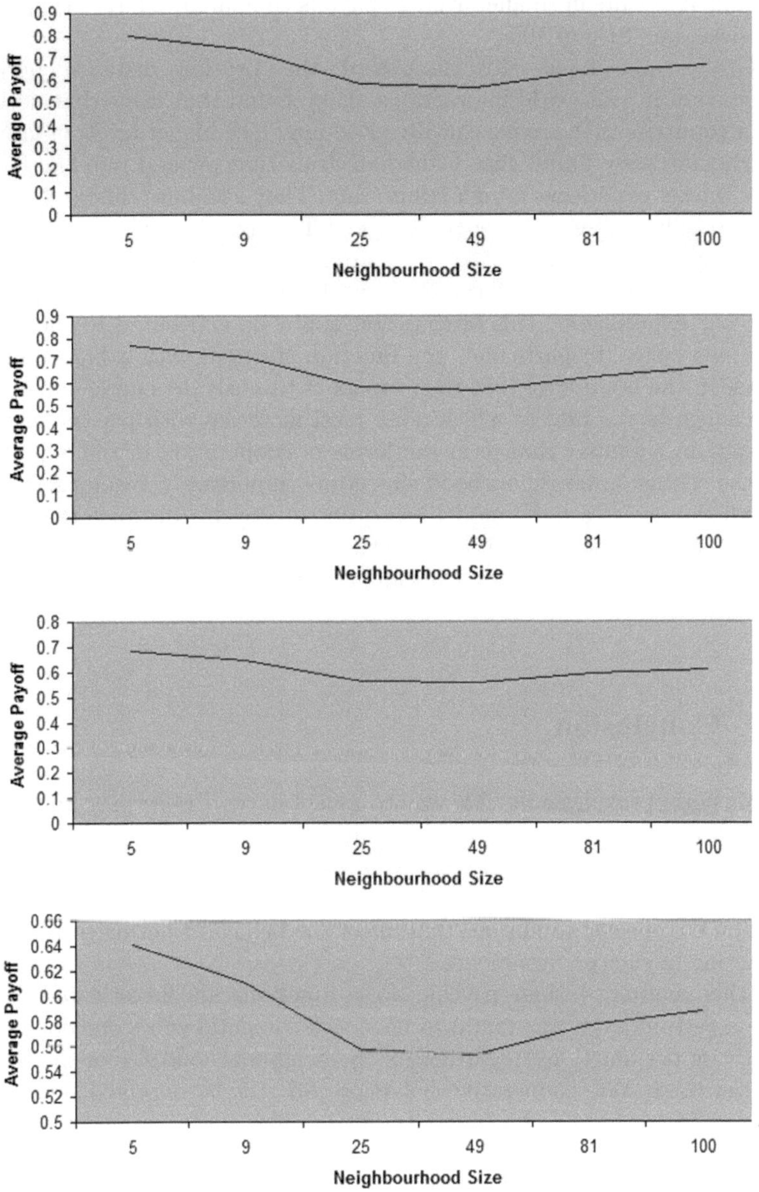

Fig. 12.9. Neighbourhood Size versus Average Payoff for the sine function at a threshold level of 1.0 for the following mutation factors (starting from the top): 0.1%, 1%, 5% and 10%.

of 25 and 49 results in smaller average payoffs compared to larger neighbourhood sizes, i.e., 81 and 100.

O'Grady and Lane [40] discovered the psychic distance paradox phenomenon in real-world interactions. They found that firms that do business in countries that are psychically close may face higher levels of failures. In particular, they found that Canadian firms that expand into the United States market experience a high failure rate. They attribute this to the firm's failure to fully realise and prepare for the differences that exist even between psychically close countries. Also, they argue that Canadian managers make incorrect assumptions about the U. S. market based on their home market.

For our experiments, this observation might be attributed to the nature of the sine curve. In particular, the function starts of with a higher rate of increase in the noise level that then tapers of towards the end (Figure 12.1). This change in the rate at which noise level increases with psychic distance may explain a similar change in the levels of cooperation in the population with the change in neighbourhood size. More importantly, results show that through the use of a noise model based on the psychic distance that better reflects real-world interactions, interesting and unexpected behavior can be produced. Furthermore, these behaviors are reminiscent of those observed in the real-world.

12.5 Conclusion

Psychic noise provides a flexible way to model noise. Unlike past approaches where all interactions involve the same amount of noise, the psychic noise is modelled after the real-world concept of psychic distance. It provides scope for further studies on interactions in more realistic environments and new insights on the specific conditions that affect the behaviors between individuals interacting in such environments.

In this chapter, various psychic noise functions are investigated. Noise, above a certain level, was found to have a detrimental effect on cooperation for most of the functions, regardless of the neighbourhood size or the threshold level used. Yet, cooperative behavior can still be observed due to the positive effects of indirect reciprocity. Introducing mutation and using various population sizes showed little difference in the observed results. However, the sine function was found to lead to behaviors in a counter-intuitive manner at very high noise threshold levels, reminiscent of the psychic distance paradox.

The psychic noise provides a more realistic modelling environment of real-world interactions compared to conventional noise models. Future studies may consider incorporating other extensions and study other noise models (e.g., noise that reduces on repeated interactions).

References

1. Alexander, R.D.: The Biology of Moral Systems. Aldine de Gruyter, New York (1987)
2. Axelrod, R.: Effective choice in the iterated prisoner's dilemma. J. Confl. Resolut. 24, 3–25 (1980)
3. Axelrod, R.: More effective choice in the prisoner's dilemma. J. Confl. Resolut. 24, 379–403 (1980)
4. Axelrod, R.: The Evolution of Cooperation. Basic Books, New York (1984)
5. Axelrod, R., Dion, D.: The further evolution of cooperation. Science 242(4884), 1385–1390 (1988)
6. Axelrod, R., Keohane, R.: Achieving cooperation under anarchy: strategies and institutions. World Politics 38(1), 226–254 (1985)
7. Axelrod, R., Hamilton, W.D.: The evolution of cooperation. Science (211), 1390–1396 (1981)
8. Back, T.: Evolutionary Algorithms in Theory and Practice, p. 117. Oxford Univ. Press (1996)
9. Beckerman, W.: Distance and the pattern of inter-european trade. Rev. Econ. Stat. 38(1), 31–40 (1956)
10. Brandt, H., Sigmund, K.: Indirect reciprocity, image scoring, and moral hazard. Proc. Natl. Academy of Science (USA) 102, 2666–2670 (2005)
11. Brauchli, K., Killingback, T., Doebeli, M.: Evolution of cooperation in spatially structured populations. J. Theor. Biol. 200, 405–417 (1999)
12. Brown, S.: The superpowers' dilemma: can game theory improve the U.S.-Soviet negotiating relationship? Negotiation J. 2(4), 371–384 (1986)
13. Chellapilla, K., Fogel, D.B.: Evolution, neural networks, games and intelligence. Proc. IEEE 87(9), 1471–1496 (1999)
14. Chong, S.Y., Humble, J., Kendall, G., Li, J., Yao, X.: The Iterated Prisoners' Dilemma: 20 Years On, World Scientific Rev., ch. 3, pp. 63–87 (2008)
15. Chong, S.Y., Yao, X.: Behavioral diversity, choices, and noise in the iterated prisoner's dilemma. IEEE Trans. Evol. Comput. 9(6), 540–551 (2005)
16. Chong, S.Y., Yao, X.: Self adapting matrices in repeated interactions. In: IEEE Symp. Comput. Intel. and Games, pp. 103–110 (2006)
17. Chong, S.Y., Xin, Y.: Multiple choices and reputation in multiagent interactions. IEEE Trans. Evol. Comput. 11(6) (2007)
18. Darwen, P.J., Yao, X.: Does extra genetic diversity maintain escalation in a co-evolutionary arms race. Intl. J. Knowledge-Based Intel. Eng. Sys. 4(3), 191–200 (2000)
19. Donninger, C.: Paradoxical effects of social behavior, p. 123. Physica-Verlag, Heidelberg (1986)
20. Eiben, A.E., Aarts, E.H., Van Hee, K.M.: Global convergence on genetic algorithms: an infinite Markov chain analysis. In: Schwefel, H.-P., Manner, R. (eds.) Parallel Problem Solving from Nature, pp. 4–12. Springer, Heidelberg (1991)
21. Flood, M.M.: Some Experimental Games, Research Memorandum RM-789. The RAND Corp., Santa Monica, California (USA) (1952)

22. Fogel, D.B.: Evolving behaviors in the iterated prisoner's dilemma. Evol. Comput. 1(1), 77–97 (1993)
23. Fogel, D.B.: An introduction to simulated evolutionary optimization. IEEE Trans. Neural Netw. 5(1), 3–14 (1994)
24. Goldstein, J.: Reciprocity in superpower relations: An empirical analysis. Intl. Stud. Q. 35, 195–209 (1991)
25. Hosseini, H.: Psychic distance, psychic distance paradox and behavioral economics: modeling MNC entry behavior in foreign markets. J. Socio-Econ. 37, 939–948 (2008)
26. Ishibuchi, H., Namikawa, N.: Evolution of iterated prisoner's dilemma game strategies in structured demes under random pairing in game playing. IEEE Transactions on Evolutionary Computation 9(6), 552–561 (2005)
27. Johanson, J., Vahlne, J.E.: The internationalization process of the firm, a model of knowledge development and increasing foreign market commitment. J. Intl. Bus. Stud. 8, 22–32 (1977)
28. Julstrom, B.A.: Effects of contest length and noise on reciprocal altruism, cooperation, and payoffs in the iterated prisoners dilemma. In: Proc. 7th Intl. Conf. Genetic Algorithms (ICGA 1997), pp. 386–392. Morgan Kaufmann, San Francisco (1997)
29. Lee, D.J.: Effect of cultural distance on the relational exchange exporters and importers. J. Glob. Market. 11(4), 7–22 (1998)
30. Leimar, O., Hammerstein, P.: Evolution of cooperation through indirect reciprocity. Proc. Royal Society of London, Series B 268, 745–753 (2001)
31. Lindgren, K., Johansson, J.: Coevolution of strategies in n-person prisoner's dilemma. In: Crutchfield, J., Schuster, P. (eds.) Evolutionary Dynamics- Exploring the Interplay of Selection, Neutrality, Accident, and Function. Addison-Wesley (2001)
32. Linnemann, H.: An International Study of International Trade Flows. North-Holland, Amsterdam (1966)
33. Lumsden, M.: The Cyprus conflict as a prisoner's dilemma. J. Confl. Resolut. 17, 7–32 (1973)
34. Maynard Smith, J.: Evolution and the Theory of Games. Cambridge University Press (1982)
35. Nakamaru, M., Matsuda, H., Iwasa, Y.: The evolution of cooperation in a lattice-structured population. J. Theor. Biol. 184, 65–81 (1996)
36. Nordstrom, K., Vahlne, J.E.: Is the globe shrinking? Psychic distance and the establishment of Swedish sales subsidies. In: Landeck, M. (ed.) Intl. Trade: Regional and Global Issues. St. Martin Press, New York (1994)
37. Nowak, M.A., Sigmund, K.: The dynamics of indirect reciprocity. J. Theor. Biol. 194, 561–574 (1998)
38. Nowak, M.A., Sigmund, K.: Evolution of indirect reciprocity by image scoring. Nature 393, 573–577 (1998)
39. Nowak, M.A., Sigmund, K.: Evolution of indirect reciprocity. Nature 437, 1291–1298 (2005)
40. O'Grady, S., Lane, W.: The psychic distance paradox. J. Intl. Bus. Stud. 27(2), 309–333 (1996)

41. Rudolph, G.: Convergence properties of canonical genetic algorithms. IEEE Trans. Neural Netw. 5(1) (1994)
42. Seo, Y.G., Cho, S.B., Yao, X.: Emergence of cooperative coalition in NIPD game with localization of interaction and learning. In: Proc. Cong. Evol. Comput. (CEC 1999), pp. 877–884 (1999)
43. Sigmund, K., Nowak, M.A.: Evolutionary game theory. Current Biology 9, R503–R505 (1999)
44. Tucker, A.W.: A Two-Person Dilemma. Stanford Univ. (USA) (1950)
45. Wolfram, S.: A New Kind of Science. Wolfram Media, Champaign (2002)
46. Wu, J., Axelrod, R.: How to cope with noise in the iterated prisoner's dilemma. J. Confl. Resolut. 39, 183–189 (1995)

Chapter 13
Intelligent Modeling and Control for Autonomous Logistics

Tobias Warden, Janusz Wojtusiak, and Otthein Herzog

Abstract. The increasing complexity of logistic networks calls for a paradigm change in their modeling and operations. Centralized control is no longer a feasible option when dealing with extremely large systems. For this reason, decentralized autonomous systems are gaining popularity in providing robustness and scalability. This chapter focuses on the use of intelligent systems in autonomous logistics. Specifically, it describes issues related to knowledge management, a machine learning-based approach to adaptability and planning, and intelligent optimization by autonomous logistics entities.

13.1 Introduction

The increasing pace of structural change in today's globalized markets bears significant implications for adequate planning and control strategies in transport logistics. Traditional supply chains are evolving into complex networks with numerous stakeholders. Three main influences that characterize these changes have been identified: the goods structure effect, the logistics effect, and the structural effect. The first effect corresponds to a shift away from mass production towards a buyers' market, which creates a trend towards individual product customization and consequently a noticeable increase in per-unit shipments. The second relates to a shift towards road freight

Janusz Wojtusiak
Machine Learning and Inference Laboratory, George Mason University
Fairfax, VA, USA
e-mail: `jwojtusi@gmu.edu`

Tobias Warden · Otthein Herzog
Center for Computing and Communication Technologies, Universität Bremen
Bremen, Germany
e-mail: `warden@tzi.de, herzog@tzi.de`

J. Kołodziej et al. (Eds.): Advances in Intelligent Modelling and Simulation, SCI 422, pp. 295–325.
springerlink.com

transport, which arises from the increasing demands for small shipments along with a high quality of service and due-date reliability. Finally, the structure effect indicates an individualization of transport on the micro-logistics level. Cooperation is needed between otherwise competing logistics service providers to satisfy today's customer needs. These three effects lead to a dramatic increase in the complexity and dynamics of today's transport logistics processes, which pose a growing challenge for the traditional paradigm of centralized process management. The widespread spatial distribution of transport nodes within the networks, as well as the link-up of competing logistic service providers in large supply-networks, restrict or even inhibit the provision of crucial information for central decision-making instances.

In recent years, the changes in many areas of logistics have led to a growing interest to create novel methods for process control, which scale naturally as managed processes grow both with regard to the number of involved stakeholders and geographic distribution. This interest has sparked multidisciplinary research that combines a variety of fields such as operations research and artificial intelligence [5]. One recurring theme, which has been widely adopted in the research community, is a gradual shift away from a hierarchical organization of control. As it becomes more and more difficult to concentrate all information and knowledge required for a top-down management of logistic processes, the idea to distribute the responsibility for the control of well-defined sub-processes has gained popularity. This idea of distribution allows for decentralized decision-making. A general term which has been proposed in [11, 10], is *autonomous cooperating logistic processes.*

Specifically, Böse and Windt have proposed a definition for the paradigm of autonomous control, stating that *"Autonomous Control describes processes of decentralized decision-making in heterarchical structures. It presumes interacting elements in non-deterministic systems, which possess the capability and possibility to render decisions independently."* [1, p. 58]. An important concept that is brought into focus, by its close ties to the definition of autonomous control, is the specific category of process owners. Traditional process control depends on skilled human controllers and managers who rely on dedicated IT-systems for decision support and their own domain expertise. Autonomous control, however, implies that processed logistics entities themselves can be augmented with decision-making capabilities. Given the preceding, the strength of the new control paradigm can hence be stated as follows: under the assumption that smart versions of logistics entities (material or immaterial) can be devised, their respective on-board decision-making capabilities should be sufficient for these entities to route themselves autonomously through complex logistic networks. This concept applies to all but few special conditions, in which human intervention is still required. Thus, instead of trying to design control schemes which seek to manage large amounts of individual processes from above, the control problem is divided into manageable sub-problems that are easier to address on a local level. The challenge then lies in ensuring that the concurrent execution and coordination of the

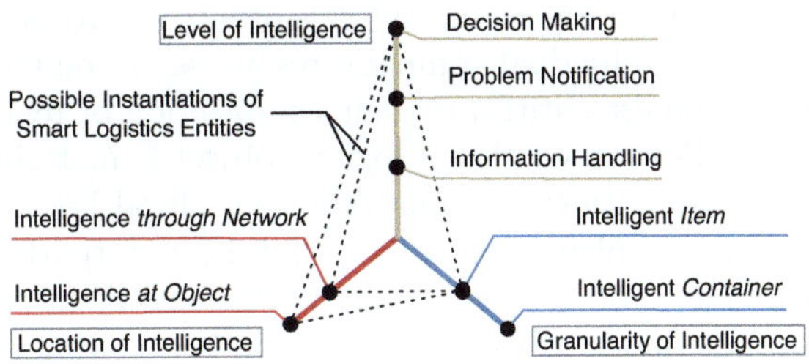

Fig. 13.1 Classification of Intelligent Products, based on Gerben et al. [16]. Note that the term *intelligent container* refers to one digital representative acting on behalf of a collection (hence: container) of physical objects.

distributed control processes by the smart logistics entities results in a satisfactory behavior of the complete system measured by an application-specific system of logistic key performance indicators. The benefits of autonomous control include local decision-making and handling of process disruptions, which in turn enable dependable and scalable process control.

The concept of smart logistics entities plays a key role in the paradigm of autonomous logistics. Therefore, it is necessary to initially provide a classification of such smart entities. Consequently, two categories of smart *physical* logistics entities have been distinguished in the literature, namely intelligent products and smart resources. McFarlane et al. define the term intelligent product as *"a physical and information based representation of an item [...] which possesses a unique identification, is capable of communicating effectively with its environment, can retain or store data about itself, deploys a language to display its features, production requirements, etc., and is capable of participating in or making decisions relevant to its own destiny."* [15] The level of intelligence an intelligent product may exhibit varies from simple data processing to complex pro-active behavior. Gerben et al. introduce a classification scheme for intelligent products, which considers location and granularity of intelligence [16] (see Figure 13.1). In production engineering, so-called holonic manufacturing systems (HMS) have been proposed as autonomous, co-operative building blocks of manufacturing systems for transformation, transport, storage and validation of information and physical objects [27]. Each HMS is understood as a composite entity (hence, a holon), which typically comprises an information-processing part and a physical processing part.

With regard to the categorization by Gerben (see Figure 13.1), smart logistics entities need to exhibit a level of intelligence which allows for autonomous decision-making. This subsumes problem notification and interaction with humans in charge as well as information handling. As for its location, it may be feasible to place an information system for autonomous control directly onto a physical represented object such as handling equipment, freight containers, or means of transport [12]. In other use cases, it suffices to have the information system access real-time information from its represented object via network connections. Finally, the decision concerning the right level of modeling granularity remains an active area of research. While some researchers, for instance in supply chain management, model complete forwarding agencies as single intelligent entity [26], other researchers employ a finer granularity down to individual company resources or commodities [24].A common trait in the smart entities is their composition of real world and control aspects. Regardless of whether a logistic object is material (e.g., enterprise resources or commodities) or immaterial (e.g., legal bodies or haulage requests), it is always complemented with a digital counterpart. Only then, can it be constituted as a smart logistics entity. One important prerequisite for these entities, besides information processing and decision-making, is the capability to co-operate and adapt to novel situations.

13.2 Multi-agent Control of Logistic Processes

Software agents are suitable for both the conceptual design, and technical implementation of real-world logistics entities. In simple form, an agent is defined by Wooldridge as: *"a computer system that is situated in some environment, and that is capable of autonomous action in this environment in order to meet its design objective."* [39, p. 15]. Wooldridge also proposed a "weak" notion of agency that is widely accepted in the agent community. It specifies the minimum criteria for intelligent agents [40, p. 116]. These criteria are: 1) autonomy, 2) reactivity, 3) pro-activeness, and 4) social ability.

These criteria meet the requirements for smart logistics entities. Autonomy and pro-activeness are mandatory to route a managed object through logistic processes. Reactivity allows for the reaction to unforeseen events, which threaten to disrupt current plans, by initiation of appropriate problem handling or re-planning. Finally, social ability is critical in situations in which agents need to cooperate or form teams to achieve their objectives.

A large number of agent architectures have been proposed in the literature [23, 3]. On one side, agents can be designed relatively simple, lightweight, and reactive. Their behavior is characterized in terms of stimulus-response patterns[1]. The other end of the spectrum is constituted by complex cognitive agents, whose architectures aim at emulating human cognitive processes. According to Wooldridge, such *intelligent agents* are conceived or implemented using concepts drawn from human context [40]. Such agents may be characterized using mental notions such as beliefs, desires, and intentions (BDI) [2], and are often implemented as BDI-agents [23]. Bounded rationality is one of the fundamental agent characteristics. Given current beliefs about its environment, and its preferences, an agent must act in accordance with its goals.

Design and development of agent-oriented approaches to autonomous logistics, especially on the level of smart logistics entities, often benefits from natural mappings of real-world entities and decision-makers onto software agents as actors and their associated roles. One paradigmatic example of

[1] See, for instance, swarm intelligence or Brooks subsumption architecture [3].

such a mapping has led to the design of the transport management agents in Section 13.3. Each such agent represents a resource (a truck), and assigned primary logistic roles including *operative container transport*, and *individual transport handling* as shown in Figure 13.4 on page 306. Once actors and roles have been mapped for a use case, the focus is then on the agent behaviors.

13.2.1 Knowledge Management in Autonomous Logistics

Intelligent agents, in autonomous logistic processes, need substantial diverse knowledge to effectively handle their primary roles in the modeled logistic processes with the desired quality of service.

Forms of Knowledge. Agents' use of knowledge ranges from declarative knowledge, e.g., ontologically modeled background knowledge [29], to operative knowledge for support of decision processes [8]. The knowledge required for the operation of a MAS can be prepared at the design stage in a knowledge engineering process, in which knowledge from human experts is acquired. The knowledge is usually represented in machine-understandable formats such as formal ontologies or rules. Models for decision support may be either devised by hand or preferably learned offline from historical data acquired from external sources or gathered specifically in the targeted application domain. The knowledge can then be made available to actors in an MAS as joint prior knowledge. In many cases, in order to accommodate for the complexity and dynamics of logistic environments, the initial provision of default knowledge alone is not sufficient. Rather it is necessary to design agents as adaptive systems, capable of agent-oriented knowledge management, e.g., autonomous knowledge revision and the compilation of additional empirical or learned knowledge. Hence, the available knowledge to an increasing degree becomes tailored to the context in which the agents are deployed.

Agent-oriented Knowledge Management. The term *Agent-oriented Knowledge Management* (AKM) refers to a collection of processes that an intelligent agent carries out throughout its activity cycle to maintain, adapt, and possibly augment its empirical and, to a lesser extent, prior knowledge through knowledge acquisition. Moreover, the technical term also comprises processes such as goal-directed information acquisition and the ascertainment of new knowledge by means of inference on using predictive or decision models. AKM bears analogies to *Personal Knowledge Management.* While the latter focuses on human decision makers, both approaches have similarities in the considered skill sets. Examples comprise self-reflection, management of individual learning, information literacy, as well as communication and collaboration. We understand AKM as a form of distributed knowledge management on the part of the individual actors within an MAS. Consequently, Langer et al. have proposed a KM framework based on the concept of both internal and external KM roles that agents can assume dynamically in addition to their primary logistic roles as the need arises [14]. In that approach, the agents within the MAS may also form knowledge networks and pursue strategies for demand-oriented knowledge transfer.

The agents, considered here, for the most part act as digital representatives on behalf of logistics entities. The agents need to perform their primary logistic activities, right from the moment of their instantiation. Therefore, the KM functions constitute *auxiliary* activities that are loosely coupled to the *primary* logistics functions. Sometimes, these auxiliary functions are executed due to independent incentives.

Knowledge Management in Decision Situations. In the face of a concrete decision situation, AKM enables an agent to gather and process raw data from approachable sensors. The sensors can be deployed on the physical object managed by the agent, as in the case of an intelligent container. Because of the context known to the agent, this raw data can then be '*elevated*' to information. Conjointly with supplemental information acquired from cooperating agents or other external information sources, such as EPC Information Services, the gathered information is filtered for relevance and integrated with the belief base of the respective agent. Subsequently, the knowledge can then be exploited. Alternatively, information may provide necessary inputs for the agent's decision support models. These are typically queried in the decision process in order to infer new knowledge. In fact, it is often the case that the demand for information acquisition arises as a direct consequence of the desire to perform inferences on decision support models. The KM functions described above are triggered by the primary logistics agent activities.

Knowledge Management with Strategic Background. The second set of knowledge management functions revolves around the initial construction of tailored decision support models at the outset of an agent activity cycle and their subsequent adaptation for new experience. These functions are typically done detached from particular decision situations. As a consequence, it is often feasible to perform these functions concurrently to the execution of an agent's primary logistic roles. The absence of trigger events out of the operational day-to-day business implies that their scheduling needs to be managed pro-actively by the agent. Figure 13.2 highlights the relationship between the primary logistic roles of an agent and auxiliary knowledge management roles. In order to perform well in a task such as the handling of container haulage, the decision-making of the agent relies on decision support models (e.g., to estimate expected trans-loading times). During its operation, the agent gathers data, that can be stored in its '*experience database*'. Thus, the data necessary to start learning updated versions of currently operationalized models is acquired on the fly. Within the scope of its knowledge management roles, the agent can then initialize learning processes, which include: complete workflow of data selection and preprocessing; the conduct of the actual machine learning scheme; comparative evaluation of the

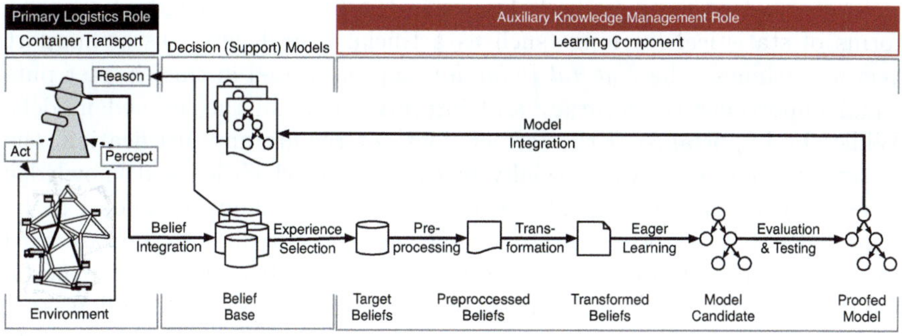

Fig. 13.2. Relationship between primary agent roles, knowledge management, and predictive modeling for decision support.

resulting model candidate; and potentially its operationalization. The process, as shown in Figure 13.2, is taking its cue from the process for knowledge discovery in large databases (KDD). In opposition to typical KDD use cases, the agents need not only search for interesting patterns in data (i.e., a model) but must implement closed-loop learning by feeding learned models back for operative use.

13.2.2 Machine Learning in Autonomous Logistics

The preceding section pointed to the role of ML schemes in the context of AKM. ML is a broad area that combines related disciplines such as artificial intelligence, statistics, data mining, robotics and cognitive science. For the purpose of this chapter it is sufficient to define learning as a process, in which an agent uses its individual experiences and potentially external knowledge to modify its behavior or increase knowledge needed to perform specific tasks.

The ML methods throughout this chapter fall into the class of supervised learning techniques. They are used to create models for accurate prediction of the future state of the environment. The agents learn and use multiple models to predict the environment in the future. Consider an example, further elaborated in Section 13.3.1, in which a transportation agent plans an optimized route to its destination. In addition to road network and current status of roads, the agent also needs to know the status of certain road segments in the near future. Current road statuses are useless for distant destinations to which the agent will be arriving several hours in the future.

The ML task considered here can be described as follows: Given a set of labeled examples (X_i, Y_i) and background knowledge, the goal is to induce a model $M(X) \rightarrow Y$, to predict the label of an output attribute Y based on an input feature vector X. If the output attribute Y is discrete the process is called classification learning. When Y is continuous the learning task is called predictive or regression learning.

In contrast to most methods known in the literature, including different forms of statistical learning (such as artificial neural networks or support vector machines), the *Natural Induction* approach used in this chapter puts equal importance on accuracy and interpretability of the learned models. While the importance of the former does not require any justification, the latter may not be clear, especially in the area in which learned knowledge is used by autonomous agents. Models learned by many methods can be regarded as a 'black box' which may give very good predictions, but are hard to understand and often impossible to validate or refine by human experts. On the other hand, models described in a language that is easy to understand, such as natural language or easy to interpret rules, can be modified by experts to reflect their background knowledge and improve predictions. Transparent models are also available on hand, when an audit trail is needed to keep track of the behavior of a system.

13.2.2.1 Natural Induction and Attributional Calculus

Natural Induction is an approach to inductive learning, whose goal is to achieve high understandability of derived knowledge [18]. It uses a highly expressive language which is called *Attributional Calculus* (AC) which combines predicate, propositional, and multi-valued logics. Because all forms of knowledge used in AC correspond to constructs in natural language, its expressions are easy to understand. The main form of knowledge in AC are attributional rules. Such rules follow the '*if* ... *then* ...' schema, but are more general than rules learned by most learning programs. Attributional rules use more expressive language, which allows for the creation of simplified descriptions as opposed to normal rules. A basic form of attributional rules is:

$$\text{CONSEQUENT} \Leftarrow \text{PREMISE} \tag{13.1}$$

Here, CONSEQUENT and PREMISE are conjunctions of attributional conditions (a.k.a. complexes). We used attributional conditions in the form [L *rel* R] where L is an attribute; R is a value, a disjunction of values, or a conjunction of values if L is a compound attribute; *rel* is a relation that applies to L and R. Other forms of attributional conditions may involve count attributes, simple arithmetical expressions, conjunctions and disjunctions of attributes, comparison of attributes, etc. [37].

A simple form of natural induction is implemented in the AQ21 system [37]. It is the newest version of the AQ family of ML systems. Given input data, problem definition, and optional background knowledge, AQ21 induces rules in the form of Equation (13.1), or in more advanced forms. These describe one or more classes in the data. A set of rules constituting a description of a given class is called a *ruleset*. By repeating learning for all classes defined by values of an output attribute, AQ21 generates a classifier.

In order to learn rules for a given class, AQ21 starts with one example, called a *seed* which belongs to the class. It generates a *star*, which is a set of maximally general rules that cover the seed and do not cover any examples from other classes. This is done by repeating an *extension-against* operation that generalizes the seed against examples not belonging to the concept being learned. Results of applying the extension-against operation are intersected, and the best rules are selected according to user-defined criteria. If selected rules do not cover all examples belonging to the class, another seed is selected (from the not covered examples) an additional rules are learned. The process is repeated until all examples of the class are covered by the learned rule set. AQ21 and other AQ implementations include several modifications to the above basic algorithm as described, for example, in [37]. Most prominent are the ability to use several seeds to protect the method against noise, employ different concept representations (attributional or relational), generate rules with different interrelationships (independent, disjoint or ordered covers), use different methods for handling data inconsistency (minimum, maximum, free and statistic-based generalization), learn rules in a batch of incremental modes, seek rules that represent the best trade-off between their consistency, coverage and simplicity, use different criteria of rule optimality, involve operators for deriving more relevant attributes (data-driven, hypothesis-driven or multistrategy constructive induction), apply prior knowledge (e.g., using knowledge-driven constructive induction), post-optimize learned descriptions, generate single or alternative descriptions, and learn rules with exceptions.

13.2.2.2 The Learnable Evolution Model

The *Learnable Evolution Model* (LEM) is an evolutionary optimization method that employs ML to guide the evolutionary process [17, 36]. Specifically, LEM creates general hypotheses identifying regions in the search space that likely contain optimal solutions and then instantiates these hypotheses to generate new candidate solutions.[2] In order to apply machine learning, LEM creates two groups of candidate solutions that are respectively high- and low-performing according to the fitness function(s) being optimized. These candidate solutions can be selected from the current population or a union of current and past populations of individuals. The group of high-performing individuals is called H-Group. The group of low-performing individuals is called L-Group. Once the groups are selected, LEM applies concept learning to create a general hypothesis describing the H-Group in contrast to the L-Group. The hypothesis is then instantiated to create new candidate solutions. In the final step, a new population is assembled from old and new candidate solutions, and the process is repeated until stopping criteria are met.

[2] In evolutionary computation, candidate solutions are often referred to as individuals.

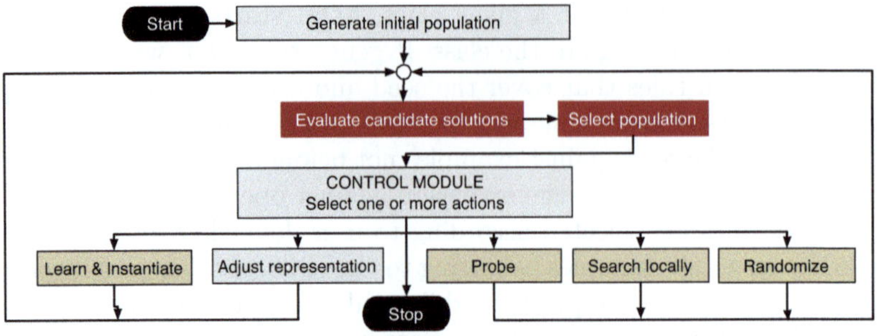

Fig. 13.3. The LEM3 top-level algorithm, adapted from [19].

The Basic LEM Algorithm. The LEM algorithm consists of elements for evolutionary computation, such as the generation of an initial population of candidate solutions, their evaluation, and selection. However, new candidate solutions are generated in a 3-step process based on machine learning:

Step 1: Example preparation, is the process whereby candidate solutions are split into H- and L-groups. This split is done based on the solution performance in terms of fitness values, constraints, and population diversity. Several methods for selecting examples have been investigated [17, 36, 35], including the combination of solutions from the current and previous populations. For simplicity, one can assume that a selected number of top solutions constitute the H-Group, and bottom ones become L-Group, although in practice more sophisticated methods can be used.

Step 2: Hypothesis formation, is the process where machine learning is applied to form a hypothesis describing why some candidate solutions perform better than others. In general, any ML method that generates easily instantiable hypotheses can be used (see next step). For this reason, LEM3 as well as previous LEM implementations use a rule-based hypothesis representation. Specifically, LEM3 applies the AQ21 rule learning system for this task.

Step 3: Hypothesis Instantiation, a.k.a. sampling, is the process during which new candidate solutions are created that satisfy the learned hypotheses. The default algorithm used in LEM3 selects with uniform distribution values (both numeric and symbolic) for attributes that are included in the PREMISE. For attributes that are not included in the PREMISE, the method chooses values from a randomly selected existing non-L-Group candidate solution. Other methods are discussed by Wojtusiak [36].

LEM3 Extensions. Successful initial implementations of the learnable evolution model sparked development of the third generation of LEM software. LEM3 extends many ideas found in the original LEM methodologies, some of which are unique in the field of evolutionary computation. The general flow diagram of the algorithm is presented in Figure 13.3. In addition to components found in standard evolutionary computation methods, such as generation of an initial population, evaluation of solutions, and selection of candidate solutions; LEM3 includes several additional components. It automatically selects one or more innovation methods to create new candidate solutions:

- *Learn & Instantiate*: to create new candidate solutions through the aforementioned machine learning mechanism;
- *Probe*: to apply traditional operators such as mutation and crossover;
- *Search locally*: to apply a user-defined local search method;
- *Randomize*: to add to the current population a number of randomly created candidate solutions, or restart the evolutionary process.

One of the novel features of LEM3 is its ability to automatically adjust the representation space through *Constructive Induction* [34, 35]. The original representation space provided to an ML, data mining, or evolutionary computation system may be inadequate for performing the desired task for concept learning, pattern discovery, optimization, etc. Constructive induction methods automatically create new representation spaces based on the original representations. The new representations allow for the determination of relationships that could not be represented in the original spaces. Since LEM relies upon machine learning to guide the evolutionary process, finding such relationships is crucial. The constructive induction process can be characterized by a function which maps the original representation space to a modified representation space. New representations are thereby created by removal of attributes that are irrelevant to the considered problem, by modification of attribute domains (e.g., by discretizing numeric attributes), and by the creation of new attributes.

13.3 Autonomous Control in Pickup and Delivery Operations

Earlier sections described the basics of AKM and decision support in complex task environments, such as autonomous logistics. Agents that are acting autonomously and over extended periods of time need the ability to learn tailored models for accurate situation assessment and prediction of future events. The agents also need adequate optimization methods that effectively exploit acquired knowledge to compute high-quality action plans. These two aspects of intelligent decision-making shall be fleshed out using the example of an MAS to represent a freight forwarder that handles container on-carriage.

The freight forwarding agency needs to a handle continuous dynamic flow of inbound haulage requests, which pertain to standard freight containers. New requests arrive to the system on a regular basis to the storage facilities that are physically distributed over large geographic areas, such as the territory of Germany. Two subsequent transport orders need to be handled for each processed container. The first comprises the transport of a full container. At this point the system does not distinguish between containers with

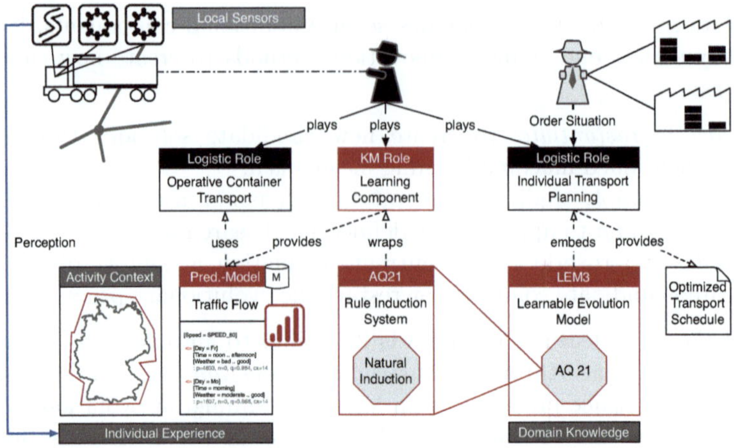

Fig. 13.4. Schematic overview of the aspects of intelligent decision support for transport management agents in a container on-carriage scenario.

respect to type[3] or enclosed contents. The second, subsequent order comprises the return transport of the empty container after unloading at its initial delivery target facility (reverse transportation).In order to handle the transport orders, which have already been successfully acquired from customers, the freight forwarding agency employs exclusively its own fleet of transport vehicles. Therefore, the scenario does not consider requests delegated to sub-contractors as an additional handling modality. The freight forwarder operates a homogeneous transport fleet of semi-trailer trucks. Each truck is equipped to carry exactly one freight container at a time. Thus, each operated truck can only be in either of the two states - empty or fully-loaded.

Informed decision-making and planning, within an MAS designed to represent the forwarding agency, assumes that the involved agents possess the ability to learn from experience, and find their best courses of action. The former has been addressed in part by Gehrke and Wojtusiak [8]. This work is summarized as a use case for individual adaption in Section 13.3.1. Specifically, it outlines an application of *Natural Induction* and AQ21 to autonomous induction of classification models to enable the prediction of traffic patterns.

The challenge of determining the optimal course of action is subsequently addressed in Section 13.3.2. The described approach is based on the learnable evolution model (see Section 13.2.2.2). The method, introduced by Warden and Wojtusiak in [38, 31], uses machine learning to guide an evolutionary optimization process to arrive more rapidly at high-performance plans.

Although both use cases for agent-oriented knowledge management and decision support have been investigated independently, they deal with complementary needs of transport management agents. As such, they can be

[3] Such as refrigerated, hazardous materials or standard container.

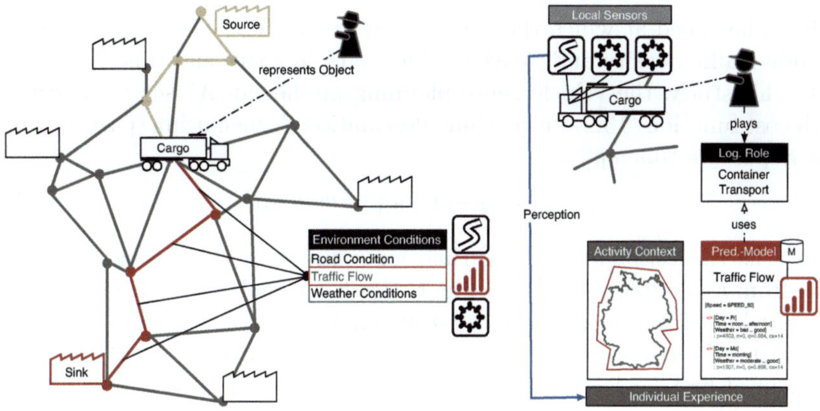

Fig. 13.5. Overview on the scenario investigated in the study on situation-aware vehicle route planning. The actual study reported upon in [8] used a synthetic road grid.

conceptualized as representing several primary logistic roles: the compilation of individual transport schedules on the strategic level of operation, and the actual handling of haulage requests on the operative level. The use cases are shown in Figure 13.4, which also highlights, in terms of the underlying technologies, the relation between learning predictive models by individual agents and optimization of individual transport schedules.

13.3.1 Learning Predictive Models for Decision Support

The performance of a planning systems depends on proper analysis of the current situation and prediction of the future. Depending on the domain, the planning scope may cover only a few hours (for regional transports) to several days (for international). Thus, the important requirements for intelligent planning systems are situational awareness and prediction.

13.3.1.1 Situation-Aware Vehicle Route Planning

In order to investigate the impact of situation assessment and prediction abilities in logistics, and to assess the importance of spatial and temporal constraints, we reduce the problem to a simplified single vehicle transportation scenario. An agent represents a truck that aims to find the fastest route in a graph-based highway grid. This becomes a complex problem because there are highly dynamic environmental conditions that may enforce or reduce the speed at which the vehicle may travel. The examined conditions include traffic density and weather which vary in space and time.

In this study, the vehicle route planning applies an A* search algorithm with cost function (13.2) for reaching destination d when using (partial) route r at departure time t_{dep}:

$$f(r,d,t_{\text{dep}}) = g(r,t_{\text{dep}}) + h(end_r,d) \tag{13.2}$$

with g as the *estimated* driving time for r and h as the estimated driving time from r's endpoint end_r to d. Heuristics h is calculated as driving time at straight line distance from end_r to d at maximum vehicle speed. The route r consists of n consecutive edges (i.e., roads) $e_i \in r$ with $0 \leq i < n$. The route segment of first k edges is denoted by r_{k-1}. The driving time g on route r is calculated by:

$$g(r,t_{\text{dep}}) = \sum_{i=0}^{n-1} \frac{length(e_i)}{v_{\text{est}}(e_i, t_{\text{dep},e_i})} \tag{13.3}$$

with v_{est} as the estimated vehicle speed on an edge which depends on the vehicle agent implementation (See [8, pp. 694]). Edge departure time for $i > 0$ is defined by

$$t_{\text{dep},e_i} = g(r_{i-1}, t_{\text{dep}}) \tag{13.4}$$

Because this setting ensures the criteria for the A* algorithm (non-negative costs and optimistic heuristics) it guarantees the optimal solution. However, the route is optimal only provided that knowledge about the environment used in the cost function is complete and correct. This is a strong assumption, because of the use of models to predict future.

13.3.1.2 Learning and Applying Traffic Models

In the presented study, traffic models were learned using simulated data collected over 15 years of simulated time. The training data consisted of 131,488 examples for each type of edge (representing a road on the map). To learn traffic models we used the AQ21 system. It was executed using a variety of different parameter settings, from which we selected the best according to predictive accuracy on testing data and simulation results. We applied the program in two modes, *theory formation* (TF) and *approximate theory formation* (ATF). The TF mode learns complete and consistent rulesets w. r. t. the training data, while the ATF mode allows partial inconsistence and incompleteness (e.g. in the presence of noise) by optimizing the $Q(w)$ quality measure [13]. In this study the weight, w, of completeness against consistency gain was 0.1, to allow for small inconsistencies in the induced rules.

Many training examples are ambiguous, meaning that for identical values of (day), (time), and (weather), different values of (speed) are assigned. Among several methods for solving ambiguity available in AQ21 we investigated two. The first method, here called *majority*, assigned the most frequent

```
[ Truck_speed = SPEED_30 ]
  # Rule 1
  <== [ Day_of_week = Fr ]
      [ Time_of_Day = Noon ]
      [ Location = ED_EE ]
      [ Weather_condition = MODERATE ]
```

```
[ Truck_speed = SPEED_40 ]
  # Rule 1
  <== [ Weather_condition = VERY_BAD ]
  # Rule 2
  <== [ Day_of_week = Fr ]
      [ Time_of_Day = Afternoon,Noon ]
      [ Location = BF_CF,EF_DF ]
```

Fig. 13.6. Excerpt of a rule set learned by AQ21.

class to all ambiguous examples, and the second method, here called *pos*, always treats ambiguous examples as positive examples. Details of the methods are in [33].

Application of learned models to classify new examples is done by executing AQ21's testing module. The program is provided with input files consisting of a testing problem description, learned models, and one or more testing examples. In the presented experiments we tested two methods of evaluating rules, namely *strict* in which an example either matches a rule or not; and *flexible* in which the program calculates the degree to which examples match rules. In the latter case the degree of match equals the number of matching conditions in the rule to the total number of conditions in the rule. Detailed description of the parameters and other rule matching schemas are in [18, 33]. Two methods of resolving imprecise classifications are tested here. The first method is pessimistic, namely it assumes the worst of the given answers (the lowest predicted speed). The second method is based on frequency of classes. For instance, if an example matches two classes "slow" and "very slow," and the former is more frequent in the training data, it is reasonable to assume that the prediction should be "slow."

Average predictive accuracies and precisions [33] of models learned and applied with different AQ21 parameters on testing data with variances of 0.001 and 0.01 for *dens*(e,t) ranged from 91% to 100% and 71% to 100%, respectively. The best results were obtained by AQ21 in the theory formation mode, with ambiguities treated as majority and flexible rule interpretation. There were 95% predictive accuracy, and 100% precision. All presented values are averaged for all types of roads in the simulation model. For example, the rules in Figure 13.6 are part of the model learned by AQ21. The first rule predicts speed 30 and the two remaining rules predict speed 40. Similar rules were created for other speeds in the simulation model.

Experimental evaluation of the method has been conducted using the simulation system PlaSMA. Comparison of three agents has been performed: ignorant (named "IA"), weather-aware ("WA"), and traffic-predicting agent ("AQ") with seven variants of parameters. All agents optimize the driving time of a 1000 km trip in the road network. Simulated results are provided as an average driving time and its standard deviation for each vehicle agent and parameter setting. For statistical significance experiments were repeated between 4200 and 4800 times with each run corresponding to one trip.

The results indicated strong advantage of the AQ agents over other two types of agents [7]. Ignorant vehicles that are dominated by the straight line heuristics and always choose such centered routes will not be affected. Weather-aware agents do not benefit from their knowledge as much as traffic predicting agents because only the latter will actually realize the traffic properties and consider them in planning.

13.3.2 Optimization of Transport Schedules

The previous section was concerned with using models for situation-assessment and prediction by means of machine learning at the local level of transport management agents. This section presents an approach for optimization of individual transport schedules that involves several agent types. The agents interact within the forwarder's MAS. The presented multiagent-based approach employs two parts that jointly determine the emerging system behavior. These are: 1) local valuation (prioritization) of pending transport orders, and 2) local selection of orders for handling deliveries by trucks in the transport fleet. The order valuation is part of order management; the operative order handling (which for the purposes of this research subsumes order selection) is part of transport management. This distinction is introduced into the agent-based implementation of the autonomous control system. Specifically, the system design comprises two primary classes of agents each handling one of the aforementioned management tasks. Each transport agent acts on behalf of a single truck and manages its intra-company order selection, acquisition, and subsequent operative order handling. These agents rely on adequate valuations of pending transport orders, which constitutes the basis for the order selection. We considered a simple greedy approach as baseline and an evolutionary-based planning approach.

(Re-)Valuation of Transport Orders. Transport agents select the most profitable orders. Therefore, orders with low *contractual order value* may never be selected for transport. For the freight forwarding agency, such a system behavior evidently violates contract agreements with customers. The forwarding agency requires an effective mechanism which is designed to ensure a timely handling of *all* pending orders. Although this research does not assume contractually fixed delivery times, the system is designed to process all orders in a sensible global time window, e. g., 48 hours. The approach used here is based on an autonomous re-valuation of transport orders, which emanates from the original real valuation of the orders in terms of monetary value. The introduced use of values as a means of calculation should be understood as a method of intra-organizational prioritization of orders. The transport management agents internally consider the order value for order selection which is calculated as follows:

$$value(order,t) = price(order) + prior^{+}(order,t) \tag{13.5}$$

value(*order*, t) is computed as sum of a first component *price*(*order*) which constitutes the initial order price agreed between customer and forwarder (See Equation (13.6)) and a second component $prior^{+}(order,t)$ for the order revaluation based on the period of time the cargo has already been waiting to establish a transport contract (See Equation (13.7)). The initial order price is computed as:

$$price(order) = d \cdot (c_0 + r \cdot distance(order)) \tag{13.6}$$

The constant c_0 denotes fixed costs for the operation of a semi-trailer truck, while the constant r denotes variable operating costs per kilometer. In order to accommodate the two considered "types" of containers which are associated with transport orders – with regular content, and empty to be returned to their original storage facility – the dampening factor d has been introduced.

To increase the value of a container depending on the time it has been waiting, the function $prior^{+}(order,t)$ is used. This function is monotonically increasing with t, thus for containers which are kept waiting for a longer period of time the term $prior^{+}(order,t)$ at some point begins to dominate the initial order price. Assuming a sufficiently large number of transport agents, each order will eventually be selected for transport as its value increases with time. While in general, the $prior^{+}(order,t)$ can be sophisticated and account for, amongst others, prediction of travel times, pickup and delivery time windows, and penalties, the basic version shown below has been used for initial experiments.

$$prior^{+}(order,t) = w_c(order) \cdot t^{\alpha} \tag{13.7}$$

Selection of Transport Orders. The transport management agents, which have been briefly introduced in the preceding sections, are routinely faced with the challenge of autonomously rendering decisions that determine their respective operative transport planning. Although this report concentrates on a particular transport planning approach, based on an evolutionary optimization method, a simple non-planning type of transport agent which effectively employs a greedy order selection strategy has been implemented as a baseline for the measurement of transport management performance (See [31], p. 7). A common assumption, which is reasonable within the bounds of a single freight forwarding agency, is that the transport management agent has access to the entire momentary order situation. This includes even the pending orders, which are waiting for processing at the distinct storage facilities. The data is organized as a lookup table indexed by transport endpoint pairs, which point to a priority queue of orders which belong to the given transport relation. The priority is measured within the respective order value, as introduced in the preceding section, which is updated in regular intervals by the order management agents.

A Planning Approach to Order Selection. Planning transport management agents are faced with the challenge of autonomously rendering decisions that determine their respective transport plans. These decisions thereby pertain to a choice of adequate action alternatives, the options being: choosing a transport order whose pickup point is the currently considered storage facility, or postponing that choice and relocating to another storage facility.

In essence, the transport agents need to decide which transport order to choose based on a specific time and location. That choice is guided by the gains and costs of transporting the container associated with the order. Therefore, by choosing orders in an optimized way, an agent can maximize its financial balance. The behavior of the transport management agents is thus the result of a series of constitutive decisions. This initial situation calls for provident planning in which a transport agent considers several steps ahead.The transport management agents seek an optimized pickup and delivery plan with a plan horizon of size n. Formally, such a plan is defined as:

$$plan^n = (action_1, action_2, \ldots, action_n) \text{ where } action_i \in Deliveries \cup EmptyRides$$

Deliveries refer to the set of possible delivery actions as determined by the pending transport orders, which have previously been acquired by the transport forwarding agency. *EmptyRides* in contrast refer to the set of possible empty journeys between storage facilities. Thus, a transport plan as defined above can blend deliveries and empty drives where the latter can often be interpreted as pickup tours. The space of *valid* transport plans is specified by constraints:

(C.1) Let $start : Deliveries \cup EmptyRides \rightarrow SF$ define a function which returns the source location of a particular plan step (i. e., in the case of proper orders, the pickup site). Further let $dest : Deliveries \cup EmptyRides \rightarrow SF$ define the complementary function which returns the target location of a plan step (for proper orders, the delivery site). In both cases, SF constitutes the set of storage facilities in the given scenario. The first constraint that on admissible transport plans is defined by:

$$\forall i = 1 \ldots n : start(action_i) \neq dest(action_i).$$

The rationale here is that both types of actions that be carried out as plan steps, i. e., empty relocation from one storage facility to another and execution of a delivery, comprise a non-circular movement of the truck in question. Thus, a single plan step may neither consist of a round-trip nor a rest or waiting period at a particular storage facility.

(C.2) Another constraint ensures that the tour specified by a valid plan is contiguous, which means that short cycles are precluded by this constraint thus acting as a sub-tour elimination constraint:

$$\forall i = 1 \ldots (n-1) : dest(action_i) = start(action_{i+1})$$

where $action_i$ denotes the tuple elements of a plan as defined in Equation (13.8). It is, however, possible for transport tours to revisit certain locations since loops are allowed by the formulation.

The value of a transport plan as defined above is determined based on Equation (13.5):

$$val(plan^n) = \sum_{i=0}^{n-1} val(act_{i+1}) \cdot (n-i)^{\alpha} \tag{13.8}$$

where $val(act) = val(order,t) - cost(start(act), dest(act))$ iff $act \in Deliveries$. For $act \in EmptyRides$, it holds that $val(order,t) = -cost(start(act), dest(act))$. Equation (13.8) shows that the value of the complete transport plan is a weighted sum of the values of the respective plan steps $action_i$. The parameter is used to determine a specific weighting scheme. For instance, if $\alpha = 0$, all plan steps are given equal weight in the calculation of the value for the complete plan.

13.3.2.1 Employing LEM3 for Transport Planning

In the presented study, transport agents use the learnable evolution model (introduced in Section 13.2.2.2) to search the plan space. The problem definition for the application of LEM3 in the presented context currently incorporates: 1) the storage facility where the truck for whom the planning is conducted is located at the time of planning, 2) a complete list of storage facilities where transport orders may be pending (i. e., this list comprises both such facilities where pending orders are momentarily in stock and those which are effectively unused at the moment), and 3) the size of the plan horizon. Due to this reduced problem definition, candidate plans constitute what has been referred to as *plan skeleton* rather than a fully-fledged transport plan. The concept can be formalized and associated to the definition of proper transport plans in Equation (13.8) as follows, assuming a plan horizon of size n:

$$\begin{aligned} planSkel(plan^n) &= (\ dest(action_1), dest(action_2), \ldots, dest(action_n)\ ) \\ &\equiv (\ storage_1, storage_2, \ldots, storage_n\ ) \end{aligned} \tag{13.9}$$

A plan skeleton is thus a n-tuple of storage facilities. However, with regard to the planning problem, only a sub-set of the set of all plan skeletons of size n are admissible in terms of compliance with the following constraints:

1. $\forall i = 1 \ldots (n-1) : storage_i \neq storage_{i+1}$. All actions in proper plans involve a relocation of the truck between two distinct storage facilities, either via an empty ride or a proper delivery of a container.
2. $pos(truck) \neq storage_1$ where $pos(truck), storage_1 \in SF$. In particular, the storage facility where the truck is located at planning time must not be identical with the first storage facility in the plan skeleton.

The aforementioned constraints have been implemented as part of the problem-specific LEM3 integration. As a consequence, after the creation of new candidate solutions, LEM3 is enabled to detect the number of constraint violations and filter inadmissible solutions.

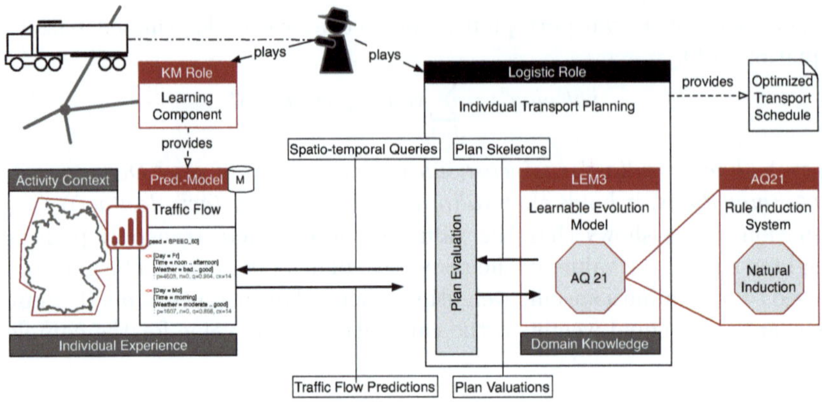

Fig. 13.7. Optimization of individual transport plans as a use case for the prediction model learnt in Section 13.3.1. The graphic provides an example for the interaction of primary logistics and knowledge management roles via models.

Since the plan skeletons without additional processing do not describe directly any particular plan which can be evaluated, an unequivocal conversion into a proper transport plan needs to be used. This conversion is possible because of two assumptions. First, transport plans always have as origin the current location of the associated truck, which corresponds to a storage facility (*pos*(*truck*)). Second, as the transport agent seeks to maximize its financial balance for each plan step, it is rational to choose the most profitable action alternative by using the following function $bestAct : SF \times SF \rightarrow Deliveries \cup EmptyRides$. This function accepts transport end points as input and returns either most profitable real transport order, or, as a fall-back if no orders with the specified transport endpoints currently exists in the system, the empty drive order. Based on a plan skeleton, the corresponding plan is:

$$plan^n = (\ bestAct(pos(truck), storage_1), \ldots, bestAct(storage_{n-1}, storage_n)\ ) \tag{13.10}$$

The transformation from plan skeletons, which constitute the plan suggestions created iteratively by LEM3, into proper candidate plans (from the point of view of the planning agent) is a mandatory prerequisite in order to apply the domain-specific weighting function. This function is used to evaluate candidates and thus drives LEM's search in the space of possible plans. The approach used, which is currently applied with the integration of LEM3, basically out-sources the weighting function from the library to the planning agent. The agent acts as user of the library, which is also equipped with the required domain-specific knowledge to execute the candidate valuation as shown in Figure 13.7. However, the route valuation is not yet integrated

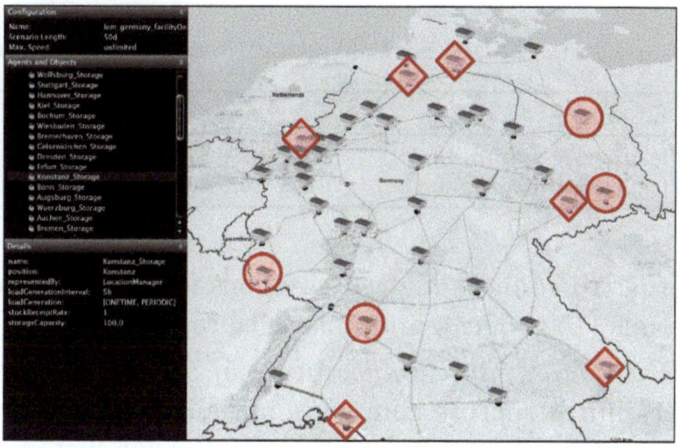

Fig. 13.8 Traffic network with storage facilities employed in experiments. Storage facilities that are not marked explicitly do not have inbound customer orders. Storages marked with a diamond have a larger, those marked with a circle a lower amount of inbound orders.

for experiments. In order to directly calculate the value of a suggested plan skeleton, the following Equation (13.11) can be applied:

$$value(planSkeleton^n) = balance(bestAct(pos(truck), storage_1)) \cdot n^\alpha + \sum_{i=1}^{n-1} [balance(bestAct(storage_i, storage_{i+1})) \cdot (n-1)^\alpha] \qquad (13.11)$$

The function $balance : SF \times SF \rightarrow$ Euro takes the current order value (which for all orders that have been waiting for some time is higher than the initial order price) and subtracts operation costs for the execution of these orders.

Once plans have been evaluated, LEM3 checks stopping criteria (reached desired value of plan or the maximum length of evolution is reached).

13.3.2.2 System Evaluation

The freight forwarder introduced in Section 13.3 has been implemented as an MAS for experiments within the PlaSMA [29] simulation environment. The goal is to evaluate the emergent system behavior. The implementation has been described in detail in [31]. It involves three distinct agent types: order management agents, which implement the order valuation strategy; transport management agents, which need to compile transport plans; and order information service agent (OIS), which maintains a company-wide database of pending haulage requests. The performance of transport agents employing either LEM3 or a baseline strategy for route planning was investigated for the freight forwarder MAS

Traffic Network: A realistic traffic network represented as a directed graph has been used as a common basis for the simulation experiments. The network covers the area of Germany, as shown in Figure 13.8. It contains 359 nodes and 1.044 edges. The nodes comprise the major cities of Germany (50 nodes). The edges constitute transport routes. These represent a significant part of the German motorway network (750 edges) and, to a lesser degree in

Table 13.1. Waiting times until a container is picked up.

Unit: *hours*	Mixed Fleet	LEM Fleet	Baseline Fleet
Pickup wait, full ($\mu_s \pm \sigma_s$)	$12,597 \pm 11,899$	$7,926 \pm 8,706$	$40,820 \pm 24,855$
Pickup wait, empty ($\mu_s \pm \sigma_s$)	$90,029 \pm 33,761$	$53,832 \pm 30,718$	$141,564 \pm 46,675$

order to connect highway sections or cities in the highway network, federal roads (152 edges), and inner-city roads (28 edges).

Distribution and Configuration of Storage Facilities: Modeling of the logistics infrastructure comprises initially stationary resources, in particular the allocation of storage facilities. For the conducted experiments, each city hosts exactly one facility whose relevant modeling parameters comprise its storage capacity (100 container units) and simulation-specific parameters to control the external intake of new orders (See [31, p. 23]). For the presented experiments, ten out of 42 storage facilities were configured with an external order intake as shown in Figure 13.8. These have been chosen because of their geographic location such that a reasonable distribution of orders could be established.

Configuration of the Forwarder Fleet: The experiments feature a homogeneous fleet of 16 trucks which can only transport a single container at a time. Three configurations of the transport agents were employed: first, a configuration where all trucks are managed by baseline transport agents; second, a configuration featuring LEM-enabled transport agents; and a third "competitive" scenario, where eight trucks respectively are managed by agents of either type. The last scenario was chosen specifically to analyze the emergent effects of intermixing transport planning strategies of different complexities.

Selected Findings. The evaluation of the proposed planning approach comprised two experiment series. These differed in the amount of orders created at high-intake storage facilities, i. e., creation of orders every 6h (series 1) or every 5h (series 2)[4]. We summarize selected results, which pertain to the system pickup times with different forwarder configurations in the high order set of experiments. The full experiment description which also considers a financial analysis and the distribution of truck operations has been published in [31].

Analysis of Pickup Times: The data in Table 13.1 and the supplementary plots in Figures 13.9 and 13.11 show that the LEM-based planning optimization leads to a significant reduction in waiting times for both the pickup of containers in regular and reverse transports. The plots of the distribution of pickup times depict vastly different characteristics. In Figure 13.9, the distribution of pickup times for the baseline approach is quite regular with only

[4] In both cases, orders are created every 12 hours for low-intake facilities. Details as well as information about the full system parameterization are provided in [31, p. 23].

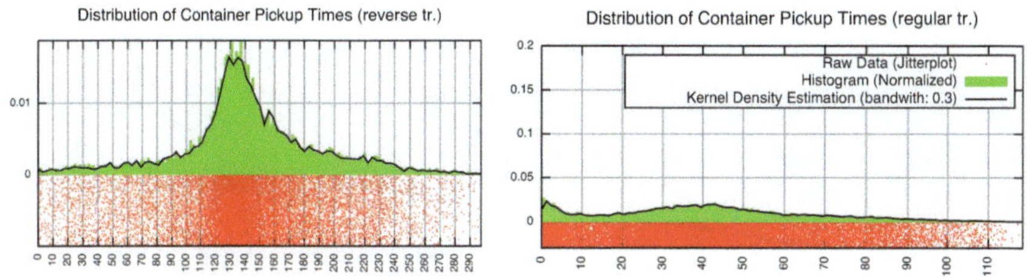

Fig. 13.9. Pickup times for 10 simulations of 60 days each with a *baseline* fleet.

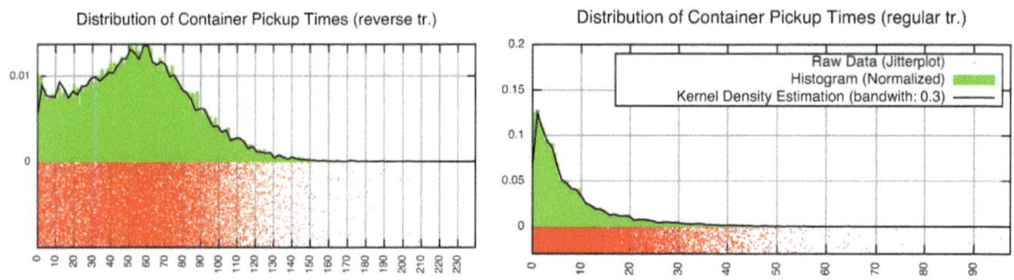

Fig. 13.10. Pickup times for 10 simulations of 60 days each with a *planning* fleet.

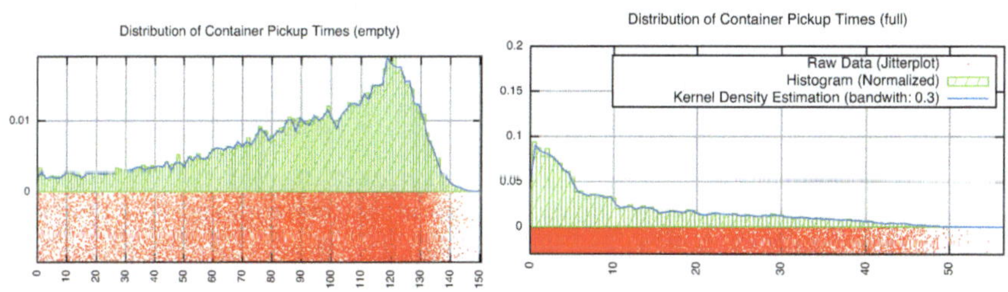

Fig. 13.11. Pickup times for 10 simulations of 60 days each with a *mixed* fleet.

two shallow peaks; the first around minimal pickup times (orders handled just-in-time) and the second peak around the measured mean. For the planning case, there is only a single peak at about 1h, followed by a very steep incline which leads to 95 % of the regular transports being picked up within 30h and thus, before the mean measured time in the baseline approach. The surplus value generated through the utilization of a planning approach is thus evident in the plots. What is also rendered obvious in the plot for the planning transport fleet is the long tail of the distribution due to a very low amount of outlier cases for which the pickup takes significantly longer than for the vast majority of cases. This effect is replicated for the case of reverse transports. It is thus a side-characteristic of the planning approach, which calls for in-depth research to enhance the approach in the future. The plot in Figure 13.11 shows for case of a mixed transport fleet that the insertion of planning agents has the effect of ablating the second peak in the baseline-only distribution. Additionally, more weight is accumulated at the initial peak, thus bringing the distribution closer to the planning-only case.

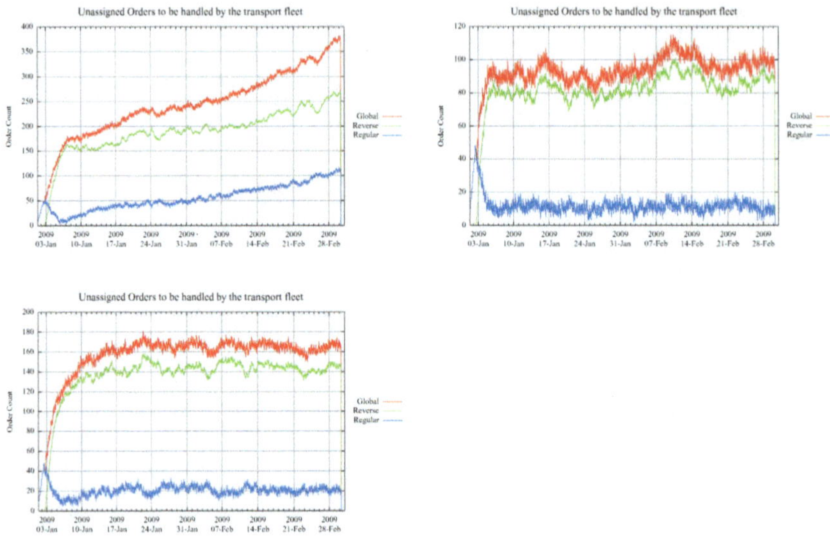

Fig. 13.12. Global development of pending orders in the system over the course of a simulation run of 60 days. *Top-left*: greedy approach; *top-right*: planning approach using LEM3; *bottom*: mixed fleet.

It is also shown, that the effective value range of the distribution is cut in half.

As a next step in the analysis process, focus is shifted from the examination of regular transport towards the reverse transports. Here, the data presented in Table 13.1 shows that in the baseline case, the mean pickup time for reverse transport is now 141,56 h – an increase of 43 % from experiments with lower order inflow as documented in [31, Section 4.4.1, pp.26], with the standard deviation growing significantly as well. The plot in Figure 13.9 also shows, that in the face of the increased order inflow, the characteristics of the distribution of pickup times for reverse transports deviates from its analog in a low order situation as examined in [31, pp. 26]. The latter was characterized by a single significant peak around 120 hours with highly uneven slopes, the left featuring a particular long outlet while the right measured only a fraction thereof. In Figure 13.9 however, the single peak of the distribution shifted to the right while the width of both slopes, left and right, amounted to more than 120 hours. What is more, the right outlet bears an even greater weight than the left. This data indicates that the number of orders which needed to be handled in this set of experiments exceeded the capacities of the baseline approach. This assessment is supported by the ever growing stock levels and pending orders within the system (cf. Figure 13.12).

The effect of combining both approaches (in equal fractions) can be described as follows (see Figure 13.11): While retaining the single peak at about 120h also found in the baseline case, the majority of the slope from that case is cleared away, and the respective weight of the distribution is transferred to the left slope. The mixture of approaches hence helps to suppress undesirable outlier cases, observed when employing only planning transport agents.

13.4 Future Challenges and Research Directions

Machine learning and guided evolutionary computation can be employed to render optimized action plans (Section 13.3.1) and provide tailored knowledge to be exploited for effective operative planning conduct (Section 13.3.2). It was shown that machine learning can be employed effectively both in knowledge-based decision-making within primary logistics roles, and within auxiliary knowledge management roles. Specifically, Section 13.3.2 suggested how both categories of roles that are played by intelligent agents can be coupled. As agent-based autonomous logistics matures with the development of more sophisticated learning approaches and planning strategies, novel challenges arise. As MAS scale from research prototypes towards industry-grade systems, size and diversity of the MAS grows. Also, the systems will have to be built for continuous operation over extended periods of time. The MAS will then have to accommodate the dynamic interchange of specific agents and/or agent types. From a KM point of view, two areas which demand attention in future research are consequently highlighted.

13.4.1 Integration of Multiple Knowledge Sources

Section 13.3.1 was concerned with the application of the AQ21 rule learning system in order to learn individually tailored classification models for situation assessment and prediction. The learned model for traffic flows pertained to a single transport agent. With respect to its application, it was tacitly assumed, that no other 'competing' models need to be accounted for. As complexity of MAS increases, it is expected that knowledge from multiple sources needs to be integrated on the fly to procure well-founded support in decision-making. The models in question may be associated with different providers and contain different scopes of applicability.

Extrinsic models an agent might want to account for include those belonging to peers within a community of practice within the MAS. In relation to the container on-carriage scenario introduced in Section 13.3, such a community of practice could be formed by all transport agents acting on behalf of the freight forwarder. More importantly, the agents operate in sufficiently similar activity contexts. Besides peer-models, an agent might also exploit models that have been built for wider, i.e., less agent-specific, employment based on accumulated experience from multiple contributing agents such as company-wide models. The least specific models may pertain to third parties, which can be interfaced from within the bounds of an MAS representing, for instance, a freight forwarder as independent knowledge and information services.

The Challenge of Model Integration. The major challenge with regards to the integration of multiple sources of knowledge is that, in general, the inferences drawn from these models will not complement each other seamlessly. Thus, the agents may have to incorporate robust methods for conflict resolution to handle contradictory predictions. While the problem of knowledge exchange is widely known, it has not been properly addressed in machine learning for existing heterogeneous models. The idea is that agents learn from the feedback they receive from other agents (when negotiating a prediction) instead of using only the outcome of the prediction. The goal is to utilize the actual evidence support different agents' decisions.

Research Problems. The integration of heterogeneous data from different applications is subject to research since the seminal work of Wiederhold [32] on the integration of heterogeneous relational databases. Formally, data integrations can be realized by mapping from the information sources ("data + semantics") onto the information of an application. The definition of this mapping must detect and remove integration conflicts where structural syntactic heterogeneities are not the biggest problem, but the actual semantic heterogeneities constitute problem that must be resolved. Therefore it is necessary to find corresponding semantic information and to consolidate the different semantic interpretations of a piece of information. This approach is called semantic mediation for heterogeneous information sources [28] and supports the acquisition of the integration mapping at the semantic level. The related problem of belief revision concerns an agent acquireing knowledge contradicting its existing beliefs [25, 20]. Costantini and Tocchio have shown that agents can exchange rule-based knowledge [4]. Prentzas et al. combined symbolic rules and neural networks in a hybrid rule base, and presented a unified approach for these models for knowledge exchange[21]. Hilario overviewed strategies for neuro-symbolic integration in which symbolic structures are described as a special case of functional hybrid approaches [9].

One area of current research concerns situations when multiple classification models that may also originate from different agents are to be applied to render a decision. Specifically, the classifiers $C_1,\ldots,C_k$ are applied to classify one specific case. Two research questions arise in these situations: 1) How to combine results from these classifiers? More importantly 2) how can individual classifiers learn from their interaction with other classifiers?
The first question has been extensively studied in the field of meta and ensemble learning [6]. Two types of approaches have been proposed that either combine outputs from classifiers (e.g. bootstrap aggregating, boosting, or stacked generalization), or select one actual output (i.e., voting). The second question is much harder and demands extensive additional research.

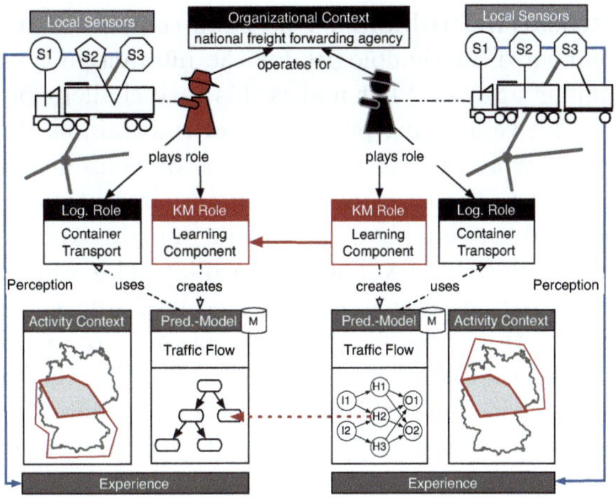

Fig. 13.13 In MAS considered in autonomous logistics, multiple agents with similar tasks act and acquire knowledge in overlapping geographic task environments. Thus, there is a potential to promote local learning efforts through collaboration and transfer of dynamic knowledge; a task which is challenging especially in the face of different knowledge representations (here: a decision tree and an artificial neural network).

13.4.2 Knowledge Transfer in Heterogeneous MAS

In Section 13.2.1, we argued that besides adequate access to decision-relevant information and background knowledge, access to tailored dynamic knowledge is required for the performance of intelligent agents in complex domains such as autonomous logistics. Dynamic knowledge thereby subsumes several categories of knowledge models that play a role in decision-making, such as decision models, prediction models, and order classification models.

The Challenge of Knowledge Transfer. In cases where the agent does not yet have access to a sufficient and representative batch of training data to construct a model using either purely inductive or, given the availability of background knowledge, hybrid analytic-inductive learning schemes which conforms designated quality standards, two action alternatives exist. The first option is to resort to the acquisition of further training data. In many real-world use cases however, such a course of action entails significant costs and might require a considerable amount of time. In the latter case, which arises for instance due to the fact that the agent is reliant on extensive monitoring of the application domain. This leads to the prolonged disadvantage of attaining a model-induced competitive advantage.

The second option builds upon a different perspective of the learning activities conducted as part of agent-oriented knowledge management and adaption activities. The core idea is to no longer treat learning as an isolated and passive process. Rather, agents are actively engaged in the learnign process, collect datam and seek advice from and collaborate with other agents as shown in Figure 13.13. The necessary domain expertise, required to enable the successful adoption of individual models, may exist in the community of agents. If relevant knowledge can be transferred effectively from peers, it may be more benefitial than learning from scratch.

Research Problems. One issue that needs to be addressed is the development of methodologies for the interactive transfer of dynamic knowledge among agents. What makes this task challenging is the heterogeneity of the respective learning problems and the training data used as input by the respective agents; as well as given or consciously introduced diversity with respect to the types of knowledge representation. When needed, agents need to be able to enable their role of a domain expert or advisor assisting advisees in learning. In the context of a bilateral learning scenario, the advisor seeks to impart learning assistance. Received advice then allows the advisee to proceed with a step-wise iterative optimization of its own initial model. Several key aspects must be addressed to realize the proposed kind of knowledge transfer in heterogeneous MAS. These relate to both to the learning agent and its interaction partner(s). The advisee needs means to assess its own learning process to identify learning deficiencies. This constitutes the basis of defining concrete learning problems that can consequently be addressed in the advisee–advisor interaction. The advisee also needs effective operations to accomodate advice from its interaction partner. Finally, the advisee needs a meta-control strategy that guides its learning progress, defines learning problems, and in engages the advisor for assistance. Such a meta-control can, for instance, be conceived as a kind of local search through model space. On the other hand, the advisor needs appropriate explanation capabilities to prepare 'helpful' advice from its own individual knowledge model. To define the structure of discrete knowledge transfer episodes among both involved actors, suitable interaction protocols as well as a content language are needed. Warden et al.[30] propose an approach for interactive individual adaption of decision support models to address the highlighted challenges.

13.5 Conclusion

The increasing complexity of logistic networks calls for a paradigm change in their modeling and operations. Centralized control is no longer a feasible option when dealing with extremely large systems. Rather, decentralized autonomous systems are gaining popularity due to their robustness and scalability. This chapter presented an overview of agent-based autonomous logistics, with specific focus on knowledge management and learning capabilities. The presented methodology assumes that an agent learns multiple models for making predictions about different aspects of the future state of the environment, and uses these models together to make decisions. The presented AQ21 rule learning system is highly configurable and robust, thus, it can be easily set up for applications involving autonomous agents. AQ21 also plays an important role in the learnable evolution model, in that it is used by transport management agents to optimize their plans. The presented experiments indicate a strong advantage of agents that use advanced learning and knowledge management techniques. This result applies to both, agents predicting the future state of the environment, and agents that optimize their plans.

The final part of the chapter outlined open challenges related to learning, and knowledge management in agent-based systems that require additional research. The highly promising obtained results, serve as a motivation for further research in the area.

Acknowledgements. This research has been supported in part by the German Research Foundation (DFG) within the Collaborative Research Centre 637 *Autonomous Cooperating Logistic Processes – A Paradigm Shift and its Limitations* at the Universität Bremen. Development of the AQ21 and LEM3 systems was partially funded by the National Science Foundation grants. Current research on the Learnable Evolution Model is supported by the National Institute of Standards and Technology grant.

References

1. Böse, F., Windt, K.: Catalogue of Criteria for Autonomous Control. In: [11], pp. 57–72 (2007)
2. Bratman, M.E.: Intentions, Plans and Practical Reason. Havard University Press, Cambridge (1987)
3. Brooks, R.A.: Intelligence without Representation. Artificial Intelligence 47(1-3), 139–159 (1991)
4. Costantini, S., Tocchio, A.: Learning by Knowledge Exchange in Logical Agents. In: Proceedings of WOA, Citeseer (2005)
5. Davidsson, P., Henesey, L., Ramstedt, L., Törnquist, J., Wernstedt, F.: An Analysis of Agent-based Approaches to Transport Logistics. Transportation Research part C: Emerging Technologies 13(4), 255–271 (2005)
6. Dietterich, T.G.: Ensemble Methods in Machine Learning. In: Kittler, J., Roli, F. (eds.) MCS 2000. LNCS, vol. 1857, pp. 1–15. Springer, Heidelberg (2000)
7. Gehrke, J.D., Wojtusiak, J.: A Natural Induction Approach to Traffic Prediction for Autonomous Agent-based Vehicle Route Planning. Reports of the Machine Learning and Inference Laboratory. George Mason University, Fairfax (2008)
8. Gehrke, J.D., Wojtusiak, J.: Traffic Prediction for Agent Route Planning. In: Proc. of the 8th Intl. Conf. on Computational Science, vol. 3, pp. 692–701. Springer, Berlin (2008)
9. Hilario, M.: An Overview of Strategies for Neurosymbolic Integration. Connectionist-Symbolic Integration: From Unified to Hybrid Approaches, pp. 13–36 (1997)
10. Hülsmann, M., Scholz-Reiter, B., Windt, K. (eds.): Autonomous Cooperation and Control in Logistics. Springer (2011)
11. Hülsmann, M., Windt, K. (eds.): Understanding Autonomous Cooperation and Control in Logistics. The Impact on Management, Information and Communication and Material Flow. Springer (2007)

12. Jedermann, R., Behrens, C., Laur, R., Lang, W.: Intelligent Containers and Sensor Networks, Approaches to Apply Autonomous Cooperation on Systems with Limited Resources. In: [11], pp. 365–392 (2007)
13. Kaufman, K., Michalski, R.S.: An Adjustable Rule Learner for Pattern Discovery Using the AQ Methodology. Journal of Intelligent Information Systems 14, 199–216 (2000)
14. Langer, H., Gehrke, J.D., Hammer, J., Lorenz, M., Timm, I.J., Herzog, O.: A Framework for Distributed Knowledge Management in Autonomous Logistic Processes. International Journal of Knowledge-Based & Intelligent Engineering Systems 10(4), 277–290 (2006)
15. McFarlane, D., Sarma, S., Chirn, J.L., Wong, C.Y., Ashton, K.: Auto ID Systems and Intelligent Manufacturing Control. Engineering Applications of Artificial Intelligence 16(4), 365–376 (2003)
16. Meyer, G.G., Främling, K., Holmström, J.: Intelligent Products: A Survey. Computers in Industry 60(3), 137–148 (2009)
17. Michalski, R.S.: Learnable Evolution Model: Evolutionary Processes Guided by Machine Learning. Machine Learning 38, 9–40 (2000)
18. Michalski, R.S.: Attributional Calculus: A Logic and Representation Language for Natural Induction. Reports of the Machine Learning and Inference Laboratory MLI 04–2, MLI. George Mason University, Fairfax (2004)
19. Michalski, R.S., Wojtusiak, J., Kaufman, K.: Progress Report on the Learnable Evolution Model. Reports of the Machine Learning and Inference Laboratory MLI 07–2. George Mason University, Fairfax (2007)
20. Peppas, P.: Belief Revision. In: Handbook of Knowledge Representation, vol. 1. Elsevier (2008)
21. Prentzas, J., Hatzilygeroudis, I.: Rule-based Update Methods for a Hybrid Rule Base. Data & Knowledge Engineering 55(2), 103–128 (2005)
22. Prentzas, J., Hatzilygeroudis, I., Tsakalidis, A.: Updating a Hybrid Rule Base with New Empirical Source Knowledge. In: Proc. of the 14th IEEE Intl. Conference on Tools with Artificial Intelligence (ICTAI 2002), pp. 9–15. IEEE (2002)
23. Rao, A.S., Georgeff, M.P.: Modeling Rational Agents within a BDI-Architecture. In: Allen, J., Fikes, R., Sandewall, E. (eds.) Proceedings of the 2nd International Conference on Principles of Knowledge Representation and Reasoning, pp. 473–484. Morgan Kaufmann Publishers Inc., San Mateo (1991)
24. Schuldt, A.: Multiagent Coordination Enabling Autonomous Logistics. Springer, Heidelberg (2011)
25. Tselekidis, G., Peppas, P., Williams, M.: Agent-Oriented Knowledge Management. In: Proceedings of the 3rd European Conference on Organizational Knowledge, Learning, and Capabilities (2002)
26. van der Putten, S., Robu, V., La Poutré, H., Jorritsma, A., Gal, M.: Automating Supply Chain Negotiations using Autonomous Agents: A Case Study in Transportation Logistics. In: Proc. of the 5th International Joint Conference on Autonomous Agents and Multiagent Systems, pp. 1506–1513. ACM (2006)
27. van Leeuwen, E.H., Norrie, D.: Holons and Holarchies. Manufacturing Engineer 76(2), 86–88 (1997)
28. Wache, H.: Semantische Mediation für heterogene Informationsquellen. DISKI. Akademische Verlagsgesellschaft Aka GmbH, Berlin (2003)

29. Warden, T., Porzel, R., Gehrke, J.D., Herzog, O., Langer, H., Malaka, R.: Towards Ontology-based Multiagent Simulations: The PlaSMA Approach. In: Proc. of the 24th European Conf. on Modelling and Simulation (ECMS 2010), pp. 50–56 (2010)
30. Warden, T., Wagner, T., Langer, H., Herzog, O.: Towards Multiagent Interactive Adaption of Individual Models for Decision Support. In: 2nd International Workshop on Semantic Interoperability, IWSI 2012 (to appear, 2012)
31. Warden, T., Wojtusiak, J.: Learnable Evolutionary Optimization in Autonomous Pickup & Delivery Planning: A Scenario, System Architecture and Initial Results. TZI Technical Report 55, Universität Bremen (September 2010)
32. Wiederhold, G.: Mediators in the Architecture of Future Information Systems. Computer 25(3), 38–49 (1992)
33. Wojtusiak, J.: AQ21 User's Guide. Reports of the Machine Learning and Inference Laboratory MLI 04–3, George Mason University (2004) (updated in September 2005)
34. Wojtusiak, J.: Handling Constrained Optimization Problems and Using Constructive Induction to Improve Representation Spaces in Learnable Evolution Model. PhD thesis, College of Science, George Mason University, Fairfax, VA (November 2007)
35. Wojtusiak, J.: Data-Driven Constructive Induction in the Learnable Evolution Model. In: Proceedings of the 16th International Conference on Intelligent Information Systems, Zakopane, Poland, June 16–18, pp. 191–200 (2008)
36. Wojtusiak, J., Michalski, R.S.: The LEM3 Implementation of Learnable Evolution Model and Its Testing on Complex Function Optimization Problems. In: Proceedings of Genetic and Evolutionary Computation Conference (GECCO 2006), Seattle, WA, July 8–12 (2006)
37. Wojtusiak, J., Michalski, R.S., Kaufman, K., Pietrzykowski, J.: The AQ21 Natural Induction Program for Pattern Discovery: Initial Version and its Novel Features. In: Proceedings of the 18th IEEE International Conference on Tools with Artificial Intelligence, Washington, D.C., November 13–15 (2006)
38. Wojtusiak, J., Warden, T., Herzog, O.: Agent-based Pickup and Delivery Planning: The Learnable Evolution Model Approach. In: Proceedings of the 5th International Conference on Complex, Intelligent, and Software Intensive Systems (CISIS 2011), Seoul, South Korea (2011)
39. Wooldridge, M.J.: An Introduction to Multiagent Systems. Wiley (2002)
40. Wooldridge, M.J., Jennings, N.R.: Intelligent Agents: Theory and Practice. Knowledge Engineering Review 10(2), 115–152 (1995)

Chapter 14
LINDSAY Virtual Human: Multi-scale, Agent-based, and Interactive

C. Jacob, S. von Mammen, T. Davison, A. Sarraf-Shirazi, V. Sarpe, A. Esmaeili, D. Phillips, I. Yazdanbod, S. Novakowski, S. Steil, C. Gingras, H. Jamniczky, B. Hallgrimsson, and B. Wright

Abstract. We are developing *LINDSAY Virtual Human*, a 3-dimensional, interactive computer model of male and female anatomy and physiology. *LINDSAY* is designed to be used for medical education. One key characteristic of *LINDSAY* is the integration of computational models across a range of spatial and temporal scales. We simulate physiological processes in an integrative fashion: from the body level to the level of organs, tissues, cells, and sub-cellular structures. For use in the classroom, we have built *LINDSAY Presenter*, a 3D slide-based visualization and exploration environment that presents different scenarios within the simulated human body. We are developing *LINDSAY Composer* to create complex scenes for demonstration, exploration and investigation of physiological scenarios. At *LINDSAY Composer*'s core is a graphical programming environment, which facilitates the composition of complex, interactive educational modules around the human body.

14.1 Motivation

Health care systems all over the world are struggling to provide affordable and comprehensive care. Consequently, the need for excellent education and training of medical staff and doctors is more important than ever. At the same time, emerging digital technologies render it possible to present complex contents faster and better to a wide range of audiences. Within this educational context, we are developing *LINDSAY Virtual Human*, a project

C. Jacob · S. von Mammen · T. Davison · A. Sarraf-Shirazi · V. Sarpe · A. Esmaeili · D. Phillips · I. Yazdanbod · S. Novakowski · S. Steil · C. Gingras · H. Jamniczky · B. Hallgrimsson · B. Wright
University of Calgary, Calgary, Alberta, Canada
e-mail: cjacob@ucalgary.ca

J. Kołodziej et al. (Eds.): Advances in Intelligent Modelling and Simulation, SCI 422, pp. 327–349.
springerlink.com

at the intersection of Computer Science, Education, and Medicine. *LINDSAY* provides a collection of computational tools for research, presentation and learning in the context of human anatomy and physiology. *LINDSAY* also serves as a testbed for interactive computer graphics, touch-based user interfaces, multi-scale modelling, and scalable computing solutions to run simulations seamlessly on a diverse range of computing platforms: from web-based systems and desktop computers to laptops and mobile devices.

14.1.1 Starting with Virtual Anatomy

As a starting point for this project, we attended and analyzed lectures on human anatomy given by senior instructors from the Faculty of Medicine at the University of Calgary. One of our conclusions was that a virtual 3-dimensional human body in combination with the ability to easily navigate and label the displayed contents could greatly facilitate and improve the learning experience for students. Consequently, such an application for interactive 3D content presentation became the first milestone of our *LINDSAY* project.

14.1.2 Bringing Virtual Physiology to Life

LINDSAY Presenter supports the visualization and exploration of static contents on human anatomical structures. Integrating human physiological processes into an anatomy model is challenging. Highly dynamic processes—resulting from a densely connected network of interacting components—have to be illustrated over time and within 3-dimensional spaces. We did not want to resort to mere animations of such contents [3], as this would not allow for a wide range of interactive exploration of the underlying networked units and their interactions. Furthermore, we want users of *LINDSAY* to be able to navigate to any part of the virtual human body and, even more importantly, be able to quickly zoom to physiological processes at (almost) any scale. More precisely, a user should be able to experience human anatomy on the whole-body level, on the level of organs, tissues, and cells, down to sub-cellular components. This requires an integration of mathematical and computational models of the various processes across multiple scales in space and time. *LINDSAY Composer* is our first prototype of such a multi-scale programming, composition, and visualization toolset. It allows us to explore and present physiological processes within a virtual 3D female and male anatomy model.

14.2 Related Work

Before we expand on our approach to building a virtual human, we give a brief overview of virtual anatomy, the use of anatomy atlas databases, and the use of component-based frameworks for building dynamic software environments.

14.2.1 Replicating Human Anatomy and Physiology

For more than two decades, scientists have been exploring ways to enhance medical research and education through computer-based renderings of human anatomy and physiology. The American National Library of Medicine started as early as 1989 with the composition of a comprehensive imagery database of human physiology, also referred to as the Visible Human [4]. Since then, these data sets have inspired a large number of virtual anatomy projects for research, patient consultation and education [26].

14.2.2 Virtual Human Anatomy and Physiology

In the context of education, atlases have been composed that promote the exploration of detailed anatomical terminology in a proper visual context [21]. Some systems have been further extended to incorporate data about actual biomedical processes in the human body, for instance processes of gene control [27]. One of the projects at the forefront of modeling physiology by integrating processes across scales is the Physiome project [14], which takes signalling and metabolic pathways into account to connect protein models with cell simulations. The objective is to build tissues and organs through this bottom-up approach. Such systems can be modelled and simulated by means of traditional mathematical methodologies or as large sets of self-organizing bio-agents, or *swarm* systems [8, 15, 34]. The latter, agent-based approach forms the foundation for our computational models to implement physiological processes across multiple scales.

14.2.3 Components as Dynamic Building Blocks

In addition to multi-scale modelling, we rely on a component-based architecture for a framework to generate and deliver anatomical and physiological contents. The required diversity of data types and computational processes for an integrative simulation and presentation tool can be addressed by a component-based software architecture [12]. More specifically, a component can be broadly defined as a unit of independent deployment, which has a persistent state [18]. The design of component-based software architectures has advantages in regard to various application domains. Frameworks for (human) collaborative work can be implemented by brokering groupware components [29]. The coordinated execution of heterogeneous components works for organizing human collaboration as well as complex code bases that comprise large sets of interoperable, reusable software components. For example, a component-based augmented reality framework could manage elements and modules for user interfaces, tracking or object modelling [6].

Computer games face a similar challenge of integrating vast numbers of software components, whether related to contents, providing networking infrastructure or user interfaces [28]. Component-based frameworks have been investigated and applied in the context of Massively Multiplayer Online games [1]. For practical reasons, multi-facetted game units are defined by aggregating distinct software components rather than by using established methodologies of object-oriented inheritance [32, 2].

An overview of component-based client/server frameworks is provided in [19]. Sets of component-based distributed embedded systems facilitate coordinated interactions [5]. Redeployment of components across a network infrastructure results in improvements regarding service availability [20]. Alternatively, mobile devices can exchange software components in peer-to-peer networks in order to address user requests [25]. The exchange of software components in heterogenous hardware infrastructures might require adaptation of the control over the respective components or of their data. In [11], such an adaptation strategy is presented for transferring components among devices of varying degrees of computational power, e.g., from a desktop/server to a set of mobile devices. In particular, adaptation is realized by specific drivers that serve as middleware to translate the broadcast software components.

14.3 The LINDSAY Virtual Human

We started developing *LINDSAY Virtual Human* in May 2009 as a collaboration between the Department of Computer Science in the Faculty of Science and the Undergraduate Medical Education program in the Faculty of Medicine. The *LINDSAY* project currently has a demonstration room that replicates a medium size class room (20-30 students) with a large rear-projection screen. With *LINDSAY* we create, use and test (1) stereoscopic display technologies, (2) software for 3D content sharing across the internet, (3) motion capture technologies to develop seamless gesture-based user interfaces, and (4) wireless remote controls and touch interfaces based on iPhones, iPod touch devices and iPads. Figure 14.1 provides a snapshot of the different components of the *LINDSAY* project that its development team is currently working on.

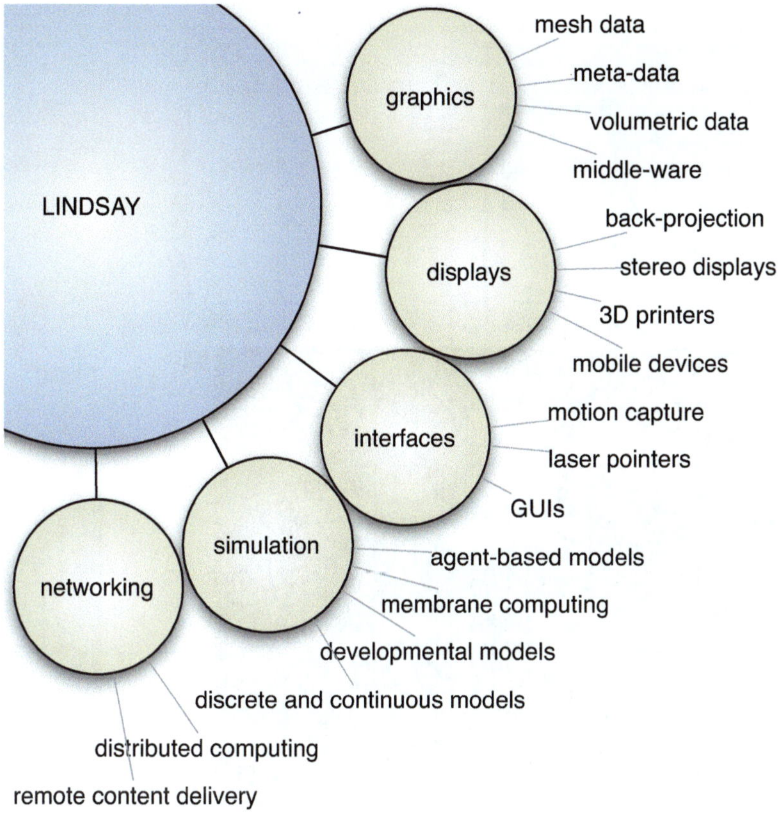

Fig. 14.1. As a truly interdisciplinary project, the *LINDSAY* system requires the integration of ideas, techniques and solutions from a wide range of fields within the disciplines of information and communication technologies.

14.4 LINDSAY Presenter

A typical scenario for an anatomy lecture for 1st year students works usually like this: The anatomy instructor is standing in front of a class of a few hundred students, elaborating on the human skeleton, muscle, and nervous systems. The instructor is using a plastic model to explain the various muscles of the hand, draws with a pen on rubber gloves to illustrate tissue connectivity, or uses projected images on a screen in front of the class in the form of a slide presentation. This is not an ideal situation for the following reasons:

- Only very few students can actually follow the instructor's explanation, as they sit close enough to see which anatomical structures are being pointed out. Most of the students in the classroom are relegated to what

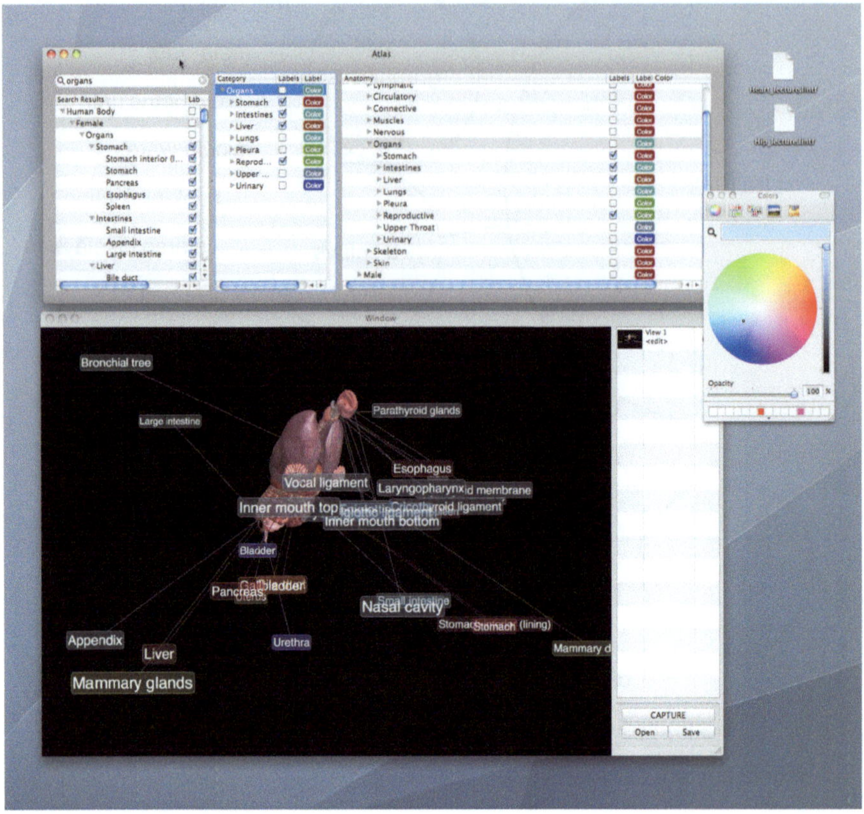

Fig. 14.2. The *LINDSAY* Presenter interface with its two main windows: (a) The top window displays the hierarchical list of the male and female anatomy atlas. (b) Below is the main display window, in which the currently selected anatomical structures are visualized.

they see on the projected slides, which they might (in many situations they don't!) have also on their laptop screens in front of them.

- The location of anatomical structures within the human body is best understood by getting a true feeling of their arrangements in 3D space. That is, it helps to be able to rotate the body, dissect to reveal hidden structures, and reassemble to get a true sense of each structure's exact position.
- The interrelationships between anatomical structures—such as organs, connective tissues, muscles, cardiovascular or lymphatic networks—is best revealed by, again, studying their points of connection and their distribution across the body.

Obviously, these aspects are hard to illustrate to students without being able to convey the true dimensionality, spatial arrangement and connectivity of anatomical structures. Furthermore, complex structures of the human body and their interrelationships reveal their functions much more easily by seeing them in action and in coordination with other components. Our *LINDSAY Virtual Human* tries to address exactly those issues.

14.4.1 Anatomy Atlas

The backbone of *LINDSAY* is an anatomical database, derived from a male and female anatomy model acquired from Zygote, Media Group, Inc. [35]. The model consists of a set of 3D surface files, organized in folders according to anatomical categories (vascular system, circulatory system, muscles, etc).

However, each file has labels which are often abbreviated and sometimes have no reference to anatomical categories of parts. Using anatomical atlases, we properly labelled and further categorized the parts, storing the data in an XML file. We converted the model to the COLLADA format, which we use as our data interchange format [10]. This facilitates the storage and transfer of not only static model data, but also rigging and information relevant to animation. We import and store the model data in our own binary format, along with the XML file containing the proper labelling and hierarchical organization of anatomical parts.

A snapshot of the user interface to the database is depicted in Figure 14.2. The atlas contains female and male anatomy, which is accessible via a hierarchical list. The selections through this interface determine which anatomical structures are displayed in the visualization window. Both single and multiple selections are possible, so that, for example, the cardiovascular system (heart and blood vessels) can be displayed in combination with the lymphatic network of the immune system.

The database is also searchable. Search terms can be entered in the field on the top of the left column of the atlas interface. Below this search field, the entries that match the query are displayed. From this result list, further selections can be made to add to the components displayed in the visualization window. The centre column serves as a common-item placeholder. Items from the atlas or search can be dragged in this middle table in order to keep only those components that are needed in a neat and arranged list.

Each entry in the atlas can be annotated with a label in the visualization window. This is illustrated in Figure 14.2, where many of the displayed inner organs are identified by their labels. The labels are arranged in the 3D space around the virtual human. Whenever the model is rotated, the positioning of the labels is adjusted to help with their readablity. As the labels are printed on a transparent background, it is usually easy to read most of them at the same time; if necessary, a small rotation of the anatomical structure would quickly reveal any hidden labels. This greatly improves legibility of the labels and gaining an intuitive sense of which label is associated with which anatomical parts. Organizing the labels and identifying the corresponding parts is also facilitated by different colours, which can be set through the colour chooser window.

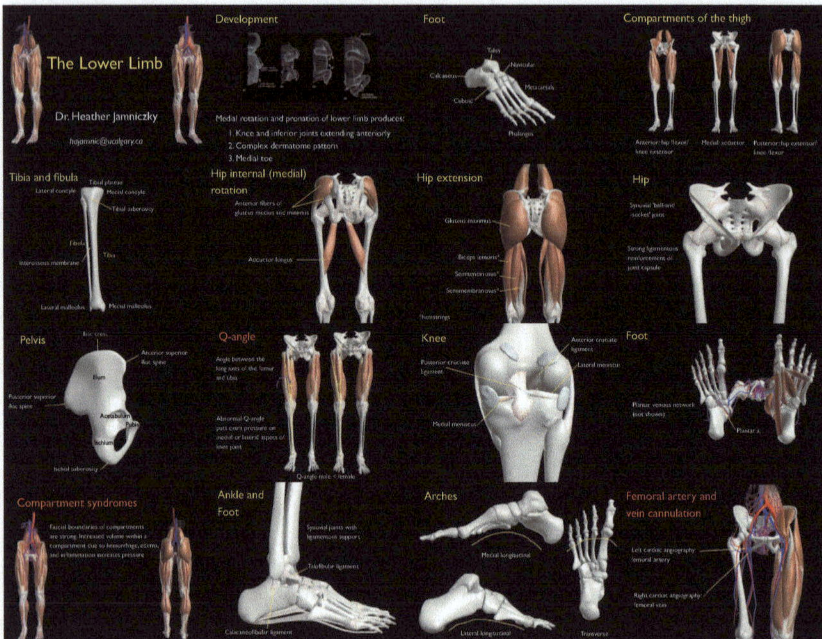

Fig. 14.3. Example slide set created with *LINDSAY Presenter.*

14.4.2 Interactivity

The 3D anatomy model can be rotated, moved, as well as scaled—so that structures from the whole body level to the organismal, cellular, and subcellular level can be inspected. These changes of perspective can be controlled using keyboard commands in combination with a normal 2D mouse and its control buttons. In addition, we have created a remote control application which runs on iPod touches, iPhones, and on a Pen-Smart pad.

We have also designed prototypical implementations of a gesture-based control interface, where gestures are identified on a video stream from a camera and then translated into navigational commands within the virtual human. Although they were generally well received by the anatomy instructors, they could not (easily) allow for a high degree of interaction, comprising numerous commands. We have conducted experiments with laser pointers that we can use as drawing devices on the projection screen. Here, again, a video camera behind the screen is identifying the position of the laser beam on the screen, translating these into the proper coordinates within the main display window. In the meantime, we have replaced the laser pointer by an iPod touch and iPad, where the drawing is now performed on the touch surface. Figure 14.4(a) shows the navigation screen of our remote control application on an iPhone. Moving a single finger across the touch interface relocates the virtual human in its x-y-plane. Moving two fingers apart or close to each other will increase or decrease the zoom level, respectively. Rotating the iPhone/iPod around its central and lateral axis will make the virtual human turn accordingly. A vertical two-finger swipe moves a customizable cutting surface in and out of the virtual body, which allows one, for example, to see through the ribs, skin and muscles into the inside of the heart. This combination of touch gestures and rotation makes the iPhone/iPod an intuitive remote control device for instructors, who should not need to use the keyboard or mouse for navigation of and around the virtual human.

The iPhone/iPod application can also be switched to atlas mode, where a replica of the hierarchical anatomy ontology is available through the touch interface. Figure 14.4(b) illustrates how this touch interface works. As the iPhone/iPod screen is rather small, it is not practical to display the hierarchical anatomy list all at once. Rather, we opted to show only a single hierarchical level per screen. Selections of anatomical structures are made by double-clicking. Any selections made through the iPhone/iPod interface are immediately replicated on the atlas interface in the main application (Fig. 14.2).

Both through the main *LINDSAY Presenter* and its wireless touch devices, the user can draw simple shapes and lines to enhance or highlight the 3D model currently on display. Custom labels can also be added, which can optionally be connected to specific anatomical structures. These labels are in addition to the already built-in annotations, which display the names of the anatomical structures in the atlas.

14.4.3 Creating 3D Slides

As *LINDSAY Presenter* is meant to be used as a presentation tool of human anatomy, we have included an easy mechanism to build sequences of 3D slides. This is similar to the way standard slide presentation software, such as Powerpoint© or Keynote© work; after having created a set of slides, one can progress from slide to slide to enhance an oral presentation. In

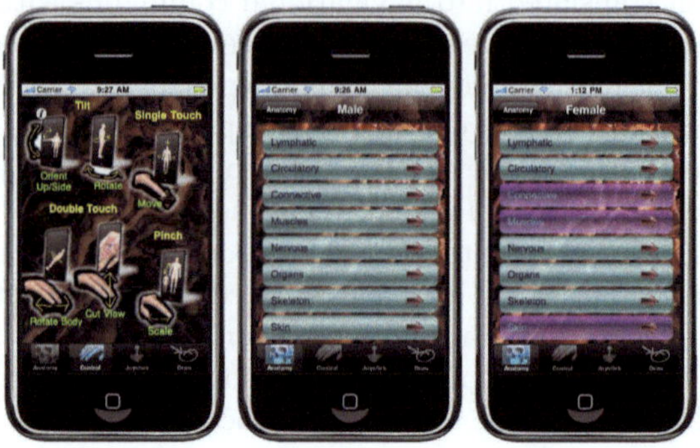

(a) (b) (c)

Fig. 14.4. The *LINDSAY Presenter* remote control applications running on iPhone and iPod touch devices: (a) The navigation interface allows the virtual body to be moved, scaled, and rotated. Different styles of cuts can 'open' the body. (b) The interface for the atlas gives access to all anatomical structures. (c) Three categories of anatomical parts are selected (indicated by the pink background), which are instantly displayed in the main window.

LINDSAY Presenter slides are actually scene descriptions of the 3-dimensional contents; therefore, we refer to these scenes as "3D slides". At any point a snapshot of the current 3D scene can be taken by selecting the `CAPTURE` button at the bottom right corner of the main display window (Fig. 14.5). This then adds a thumbnail in the 3D slide column to the right of the display window. Flipping between 3D slides creates a smooth, animated transition between the associated scenes. A set of such 3D slides can be saved (`Save`) and later loaded (`Open`) into the *LINDSAY Presenter* for subsequent presentation, where the 3D slides can be used as guidance. Figure 14.6 gives examples of such a sequence of 3D slides, here illustrating the anatomy of the female hip. It is worth mentioning, that each 3D slide is, of course, completely interactive, as the "slides" are not simple screen captures of the current 3D contents, but contain the actual scene graph of the current display.

14.4.4 Volumetric Data Integration

We acquired image slice data (Fig. 14.7(a)) of the human body from the Visible Human Project [22]. Using the OpenGL Shading Language, we can render a cut plane of the head from any orientation (Fig. 14.7(b)). Combining the shape of the head with a cut plane, an arbitrary cross-sectional view from any position or orientation can be achieved as demonstrated on the iPhone (Fig. 14.7(c)). We also render using multiple layers to visualize all

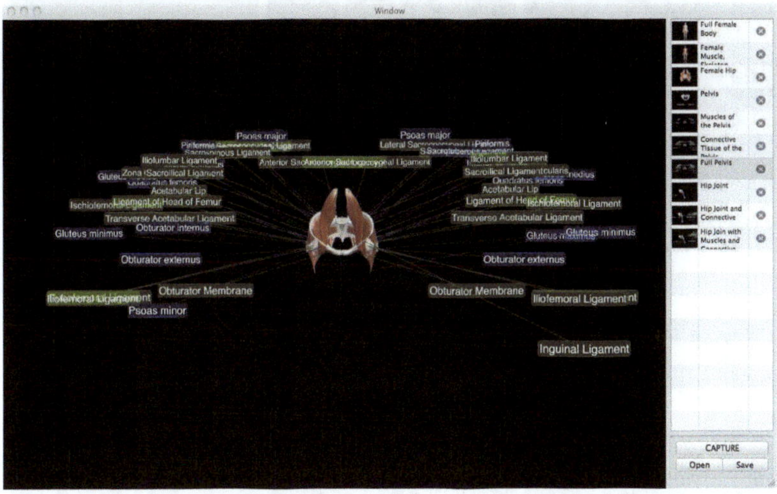

Fig. 14.5. 3D Slides in *LINDSAY Presenter*: Demonstration of a typical setup for a lecture, here, as an example, on muscle anatomy. See Figure 14.6 for close-ups of selected 3D slides.

the slices simultaneously to achieve a volumetric visualization; for this we use the shader again to allow for an arbitrary viewing orientation(14.7(d)). With the application of transparency and colour filters, inner structures can be revealed or isolated (Fig. 14.7(e), (f)).

The Zygote model [35] is stylized and representational, designed to be illustrative of the major structures and well suited for educational purposes. For example, arteries are coloured solid red and veins are coloured solid blue. The Visible Human slice data on the other hand is from photographs taken of actual slices of a cadaver, and often individual organs aren't as readily discernible and surface boundaries not always obvious.

Rendering the data sets together allows for further exploratory abilities. Figures 14.7(d-h) show the slice data superimposed over the 3D model data. Thus, direct comparisons can be made between the representational visualization and the realistic and physically accurate volumetric datasets. In Figures 14.7(g) and 14.7(h) the cut plane reveals the inner structures of the 3D model, while the slice data remains simultaneously overlaid.

14.5 LINDSAY Composer

One of the key objectives for *LINDSAY* is to support the creation of computational models in real-time and combine this with 3-dimensional, highly

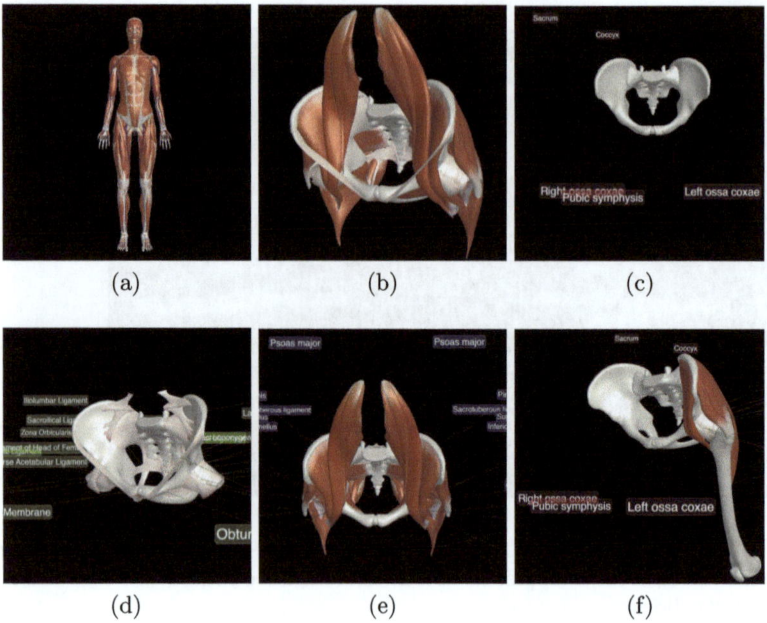

Fig. 14.6. Examples of 3D Slides in *LINDSAY Presenter*: A typical sequence of 3D scenes that would mark keypoints of navigation through the *LINDSAY* anatomy. Smooth, animated transitions are generated to blend between slides: (a) Female muscle skeleton, (b) female hip, (c) pelvis, (d) connective tissue of pelvis, (e) pelvis muscles, (f) hip joint with muscles.

interactive visualizations of physiological processes related to the human body. *LINDSAY Composer* is a primarily graphics-based programming environment, through which complex biological processes can be compiled using simple drag-and-drop interfaces. Available "building blocks" for such compositions are, for example, fluid fields, cameras, anatomical objects, protein shapes, or cell types. A time line helps to trigger and keep track of events within the simulation, which then allows one to run and replay simulations as needed (compare Fig. 14.12).

Similar to first-person 3D computer games, the *LINDSAY Composer* creates an environment that immerses the user into a virtual 3-dimensional world, with the context of anatomical structures and processes. Different from prerecorded contents, the underlying computational engines, in combination with the dynamic scene composition, allow for interactive explorations that are guided by the user's intuition and curiosity. Software engineering methodologies, hardware advancements and agent-based models turn simulations into a highly interactive multimedia experience.

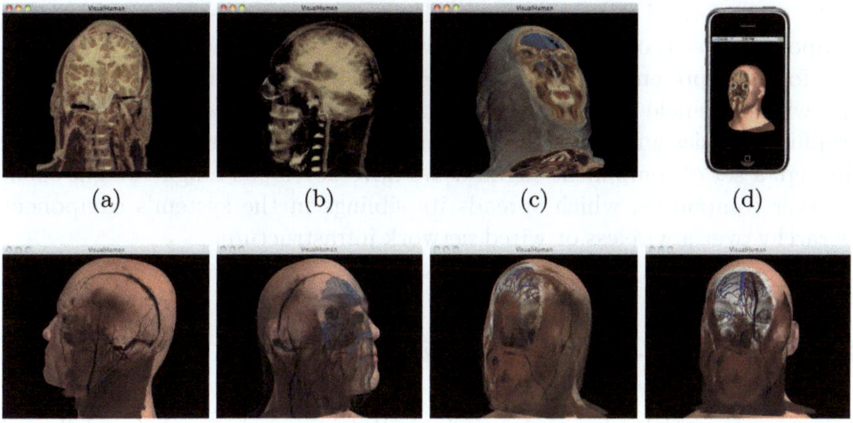

(a) (b) (c) (d)

(e) (f) (g) (h)

Fig. 14.7. Integration of volumetric data from the Visible Human with surface model data. (a-c) Renderings of the volumetric data from the Visible Human data set within *LINDSAY Presenter*; (e-h) the anatomical model data is superimposed on the volumetric data set. (d) The renderings work on both desktop and mobile devices.

14.5.1 The Computational Framework

LINDSAY is being built as a fully functional, 3-dimensional model of male and female anatomy and physiology. Our objective is to use the virtual human to illustrate, for example, which muscles are involved when we run or when we hold a pen. Similarly, we can illustrate and investigate the functional structures and defense processes involved in the human immune system. The immune system, in particular, encompasses complex processes across a wide range of scales: from the organismal level (e.g., the thymus) to the networked lymphatic system (including lymph nodes) down to the cellular (B-, T-cells) and the sub-cellular levels (gene regulation in response to pathogens).

Consequently, *LINDSAY*'s modeling framework aims to integrate mathematical and computational models across scales. For example, we use fluid dynamics to simulate blood flow at the whole body level. Using interactive zooming, we can get inside a blood vessel, dive into a capillary or hop onto a blood cell to carry us through the blood stream.

At this level we switch to agent-based simulations, where "bio-agents" (i.e., cells, proteins and other molecular structures) are being tracked within the simulated 3D virtual body space. Physics engines, mainly used for computer games, have been adjusted to provide realistic and yet quickly computable scenarios that replicate physiological processes within the human body.

As part of the LINDSAY Virtual Human project, we have developed a component-based computational framework that allows the utilization of various formal representations, computation engines and visualization technologies within a single simulation context. For our agent-based simulations, the graphics, physics and behaviours of our interacting entities are implemented through a set of component engines. We have developed a light-weight client/server component, which spreads its siblings in the system's component hierarchy over a wireless or wired network infrastructure.

14.5.2 Agent-based Modelling

Agent-based models play an increasingly dominant role for the modeling of biological and social systems [17, 33, 24]. We see the most promising potential in agent models that incorporate swarm intelligence techniques [7, 9], as these result in more accurate and realistic models. The use of agents as independent and interacting entities is particularly crucial when spatial aspects play a key role in defining patterns of interaction, in understanding their emergent properties, and in helping to shed light on the inner workings of physiological processes.

All biological systems, such as the human body, are inherently driven by interaction processes in a 3-dimensional world. Therefore, in order to capture physiological processes within our *LINDSAY Virtual Human*, we utilize swarm-based, 3-D simulations which exhibit a high degree of self-organization, triggered by relatively simple interactions of a large number of 'bio-agents' of different types.

The *LINDSAY* human body is a perfect example that allows for middle-out modeling [31]. Other models, which we have worked on before and are currently incorporating into *LINDSAY*, include the study of chemotaxis within a colony of evolving bacteria [13, 23], the simulation of transcription, translation, and specific gene regulatory processes [16], as well as studies of affinity and cooperation among gene regulatory agents for the λ switch in *E. coli* [15].

14.5.3 Component Architecture

LINDSAY Composer is written in Objective-C, but is also compatible with C, C++, Python and any other scripting languages that can interface with the Objective-C runtime system. The core framework of *LINDSAY Composer* is as minimal as possible. Having such a simple framework makes it very easy for developers to extend the system with plugins that add new component types, new component engines, and even new user interfaces. Without great

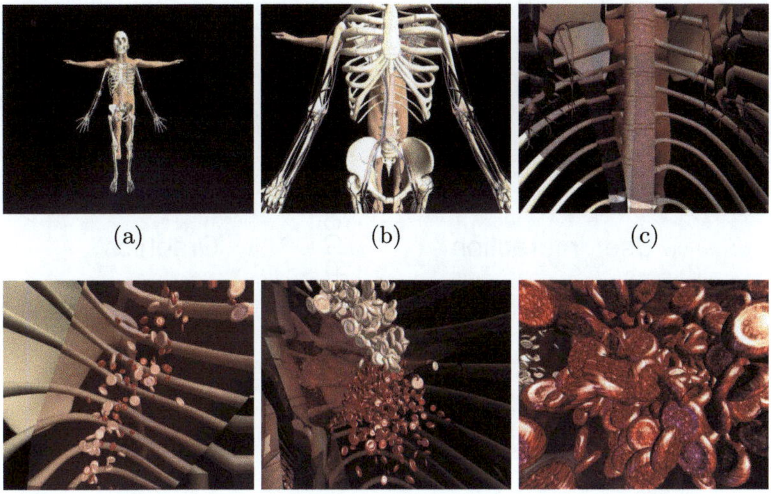

(a) (b) (c) (d) (e) (f)

Fig. 14.8. Computational modeling of physiological processes across scales by example of the circulatory system: (a) full body view, (b) close-up with upper skeleton, (c) inside the rib cage, (d) approaching the main artery, (e) inside the main artery with red and white lymphocytes, (f) close-up of a cluster of lymphocytes.

efforts, components can collaborate with each other due to the simplicity of a component's API.

In *LINDSAY Composer* a simulation is represented as a hierarchy of components that encapsulate state and behaviour. A component may be registered with a computational engine (Fig. 14.9). For instance, one might have a physics engine managing rigid body interactions, those rigid bodies would be represented in the simulation by physics components. It is not necessary for a component to be registered with a component engine, as is the case with a component that holds only position and orientation data, which we call a transform component.

Components can query their parents for siblings of a given type. Components can also query the entire simulation for components that match a given expression. A common use for this mechanism is for sibling components to share a transform component. For example, one might have a physics component updating the transform, while a rendering component draws a 3D mesh relying on the very same transformation information. In another instance, a camera may be manipulated by a user interaction component that updates a transform component's orientation and position. A graphics component would then adjust the view in the direction indicated.

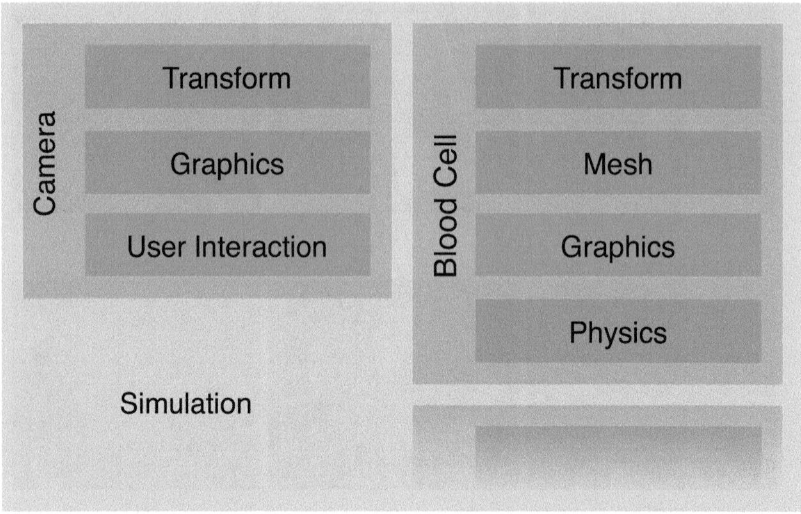

Fig. 14.9. A component hierarchy typical of most simulations. Boxes represent components, the embedding of boxes within each other is representative of the component hierarchy, shading is used to exemplify this.

LINDSAY Composer uses a number of engines to form the core computational framework (Fig. 14.10). For instance, a *physics* engine manages rigid body interactions and dynamics, a *graphics* engine renders the simulation to the screen, a *networking* engine handles the distribution of components to other nodes in a network, and an *interaction* engine allows the user to interact with the simulation via mouse and keyboard or multi-touch enabled devices. Each simulation engine iterates through the components it manages to update them. Figure 14.10 shows how components are associated with different component engines while still being part of a component hierarchy.

Especially important for the interactive simulation control as well as for collaborative classroom experiences are the networking components of the *LINDSAY Composer* [30]. A server component is anchored into the simulation's component hierarchy which replicates its siblings and sends them over a network to one or more client components. There can be multiple servers and clients within a single simulation. This allows for distribution of a shared simulation to a number of client devices, which is a typical scenario in a classroom setting. It also allows for interaction with a simulation via mobile devices such as the iPod Touch, iPhone or iPad.

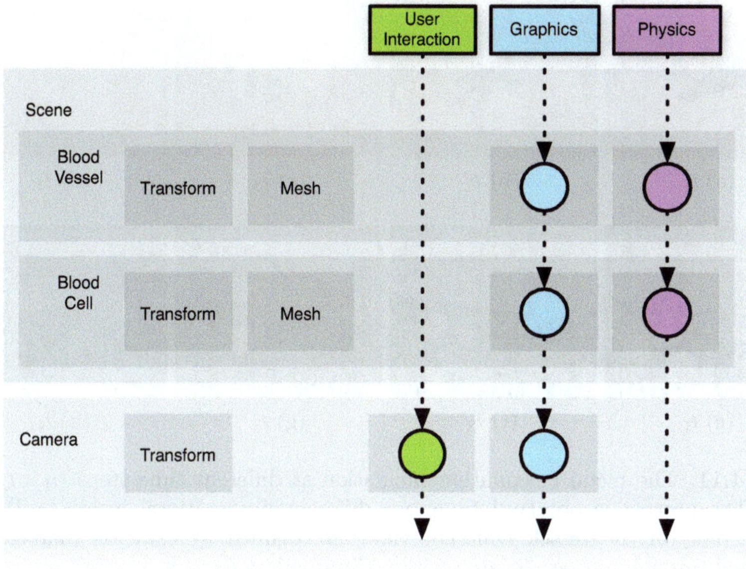

Fig. 14.10. Example architecture of component engines: The square boxes at the top are simulation engines for user interaction, graphics, and physics. Elements in the simulation, such as blood vessels and blood cells, contain components for their rendering, interaction, and physical properties. Those components register with the respective engines (denoted by the arrows). Other components, such as transform and mesh, are not registered with a simulation engine. As in Figure 14.9, the nesting of boxes (and circles) corresponds to a components place within the component hierarchy.

14.5.4 Graphical Programming Interface

In this section we explore the *LINDSAY Composer* user interface. Figure 14.11 shows a blood clotting simulation that we built with *LINDSAY Composer*. The simulation displays two separate views of the same model at different time steps [34]. At the top one can see red and white blood cells, fibrinogens, fibrins, platelets, and other messenger molecules. All of these are implemented as separate agents, that interact with one another and with the blood vessel wall, which is lined by endothelial cells. Over time, one can see that the wound site is eventually filled by platelets and fibrins, which stops the simulated bleeding. The bottom row shows the same simulation from a perspective outside of the blood vessel, which is embedded in the circulatory system of the virtual body (compare Fig. 14.8).

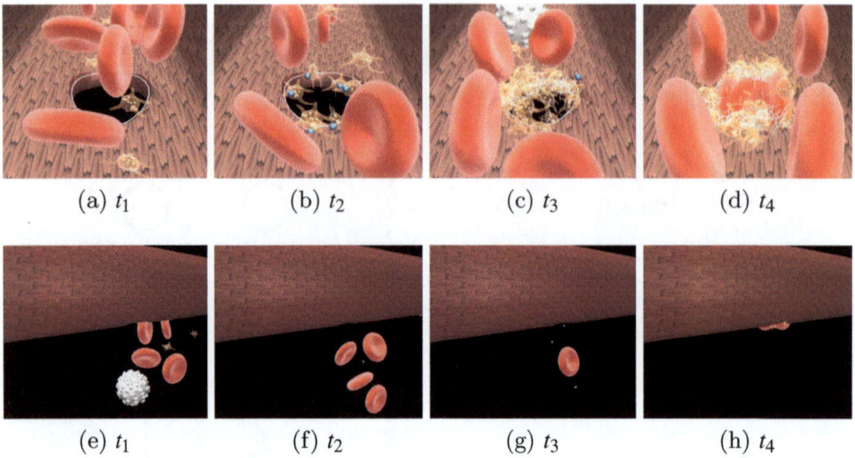

Fig. 14.11. The blood coagulation simulation at different time steps ($t_1 < t_2 < t_3 < t_4$). The process is observed from two different perspectives: inside (a-d) and outside (e-h) of the vessel. Different views are defined by cameras that can be navigated through and are located within the simulations.

Building a simulation in *LINDSAY Composer* is achieved via a drag and drop interface. Users can drag components from a library of component types into the simulation view, where they will appear immediately if they have a graphical representation. Figure 14.12 shows a typical screen display during the modelling process of a simulation.

In the toolbar, at the top of the window, we see controls for the camera, for bringing up the scene inspector (Fig. 14.13), for the frame rate, and for hiding the template library and timeline views. The user may directly interact with the scene by way of a 3D navigation and selection interface. Navigation and selection is performed via mouse and keyboard, or via networking components through mobile devices with touch interfaces.

In Figure 14.12, the component library is displayed on the left-hand side. From here, components are dragged into the simulation view on the right. Directly below the simulation view is the timeline view. We use a timeline to control the lifetime and properties of components within a simulation. This allows for the dynamic control of a simulation. The horizontal bars in the timeline correlate with the objects that exist in the simulation.

Overlaying the topmost time-bar we see an interpolation graph, which is represented as a component within the simulation. In our blood clotting simulation this graph corresponds to the velocity of the blood within the blood vessel, hence we can simulate the pulsing action of the heart on blood within a blood vessel. These interpolation components can be applied to any compatible property of any component, or set of components within the simulation.

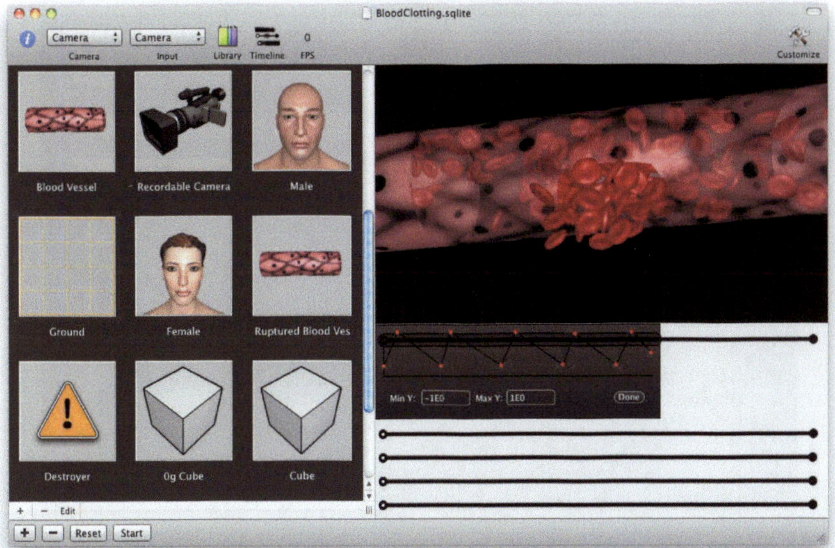

Fig. 14.12. Components are dragged from the left into the rendering view on the right to compose a simulation. An overview of the interaction processes is provided in a configurable timeline (bottom-right). The centred overlay window shows an interpolation component that directs the blood flow.

Another important part of the *LINDSAY Composer* interface is the inspection of templates and of the hierarchy of a simulation (Fig. 14.13). We dynamically generate the interaction dialogs for a component at runtime. This user interface allows for the realtime inspection and editing of properties as the simulation runs. We also allow for preprogrammed interfaces by checking for their existence at runtime.

14.6 The Educational Perspective

Medicine is replete with complex information spaces. Integrating physiology and pathology, as well as functional anatomy, within true anatomical displays—as opposed to full cartoon images of the body—gives users the opportunity to learn multiple content pieces at the same time. From an educational perspective, we are approaching an application of cognitive flexibility theory and constructivism within a simulation environment that most closely represents true (not cartoon) anatomy.

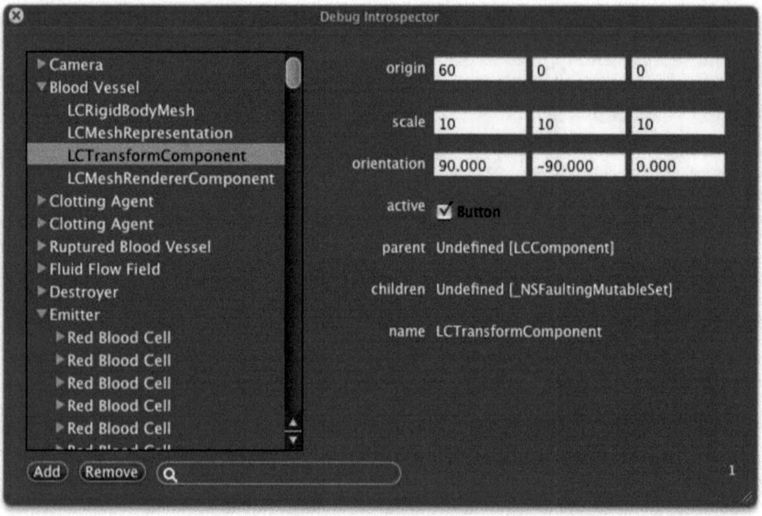

Fig. 14.13. (a) The *LINDSAY Composer* GUI for inspecting the scene, its hierarchy, and its components.

Putting the tools in the hands of the users allows them to build their knowledge and to construct pathways for their own learning. Each learner is able to pace their learning, and/or create a physiological or anatomical environment that suits their personal learning needs. Faculty is able to construct, play, pause, rewind, re-create and deconstruct the learning environment in order to emphasize, assess, or modify learning tempo and goals to meet learners' needs.

14.7 Current and Future Work

Interactive Classroom

After anatomical and/or physiological scenarios have been created using *LINDSAY Composer*, one can use *LINDSAY Presenter*to run, demonstrate and explore these compositions. We are currently working with instructors from medicine, nursing, kinesiology, and veterinary medicine to turn our software into an effective presentation toolset for the classroom. Using the described client/server architecture, we are now able to create multiple visualization and remote control components that are shared on a wireless network. This lays the foundations to create innovative scenarios for inquiry-based learning and teaching. Using iPods and iPads, we deliver visualizations to wireless devices, which, in turn, can be used to inspect and control different aspects of a shared physiology model, which runs on a powerful simulation server.

Remote Learning and Content Sharing

Building an infrastructure for immersive visualization revolutionizes the way human anatomy and physiology is being taught. Such 3D contents can be shared worldwide across the internet by using 3D content sharing technologies, where multiple users can see, "grasp" and interact with 3-dimensional anatomical objects—from organs to cells—and observe the unfolding of simulated physiological processes. This means, contents produced in one location can be distributed to other educational and research institutions across the globe.

14.8 Conclusions

We described the *LINDSAY Virtual Human* project, gave details on its programming frameworks, as well as its display, visualization, and interaction technologies. We also shared our initial experiences about how to bring computer modeling, 3D visualization and human-computer interaction technologies into medical classrooms. More information on the *LINDSAY Virtual Human* project is available on our website:

`http://lindsayvirtualhuman.org`.

Acknowledgments. We would like to thank the Department of Undergraduate Medical Education, Faculty of Medicine, at the University of Calgary for supporting the *LINDSAY Virtual Human* project. We also greatly acknowledge financial support for the *LINDSAY* project by the Lindsay Leigh Kimmett Memorial Foundation (lindsaykimmett.org). Part of this project is supported by the Natural Sciences and Engineering Research Council of Canada. The anatomy model data sets are provided by Zygote Media Group, Inc. (www.zygote.com).

References

1. Entity systems are the future of MMOG development (September 2007), `http://t-machine.org/index.php/2007/09/03/entity-systems-are-the-future-of-mmog-development-part-1`
2. Evolve your hierarchy (January 2007), `http://cowboyprogramming.com/2007/01/05/evolve-your-heirachy`
3. Harvard BioVisions (January 2012), `http://multimedia.mcb.harvard.edu`
4. Ackerman, M.J., Yoo, T., Jenkins, D.: From data to knowledge - the visible human project® continues. In: MEDINFO 2001, pp. 887–890 (2001)

5. Årzén, K., Bicchi, A., Dini, G., Hailes, S., Johanesson, K., Lygeros, J., Tzes, A.: A component-based approach to the design of networked control systems. European Journal of Control 13(2), 261–279 (2007)
6. Bauer, M., Bruegge, B., Klinker, G., MacWilliams, A., Reicher, T., Riß, S., Sandor, C., Wagner, M.: Design of a component-based augmented reality framework. In: International Symposium on Augmented Reality, Los Alamitos, CA, USA, p. 45 (2001)
7. Bonabeau, E., Dorigo, M., Theraulaz, G.: Swarm Intelligence: From Natural to Artificial Systems. Santa Fe Institute Studies in the Sciences of Complexity. Oxford University Press, New York (1999)
8. Burleigh, I., Suen, G., Jacob, C.: Dna in action! a 3d swarm-based model of a gene regulatory system. In: First Australian Conference on Artificial Life, ACAL 2003, Canberra, Australia (2003)
9. Camazine, S., Deneubourg, J.-L., Franks, N.R., Sneyd, J., Theraulaz, G., Bonabeau, E.: Self-Organization in Biological Systems. Princeton Studies in Complexity. Princeton University Press, Princeton (2003)
10. COLLADA.org. Collada - digital asset and fx exchange schema (April 2010), `http://www.collada.org`
11. de Lara, E., Wallach, D.S., Zwaenepoel, W.: Puppeteer: Component-based adaptation for mobile computing. In: USENIX Symposium on Internet Technologies and Systems (2000)
12. Heineman, G.T., Councill, W.T.: Component-Based Software Engineering. ACM Press (2001)
13. Hoar, R., Penner, J., Jacob, C.: Transcription and evolution of a virtual bacteria culture. In: IEEE Congress on Evolutionary Computation. IEEE Press, Canberra (2003)
14. Hunter, P.J., Borg, T.K.: Integration from proteins to organs: the physiome project. Nature Reviews: Molecular Cell Biology 4, 237–243 (2003)
15. Jacob, C., Barbasiewicz, A., Tsui, G.: Swarms and genes: Exploring λ-switch gene regulation through swarm intelligence. In: IEEE Congress on Evolutionary Computation, CEC 2006 (2006)
16. Jacob, C., Burleigh, I.: Biomolecular swarms: An agent-based model of the lactose operon. Natural Computing 3(4), 361–376 (2004)
17. Johnson, S.: Emergence: The Connected Lives of Ants, Brains, Cities, and Software. Scribner, New York (2001)
18. Leavens, G.T., Sitaraman, M. (eds.): Foundations of component-based systems. Cambridge University Press, New York (2000)
19. Lewandowski, S.M.: Frameworks for component-based client server computing. ACM Comput. Surv. 30(1), 3–27 (1998)
20. Mikic-Rakic, M., Malek, S., Medvidovic, N.: Improving availability in large, distributed component-based systems via redeployment. Component Deployment, 83–98 (2005)
21. Mitsuhashi, N., Fujieda, K., Tamura, T., Kawamoto, S., Takagi, T., Okubo, K.: BodyParts 3D: 3D structure database for anatomical concepts. Nucl. Acids Res. (2008), gkn613
22. N. L. of Medicine. National library of medicine: The visible human project (April 2010), `http://www.nlm.nih.gov/research/visible/visible_human.html`
23. Penner, J., Hoar, R., Jacob, C.: Bacterial chemotaxis in silico. In: First Australian Conference on Artificial Life, ACAL 2003, Canberra, Australia (2003)

24. Pogson, M., Smallwood, R., Qwarnstrom, E., Holcomb, M.: Formal agent-based modelling of intracellular chemical interactions. BioSystems (2006)
25. Roussain, H., Guidec, F.: Cooperative component-based software deployment in wireless ad hoc networks. Component Deployment, 1–16 (2005)
26. Seifert, M.F.: Visible human projects special issue. Clinical Anatomy 19(3), 191–289 (2010)
27. Sensen, C.W., Soh, J.: CAVEman, An Object-Oriented Model of the Human Body. In: Sensen, C.W., Soh, J. (eds.) Advanced Imaging in Biology and Medicine, vol. II, pp. 289–300. Springer, Heidelberg (2009)
28. Stoy, C.: Game Object Component System. In: Game Programming Gems 6, pp. 393–403. Charles River Media (2006)
29. Tietze, D. A.: A Framework for Developing Component-based Co-operative Applications. PhD thesis, Technische Universität Darmstadt, Darmstadt, Germany (2000)
30. von Mammen, S., Davison, T., Baghi, H., Jacob, C.: Component-based networking for simulations in medical education. In: MediaWiN 2010: IEEE Workshop on Multimedia Applications Over Wireless Networks, Riccione, Italy. IEEE Computer Society, IEEE Press, Washington (2010)
31. Walker, D.C., Southgate, J.: The virtual cell–a candidate co-ordinator for 'middle-out' modelling of biological systems. Briefings in Bioinformatics 10(4), 450–461 (2009)
32. Wilson, K.: Game object structure: Inheritance vs aggregation (July 2002)
33. Wolfram, S.: A new kind of science. Wolfram Media Inc., Champaign (2002)
34. Yazdanbod, I., Marcus, S.: An agent-based simulation of blood coagulation processes. The Journal of Undergraduate Research in Alberta 1(1), 13–18 (2010)
35. Zygote Inc. Zygote human anatomy 3d model (April 2010), http://www.zygote.com

Chapter 15
Comparison of Enterprise Integration Modelling Concepts Based on Intelligent Multi-Agent System

Paul Eric Dossou, Paweł Pawlewski, and Philip Mitchell

Abstract. CIMOSA and GRAI methodologies are mainly used for enterprise modeling. Two supporting tools, respectively VLPROGRAPH AND GRAIMOD, are developed for supporting these methodologies by using multi-agent concepts. The architecture of VLPRO-GRAPH (Very Long Process Graph) is based on the assumption that the system will support the MPS creation in the ERP system and will be plugged in to the ERP system database by, for example java connector. The concepts of GRAIMOD are detailed. This tool combines the theory of Multi-Agent systems, Artificial Intelligence and Metaphors of Mind. The tools developed are compared in order to defined opportunities for future research. Both tools are developed by using Java language and Jade platform.

15.1 Introduction

The different economic and financial crises existing today have increased the necessity of enterprises to be prepared and well-organised. Enterprise modelling is one way for restructuring them in order to improve their performance and being more efficient. Three methodologies are mainly used for modelling enterprises: PERA, CIMOSA and GRAI. Enterprise modelling involves not only global enterprise performance improvement but also local improvements.

Paul Eric Dossou · Philip Mitchell
Icam Group, Icam Vendee, 28 Boulevard d'Angleterre,
85000 La Roche-Sur-Yon, France
e-mail: Paul-eric.dossou@icam.fr

Paweł Pawlewski
Poznań University of Technology, ul.Strzelecka 11 60-965 Poznań; Poland
e-mail: pawel.pawlewski@put.poznan.pl, philip.mitchell@icam.fr

J. Kołodziej et al. (Eds.): Advances in Intelligent Modelling and Simulation, SCI 422, pp. 351–380.
springerlink.com

The first part of this chapter will present how to use CIMOSA for improving enterprises locally. The chapter presents the results of the authors' studies which consisted in planning the process taking place in an enterprise characterized by the manufacture of complex products (machine building). The idea of the so called domains of the CIMOSA concept, which has been used for the modelling, has been presented. A planning process based on the multi-agent architecture called VLPRO-GRAPH has been extended by the concept of a special agent which checks the material flow in the production system. The formal model of this flow has been introduced. Finally results of the numerical experiment of analysis and checking process graph are presented.

The second part of this chapter is about GRAI Methodology. GRAI Methodology is used for global performance improvements. GRAI Methodology is designed and defined for managing this modelling. This method is used for example to choose and implement a computer tool (Supply Chain management and ERP) which meets the real market needs (globalisation, relocation, capacity to be proactive, cost optimisation, lead time, quality, flexibility, etc....). GRAIMOD is a tool being developed for supporting the methodology. This chapter proposes to present in detail the concepts of GRAIMOD. Different concepts have been presented relating to modules of the global tool. The general structure of the tool is based on Case Based Reasoning (CBR). Indeed, a Problem Solving Method has been developed in this context for managing the different possibilities of modelling and improvement. These concepts are combined with Multi-agent systems for developing the new tool. Case Based Reasoning (CBR) remains widely used for the definition of the needs for the design and development of expert systems. The originality of this article is that it shows how the reasoning is combined with the theory of Multi-Agent systems and Artificial Intelligence and Metaphors of Mind. The stages leading to the achievement of the scenarios will be presented together with initial development difficulties.

The BDI architecture could be used but we have chosen the training agents because of their particularity to learn. This system is used for defining how to improve the enterprise performance and for capitalizing the process followed. According to CBR concepts the new case studied could be capitalized, but the parameters would also improve the reference model of the enterprise domain. Java is chosen for developing GRAIMOD. The Jade platform is being used in relation with FIPA-ACL for developing the different modules of GRAIMOD.

The final part of this chapter will look at similarities and differences between the two methodologies and the concepts presented for the supporting tools.

15.2 Cimosa and VLProGraph Architectures

15.2.1 Cimosa

CIMOSA (Computer Integrated Manufacturing Open System Architecture) is a methodology used for designing CIM open system architecture, for defining a set of concepts and rules for facilitating the realisation of future manufacturing systems and for aiding the reimplementation of existing systems. This methodology has been developed within the framework of the European strategic programme ESPRIT. It is an architecture for describing enterprises by giving characteristics for modelling production systems on three levels: definition of requirements, design specification and implementation description [6]. CIMOSA is a reference architecture which supports the description of the enterprise from the management level to the shop floor level. CIMOSA consists of an enterprise modelling framework, an enterprise modelling language and an integrated infrastructure. So, it is normal to use this methodology for the improvement of enterprise performance at the local level.

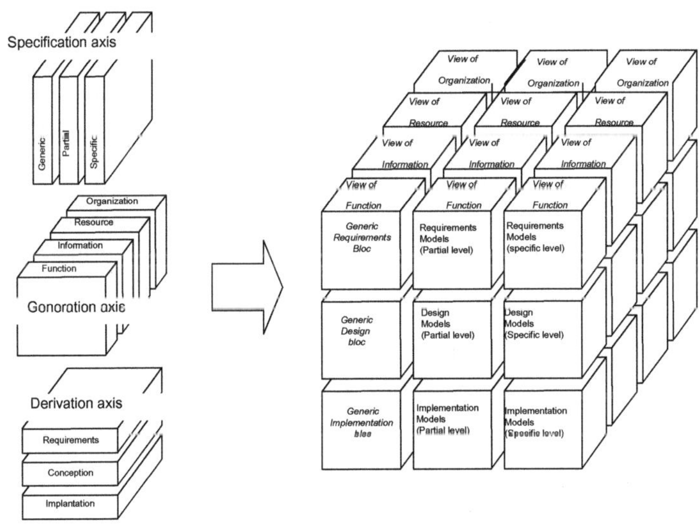

Fig. 15.1. Architecture of CIMOSA

15.2.2 Multi-agent Approaches to Production Planning and Scheduling

Traditional approaches to production planning and scheduling in MRP based logic do not consider real-time machine workloads and shop floor dynamics.

Therefore, there is a need for the integration of manufacturing process planning and control systems for generating more realistic and effective plans. Traditional approaches to production planning and scheduling are based on Centralized Optimization Algorithms, Close Loop Optimization and Distributed Process-Planning (DPP) Approaches. The overview of the approaches to agent-based production planning and scheduling can be found in [9]. Agent-based approaches provide a distributed intelligent solution by multi-agent negotiation, coordination, and cooperation. The following researches refer to the application of multi-agent systems for production planning purpose [9]:

- bidding based approach - the process routes and schedules of a part are accomplished through the contract net bids;
- a multi-agent architecture based on separation of the rough process-planning task as a centralized shop floor planner from the detailed process planning conducted through agent negotiations;
- based on cascading auction protocol, provides a framework for integrating process planning and hierarchical shop floor control.

Agent-based approaches provide a natural way to integrate planning and control, activate and make it possible to simultaneously optimize these functions. Some detailed studies have inspired our research. Meta-Morph is an architecture proposed by Maturana et al. [4]. This is an adaptive multi-agent architecture whose purpose is to integrate distributed manufacturing resources for intelligent manufacturing systems. Distributed manufacturing resources are encapsulated as intelligent agents. They are grouped according to the functionalities such as design, marketing, scheduling and material supply. Mediators are special coordinators that are attached to every group. These Mediators connect agents inside a group. An enterprise Mediator is assigned to connect with the Mediators of each group. Preferably, the main agent is a principle Mediator that connects the enterprise Mediators together and forms a Mediator-centric hierarchical architecture, in which scheduling Mediators perform resource scheduling. An agent-based framework to integrate enterprise resources in order to timely plan and re-plan manufacturing processes is proposed by Peng et al. [8]. In this framework, different functional agents such as a monitoring agent, an analysis agent and a data mining agent to deal with the process planning or fault resolving, are included. A Holonic concept is used to build an agent-based system for logistics and production planning by Karageorgos et al. [3]. They encapsulate manufacturing units at distributed locations as intelligent agents and use agent negotiation mechanisms to optimize logistics and production planning but this solution does not provide effective resource of alternative strategies. Works of Sikora and Shaw [10] focus on the agent coordination mechanisms. They encapsulate manufacturing information system components such as decision software modules, process controllers and database as intelligent agents and integrate the agents as a multiagent framework for a shop floor scheduling.

15.2.3 Movable Resources

According to APICS terminology [1], resources are anything that adds value to a good or service in its creation, production or delivery. Resources are classified as non-renewable and renewable, from the viewpoint of their use, creation and consumption. Non-renewable consumable resources are those which are completely consumed in the production process. Among such resources there are different kinds of materials, forms of energy, products, raw materials, components and auxiliary materials, which are either manufactured in a given company or brought from contractors. As renewable consumable resources we consider machines and machinery, control and instrumentation equipment, human resources, company departments. All these resources are used in the manufacturing process; after use that are released and await another application.

The resources used in the implementation of the process are renewable resources. Renewable resources constitute a system that implements the processes. They are organized as the so-called production structures [2]. That is why, they can be called structural resources. They can be classified as [11]:

- stationary, which are permanently connected with company departments (e.g. machines);
- exchangeable, which in certain time intervals are permanently connected with company departments; yet it takes time to exchange them (i.e. single-purpose tools for machine tools, stamping dies in presses);
- free, which at any time and at negligible costs can be assigned to particular departments (e.g. workers with specific skills);
- mobile, which are assigned to the processed objects in certain departments and released in others (e.g. means of transport: carriages, pallets).

Structural resources are the environment of the process implementation and they create the implementation system of the process. Such a system is a coherent one, which acts as a single entity. Structural resources make a certain supply of resources; among them movable resources can be distinguished. A movable resource is divisible, and either a part of it or the whole resource can be moved between structural units. Examples of movable resources are as follows:

- surface occupied by the implementation system of the process,
- employment measured by the number of employees,
- stocks considered as a full supply of stock in the system (according to the "four wall inventories" concept [12]),
- means of transport assigned to the system.

15.2.4 Implementation

An agent-based system is defined in the present paper as a multi-agent system that acts as a support tool and utilizes databases of main system (ERP system). A multi-agent system is a collection of heterogeneous, encapsulated applications (agents) that participate in the decision making process [5]. The architecture of the tool (VLPRO-GRAPH - Very Long Process Graph) is based on the assumption that the system will support the MPS creation in ERP system and will be plugged into an ERP system database by, for example, java connector. This architecture was introduced in [7]. A production process based on CIMOSA domains is presented in Figure 1 where domains are [7]:

- *D*1 - order and management planning,
- *D*2 - license documentation,
- *D*3 - Loads planning,
- *D*4 - Sales and production planning, scheduling for strategic machines,
- *D*6 - procurement,
- *D*7 - Quality, process and costs control,
- *D*8 - production sub-process 1
- *D*9 - production sub-process 2
- *D*10 - production sub-process 3

Figure 15.2 illustrates the structure and the amount of agents which can be found in each layer. This is an extended structure from [7]. A new agent has been introduced, i.e. *MR*-Agent (Movable Resource agent). The main task of this agent is checking the material domains - *D*6, *D*7, *D*8, *D*9, *D*10 for movable resources allocation and limits. The task of this agent is integrated with the planning process which is described at three layers reflecting to [6]: *A* - the whole process perspective, the so-called whole process planning; *B* - the entity level where the whole process plan is divided into sub-plans which are executed by each subprocess and transformed for individual production schedule at the domain level and where local re-planning activities take place; *C* - domain sub-layer where production control activities are executed and information about disturbances is gathered and passed to upper levels.

The graphical user interface agent creates a graphical user interface (GUI) for the interaction of the MAS (Multi-Agent System) with a production manager (direct users). The *GUI*-Agent is able to initialize and send behaviour parameters and messages to the Super-Visor Agent (*SV*-Agent). The *SV*-Agent is exactly one in the system because the data from all the Domain agents (*D*-Agent) are fused at this agent to generate re-planning schedules for the production. The *SV*-Agent is responsible for the control of the logic of all agents and creates the plans for the *D*-Agent. The *SV*-Agent starts the work of *MR*-Agent which checks the movable resources, allocations and limits. According to the results of *MR*-agent work, the decision of plans execution is prepared. The planning process is presented in Figure 15.3. *D*-Agents are

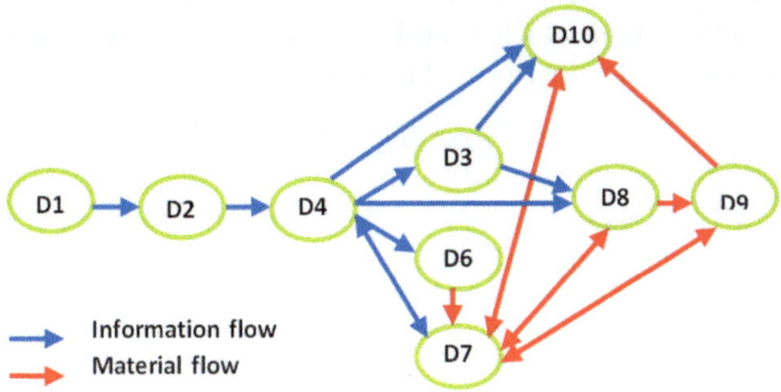

Fig. 15.2. Production process based on CIMOSA domains (based on [7])

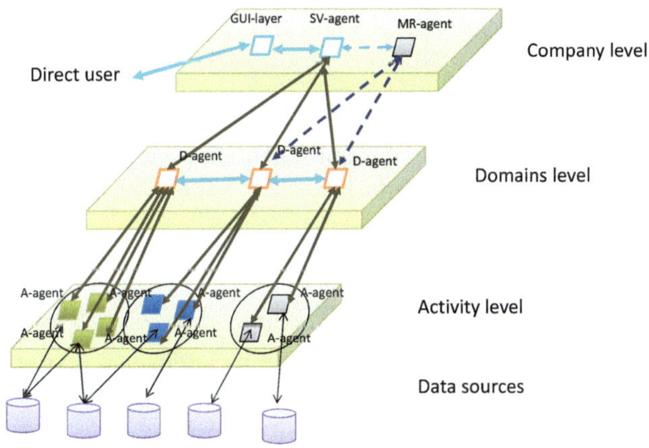

Fig. 15.3. Extended VLPRO-GRAPH agent model (based on [7])

initialized by the SV-Agent and they are responsible for the translation of the long process plan into detailed schedules. The agent is allowed to prepare the number of alternative (contingency) local plans as long as they do not conflict with the long MPS process. The local re-planning activities are allowed as long as they do not influence the long MPS process. When a re-planning activity affects the long MPS process, it has to be passed to the SV-Agent. The A-Agent is responsible for the control of plans execution within sub-process based on the given performance indicators. It reports to the D-Agent in the upper layer if production plans are executed according to the given MPS.

The work of the *MR*-agent is based on the operation of adding (shifting) movable resources between structural units called *RO*. *RO* (Resource Operation) is a structural unit which contains elements of a technological process and resources required to perform this process.

$$RO_i = (F', R'), \tag{15.1}$$

where F' is a technological process, which consists of a set of the threes describing each technological operation and can be described as follows:

$$F' = (f_i, \ldots, f_{ml}) \tag{15.2}$$

a and each threesome

$$f_i = (tpz, tj, \tau) \tag{15.3}$$

where:

- tpz - setup time,
- tj - operation unit time,
- τ - inter-operation break;
- ml - number of operations for 1 unit of assortments.

The parameter R' denotes resource requirements for each product offer as an assortment and is defined as follows:

$$R' = (r_m, \ldots, r_n), \tag{15.4}$$

with the following two main components:

- R_1 - defines the resources requirements of technological operations;

$$R_1 = [r_1, \ldots, (\ldots, r_i, \ldots), \ldots, m], \tag{15.5}$$

  where:

  –

$$r_i = \{z_i, \omega_i\}; \tag{15.6}$$

  where z_i - type of resource, $i \in [1, n]$, n - size of a resource set, ω_i - unit average usage of a resource,
  – m - number of operations;

- R_2 - defines other resources requirements;

$$R_2 = [\ldots, \{z_i, \omega_i\}, \ldots,] : \forall f_i \in RO_j, \forall k \neq j\ f_i \in RO_k \tag{15.7}$$

If triplets of f_i belong to a particular RO_j then they cannot belong to any other RO_k for each j which is different from k. A unit of assortment in the present paper is defined as a product of different complexity being manufactured within a production system.

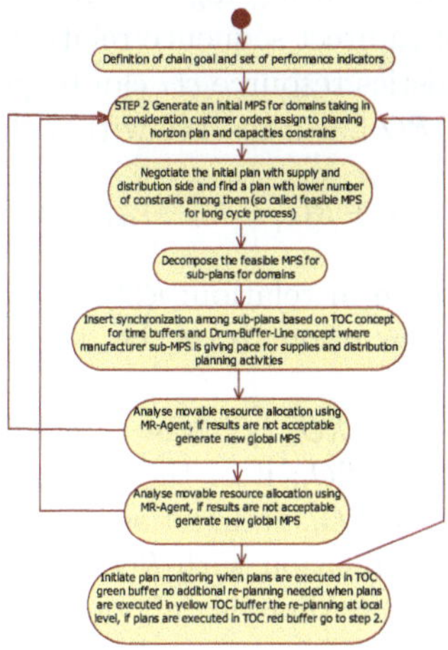

Fig. 15.4. Extended Planning algorithm for VLPRO-GRAPH agent model

Each element of resource requirements R' is connected with a parameter r_r - a requested resource to implement the production program P in the considered time interval Δt. It is a parameter determining the number of uses of a given resource required to implement the program P. Each element of resource requirements R' is connected with the parameter d_r - a disposed resource to implement production of the program P in the considered time interval Δt. It is a parameter determining the availability of a given resource while the program P is determined through the size of the resource expressed by appropriate units of measure.

The total demand for a resource for a given RO is a sum of individual demands for this resource, requested to produce all the details according to the production program and time per unit:

$$ZC_i = (\omega_l \cdot P_l \cdot t_l) \tag{15.8}$$

where: ω_l - denotes the usage of the resource, P_l - is the production program, t_l - denotes the time per unit, and $l \in [1, n]$.

The definition of a flow in the production structure is based on the requested and disposed resources. A flow with reserve occurs when RO has a reserve for any element of resource requirements R', which means that the difference between the required resource and the disposed resource is less than zero. A balanced flow occurs when RO is balanced for any element of resource requirements R', which means there is no difference between the required resource and the disposed resource. A flow with deficit occurs when RO has a deficit for any element of resource requirements R', which means that the difference between the required resource and the disposed resource is more than zero. The operation of adding (shifting) logistics resources symbol: '+'. For *RO*, which are in the direct or indirect sequence relation, an operation of adding (shifting) an arbitrary logistics resource *rm* can be performed only in the situation when the preceding *RO* has a reserve within the resource:

$$R_m \in RO_j = rm \in RO_i + rm \in RO_j : \; d_{rm} - r_{rm} \in RO_i > 0 \qquad (15.9)$$

where: s - size of *RO* files, which are in relation: $RO_j \rightarrow RO_i$ or $RO_j \Rightarrow RO_i$, and have a reserve within the given resource rm, i.e. the condition is met: $d_{rm} - r_{rm} > 0$.

Direct sequence relation symbol: $\rightarrow RO_j$ is in the direct sequence relation with RO_i, which is written as $RO_j \rightarrow RO_i$; if in the production process RO_j is a successive element in relation to RO_i. Indirect sequence relation symbol : $\Rightarrow RO_k$ is in the indirect sequence relation with RO_i, which is written as : $RO_k \Rightarrow RO_i$; if in the production process there is only one RO_j, which is in the relation: $RO_k \rightarrow RO_i$ and $RO_j \rightarrow RO_i$.

15.2.5 Numerical Experiment

The research was done based on the data collected for the sake of projects developed by the Institute of Management Technology at Poznań University of Technology and ordered by one of machine-building industry enterprises. The investigated object is a typical cooperation company, since all the finished products are units, components and parts which are passed on for further assembly. The company specializes in the production of units and components for agricultural machinery, especially:

- three-node suspension systems for tractors,
- straw shredders as parts of harvesters,
- other parts and units (e.g. frames for cultivators).

The machinery in the enterprise is typical for machine-building companies and it mainly consists of: turning lathes, milling machines, rolling mills, drilling machines, presses, frame saws and guillotine shears, fettling and cleaning machines, semi-automatic welder and automatic machine for oxygen-cutting and assembly, locksmith's and painting stands. A full list of machines grouped into RO (groups of uniform stands according to MRP II concept) is shown in Table 15.1. Table 15.2 presents input data for numerical experiment - technological routines for every produced detail and process goal.

Based on the data presented in Table 15.2 the process model was built as shown in Fig 15.4. To prepare the numerical experiment, special Java application (called ROgraph) was written. The main reason to choose the Java environment was that this environment is the same for Jade (Java Agent Development Framework). Jade is the framework which is used to implement VLPRO-GRAPH [7].

Table 15.1. List of RO used in numerical experiment

RO number	RO Name and Type
1	Semi-automatic welders (CO2) 10 pieces
2	Fettling machine OWD 1000, OPK 1000
3	Paint shop
4	Guillotine shears ScTP 16
5	Hydraulic press
6	Tumbling barrel OB1000 2 items
7	Abrasive cut-off machine BSA 5
8	Drilling machine SKT, BKR
9	Loftsman's stand
10	Turning lathes SU-50A, TPC-40x1500, TPC40x1000
11	Centreless grinder S88 75 2 pieces
12	Rolling mill UPW 12,5x70 3pieces
13	Multiple-spindle drilling machine BKG63
14	Drilling machine WKA40
15	Horizontal milling machine NRD
16	Milling machine FSS 400V
17	Oxygen-cutting stand
18	Automatic machine for oxygen-cutting
19	Locksmith's stand
20	Automatic lathe APA25 4 pieces
21	Hardening furnace PEK 3
22	Universal shears NU 25
23	Friction screw press OPC 120
24	Assembly stand

Table 15.2. Technological routines of production process

Details	**A**	**B**	**C**	**D**	**E**	**F**	**G**	**H**	**I**	**J**
Process	1 27	1 62	1 75	1 78	1 79	219	223	234	2 35	26 9
Goal	1 2000	1 1000	1 1000	1 2000	1 2000	110 00	11000	11000	1 1000	12 000
Operations	4	9	4	4	6	4	6	4	5	5
RO	1	1	1 0	6	2 1	2	1	2	2	2
RO	5	1 9	1 9	1 0	2	10	10	10	2 1	24
RO	2	2	2	1 4	1 0	19	25	24	1 0	10
RO	3	1 0	3	3	1 4	3	8	3	2 4	3
RO		9			1 2		14		3	21
RO		1 5			3		3			
RO		1 9								
RO		3								
RO		2 4								

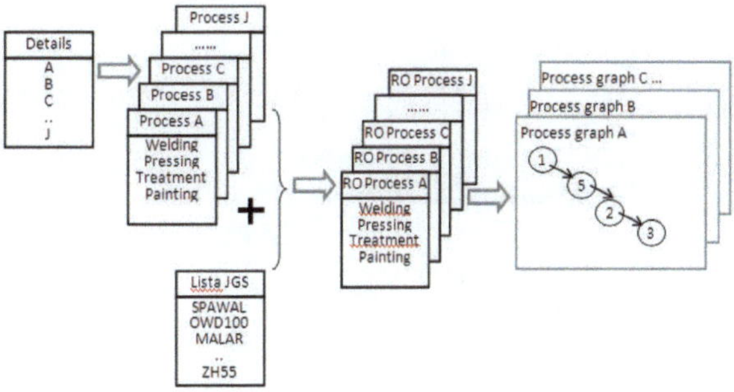

Fig. 15.5. Process model built from input data for numerical experiment

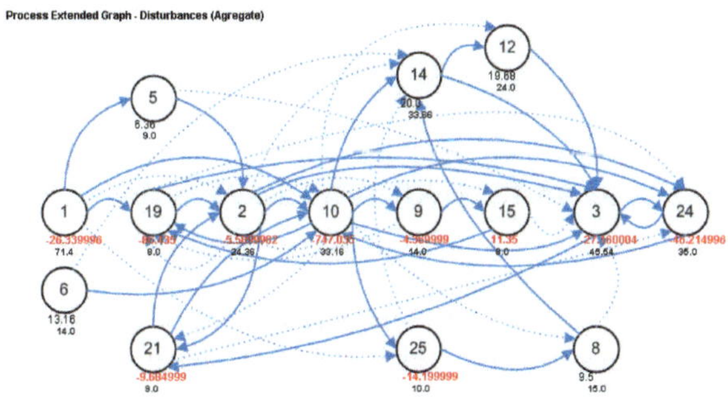

Fig. 15.6. Extended Planning algorithm for VLPRO-GRAPH agent model

Results of the numerical experiment are presented in Figure 15.5. ROgraph application generates a full process graph (solid lines present direct sequence relations and dotted lines present indirect relations) and writes numerical results on this graph. Vertex with red numbers indicates that there are flows with deficit and global MPS must be corrected; that means that a new global MPS will be generated.

15.3 GRAI METHODOLOGY and GRAIMOD

15.3.1 GRAI Methodology

GRAI Methodology is used for analyzing and designing enterprises [4]. GRAI approach is composed of four phases:

- An initialization phase to start the study,
- A modelling phase where the existing system is described,

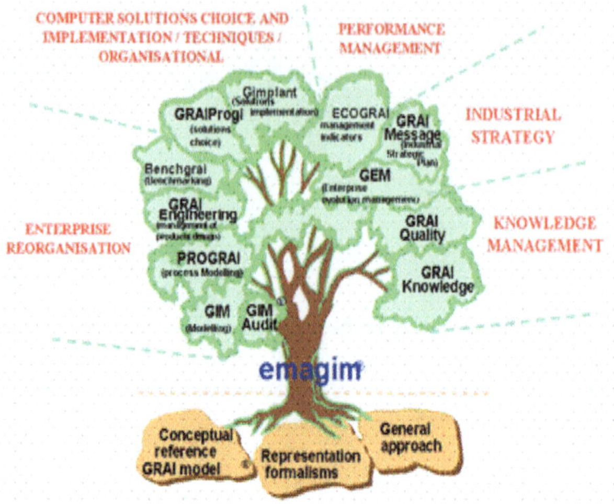

Fig. 15.7. The GRAI Methodology tree

- An analysis phase to detect the inconsistencies of the studied system,
- And a design phase during which the detected inconsistencies are corrected, and a new system is proposed.

The GRAI methodological tree is composed of five domains as clearly shown in Figure 15.2. Each domain contains modules used for modelling, analyzing and improving enterprises. GRAIPROGI and GRAI Quality can be translated into software forms, thereby, enabling enterprises to select the tools necessary for the optimization of the supply chain and managing their approach to quality. Then GRAISUC and GRAIQUAL are being developed.

This method is essentially used to choose and implement a computer tool (Supply Chain management and ERP) which meets the real market needs (globalisation, relocation, capacity to be proactive, cost optimisation, lead time, quality, flexibility, etc.).

The 'AS IS' models describe the existing system (see Fig. 15.3). The components of the system are already known, described and formalised. It is therefore possible to easily understand the system and pinpoint the strengths and weaknesses. The 'TO BE' corresponds to models for the future originating from the design phase of the GIM (GRAI Integrated Methodology) approach.

In the GRAIPROGI approach the 'SHOULD BE' illustrates the 'vision' of the future system and is considered as a key goal in the trajectory of the company development. The elaboration of 'SHOULD BE' consists in defining a very long term business plan including strategic objectives, from which development actions would be established.

Because the environment of the enterprise is in constant evolution, it is necessary to regularly update the 'SHOULD BE' in order to ensure the coherence of the development process. The definition of 'SHOULD BE' must allow to determine the key factors for the success of the enterprise. It is precisely these elements which will enable the enterprise to achieve its global objectives.

The 'AS IS' and 'TO BE' models have the same structures (physical, decisional, informational, functional and process models); The 'TO BE' model is therefore the result of a combination of the ambitions of the enterprise, the constraints of the existing system and the realistic aspirations taking into consideration the economic environment. An action plan is defined. The next stage is the execution of the action plan over the short, medium and long terms. At the same time, the 'TO BE' models are transformed into specifications needed to obtain the road book so as to, on the one hand, reorganise the enterprise and improve the performance and, on the other hand, to select the most suitable tool.

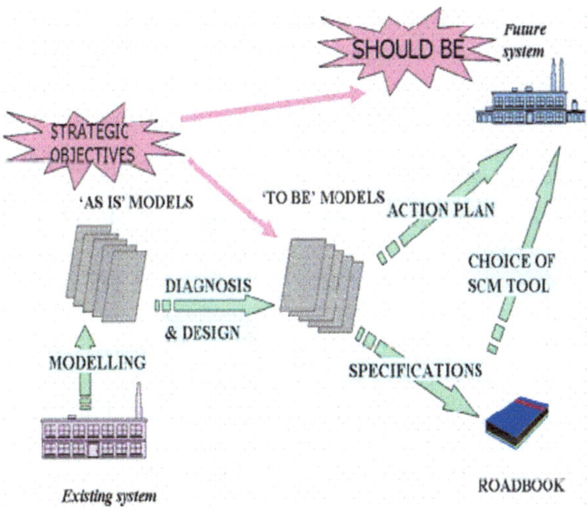

Fig. 15.8. Use of enterprise modelling for choosing an SCM tool

From the specifications we deduce:

- The global architecture contains the desired computer features (client/server architecture, data base SQL server...). This architecture can lead to the development of SCM software. Most frequently, however, enterprises normally choose already existing software.

- An action plan determines the evolution of the project with the choice and implementation of software. The specifications should also consider social, technical and human factors.
- The design phase is where the future solutions corresponding to the global architecture and the different projects are constructed (development), whereas the implementation phase is where the solutions are integrated into the system.
- Finally, the exploitation phase is where the system is exploited with its new components. During this period it is important to pay attention to the performance of the system.

To determine the future architecture, GRAIPROGI proposes the transformation of the existing architecture of the enterprise into specifications [9]. GRAIMOD is a tool developed for supporting the different phases of GRAI approach. This would lead to a new architecture for the enterprise and also the development in the computer domain. At present, no single solution exists to represent the architecture. However, the performance of the future tool is closely linked to the choice of architecture.

The 'TO BE' is used to define the specifications, allowing to take into account all the functionalities of the enterprise. The action plan classifies in terms of time the different simultaneous and sequential actions (potential solutions).This makes it possible to clearly identify the future architecture and its functionalities (see Fig. 15.4).The different functionalities are presented in a questionnaire form to the enterprise developing the software. Therefore, the software most adapted to the future functionalities of the system can be selected. The implementation stage is completed in a project form and leads to the progressive use of the chosen software. There is cohabitation between the old software and the new architecture. At this stage, the system of performance indicators must also be implemented in order to follow the development of the system and correct any deviations.

The exploitation phase refers to putting the software into operation and the correcting any difficulties which occur during use. This enables to define a number of points crucial for the future evolution of the system (i.e. access to the future installation of the aid to decision making - Data Mining.).

15.3.2 GRAIMOD

15.3.2.1 Concepts of GRAIMOD Modules

Different concepts were defined for developing GRAIMOD. For instance, we define knowledge as the process which transforms the whole set of known information C_i (stable state) into another C_{i+1}. Knowledge C_{i+1} can be therefore defined as a sum of disjointed information or as a progressive improvement of the whole - C_1 implying a restructuring of already acquired information. How

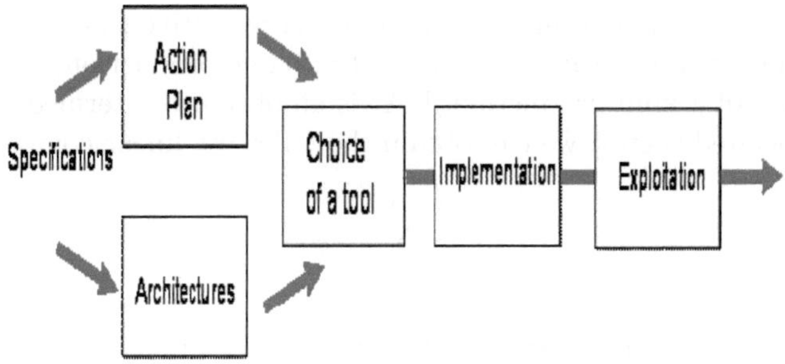

Fig. 15.9. Choice of an ERP or a SCM Software

does a child obtain understanding of the world around him? He integrates the new element with his already acquired knowledge and he structures his learning by employing actively what he has just experienced.

Both cases correspond to a refinement of knowledge by the addition of distinct (new) knowledge or the improvement of existing knowledge [7]. They are therefore used during the different phases leading to the choice and the implementation of a SCM tool.

This concept is used for defining three modes of knowledge representation:

- rules
- cases
- reference models

Then mechanisms of a problem solving method are associated to different types of reasoning. They are called upon by the expert system (GRAIXPERT) of GRAIMOD according to the considered design step, the acquired knowledge on the design domain and the nature of the design (innovative, etc.) [7], [8]. Therefore, we have two problems to solve: the enterprise typology and, at the same time, the creation of pertinent generic models. The use of generalisation and classification reasoning not only allows to solve these problems, but also to carry out the design of the new system. The classification reasoning allows to elaborate enterprise classes C_i from pertinent criteria c_p. The first criterion transforms the set of enterprises into intermediate subsets. The other one divides these subsets into reduced elements until the obtaining of more elaborated classes. This organisation is progressive according to changes of the pertinent criteria of the enterprise. An increase in the amount of criteria improves the typology and increases the number of classes. We define the set S_p of enterprises which is characterised by the vector s_p. We also define for each criterion c_p a vector α as the number of columns m corresponding to different attributes (values) associated to the criteria.

$$\alpha = (\alpha_1, \ldots, \alpha_j, \ldots, \alpha_m)^T \tag{15.10}$$

For example, the application of the 'production repetitiveness' criteria implies the obtaining of three production attributes: batch production, mass production and one of a kind production. If X_j is an obtained intermediate set and x_j the associated vector, we can obtain the following linear combination:

$$S_p = \sum_{j=1}^{m} x_j \cdot \alpha_j \tag{15.11}$$

The union of the intermediate sets X_j is the global set S_p of enterprises. The following linear combination is deduced for three criteria corresponding to the associated vectors α, β and γ:

$$S_p = \sum_{j=1}^{m} \alpha_j \cdot \left\{ \sum_{k=1}^{r} \beta_k \cdot \left\{ \sum_{i=1}^{n} \gamma_i \right\} \right\} \tag{15.12}$$

where m, r and n correspond to the number of column (attributes) of α, β and γ and c_i a class of enterprise. The number of enterprise classes not only varies according to the chosen criteria, but also according to the number of columns. The classification could be changed dynamically according to enterprise typology modifications or changes in criteria.

According to this typology, the structure of rules and reference models corresponding to each part were defined. Then, for each part a base containing studied cases were elaborated. These kinds of knowledge were used for designing a new enterprise model.

15.3.2.2 The Use of CBR

Case Based Reasoning allows to solve new problems by using the knowledge acquired from previous cases [2], [4], [5]. This reasoning is used, for example, in GRAIMOD for defining the structure of GRAIXPERT, a hybrid expert system (see Fig. 15.11).

The architecture of this system contains three different bases for managing the study of a new case. A model base is used for managing elaborated reference models. A rule base is used for analyzing the models of the system in question. And a case base is defined for capitalizing different studies for future use.

This system also contains different modules. For each new case, an old case with a similar context will be extracted from the case base, adapted in order to improve the new system (adaptation module and case builder), according to CBR rules.

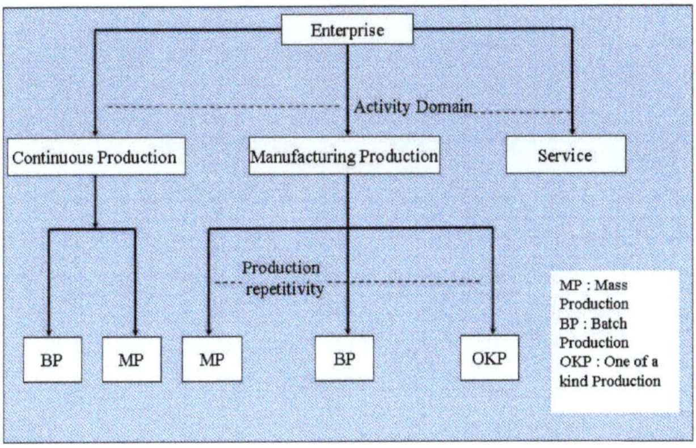

Fig. 15.10. Typology of enterprises

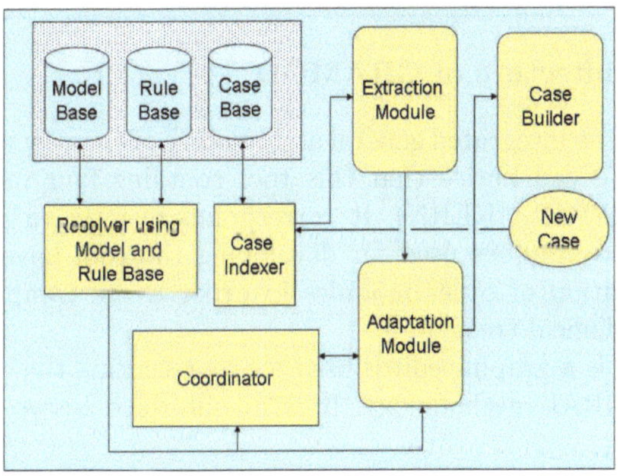

Fig. 15.11. Architecture of a Hybrid system

This architecture implies the development of a problem solving method associated to the different phases of the methodology [12].

This problem solving method (see Fig. 15.12) combines CBR with different types of reasoning (decomposition, direct correspondence, transforming reasoning). The next step after the definition of the tool concepts is the development. The combining of CBR approach with Multi-Agents system theories is proposed. The objective is to obtain an efficient tool for improving enterprise performance and particularly the supply chain performance.

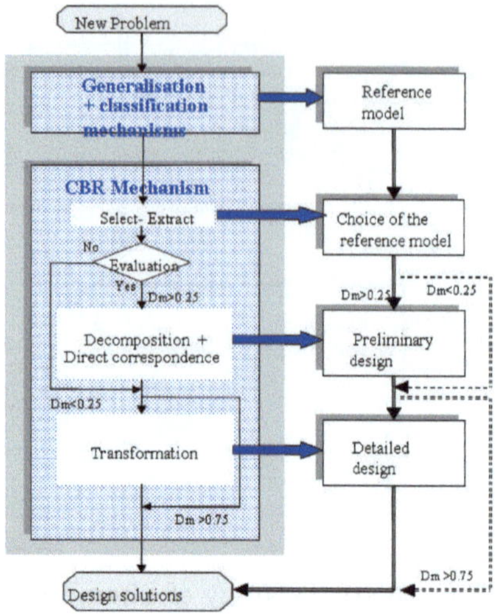

Fig. 15.12. Problem solving method

15.3.2.3 Architecture of GRAMOD

In Figure 15.13 the integrated general architecture of the new tool GRAIMOD is presented. We can notice that this tool contains four modules working around a kernel (GRAIKERN). It corresponds exactly to the actual level of design. The techniques used for developing the tool have to integrate a progressive addition of other modules in order to be compatible with the GRAI methodological tree.

GRAIKERN is a graphic editor used for representing the different models associated to GRAI methodology. It is an interface between the different modules.

GRAIXPERT is a hybrid expert system for managing the analysis of the existing system and proposing a new system. It is composed of two sub-modules in interaction with GRAIKERN: the Knowledge Capitalization (KCM) and the Knowledge Based System (XPERTKBM) [3], [4], [5].

The knowledge based system contains a rule base used for analyzing the different models in order to detect inconsistencies and propose corrections. A dictionary is used to translate the user's expressions into standard expressions provided by the GRAI methodology.

The knowledge capitalization process needs some aptitudes to manage different know-how and points of view. It must integrate this knowledge in an accessible, usable and maintainable form. It offers an expertise model based on the knowledge of the experts but also on previously carried out studies. The capitalization module is composed of an acquisition module for integrating other expert knowledge, a case base for capitalizing cases and reusing them during a new modelling and a reference model base containing models according to different types of enterprise domain. For elaborating the reference

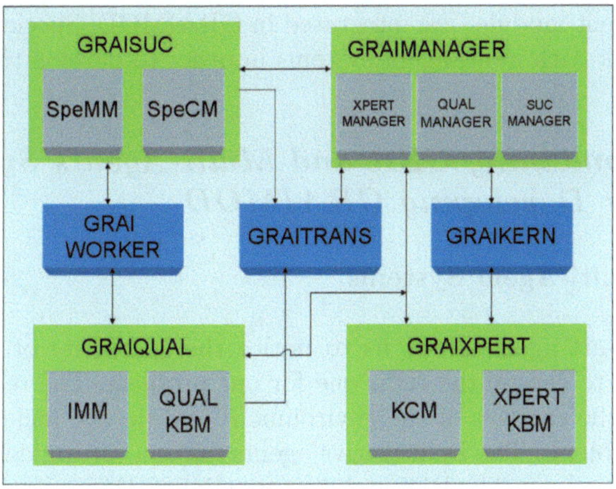

Fig. 15.13. Architecture of GRAIMOD

models, a production typology is developed. This typology is improved by the addition of new criteria.

GRAIMANAGER is a management module used for organising the different interactions between the modules of GRAIMOD. It controls and manages the system's interactions with the users. It presents the users with appropriate questions and choices together with the necessary information about the characteristic of the studied enterprise. It also manages the rules classified according to a typology of production systems. Its main tasks are the modification, suppression or selection of the applicable rules in a given context. It is also used for the loading and the saving of rule files. Finally, it controls the design process, different actions of the sub-modules and their interactions. The GRAIWORKER is the work base elaborated for managing, modifying and capitalising knowledge about the studied case. GRAITRANS is a Transfer Interface used for putting the new case in GRAIXPERT in order to improve its Cases Base. The reference model elaborated for each enterprise domain will be improved by the acquisition of this new model in GRAIXPERT [6], [7], [8].

GRAISUC is a module used for managing the choice of an ERP or SCM tool for an enterprise. It is composed of two sub-modules SpeMM and SpeCM. The Specification Management Module (SpeMM) is used for choosing the appropriate ERP or SCM Tool of an enterprise. The specifications obtained are capitalised in the Specification Capitalisation Module (SpeCM).

GRAIQUAL is a module used for managing quality approach implementation or quality improvement in an enterprise. It contains two sub-modules IMM and QUALKBM. The Improvements Management Module (IMM) is used for managing the different quality action plans of the enterprise. It contains different quality tools. The Quality Knowledge Base Module (QUALKBM) is elaborated for containing the rules related to quality certifications in order to use them for improving or elaborating quality in an enterprise.

These different modules are integrated in GRAIMOD. Special attention is paid to the link with multi-agent systems in order to develop this new tool.

15.3.3 *Combining CBR and Multi-Agents Systems for Developing GRAIMOD*

15.3.3.1 Multi-Agent Systems

A state-of-the-art agent allows us to notice the specificity of each type of agent in order to choose the right one for our problem. The reactive agents react only to the changes of the environment and do not include reasoning for solving problems. The Deliberative agents make a certain deliberation for choosing their actions. The BDI architecture (Belief, Desire, Intentions) is an approach used for the design of deliberative agents. A set of beliefs represents information of the agent about its environment. A beliefs revision function takes the sensor inputs and actual beliefs of the agent and determines the new set of beliefs [9]. The options generation function determines the available options for the agent (i.e. desires) by using actual beliefs on the agent about its environment and its actual intentions. A set of desires represents the options available to the agent. A filter function represents the deliberation process of the agent and determines the intentions of the agent by using its desires, its beliefs, and actual intentions. A set of actual intentions represents the actual attention centre of the agent. A selection function of actions determines the action to be taken by using the actual intentions of the agent.

The training agent is composed of four parts. The first part is the critical one which informs the training module how good the action of the agent is (see Fig. 15.14). The training module uses a certain retroaction on the actions

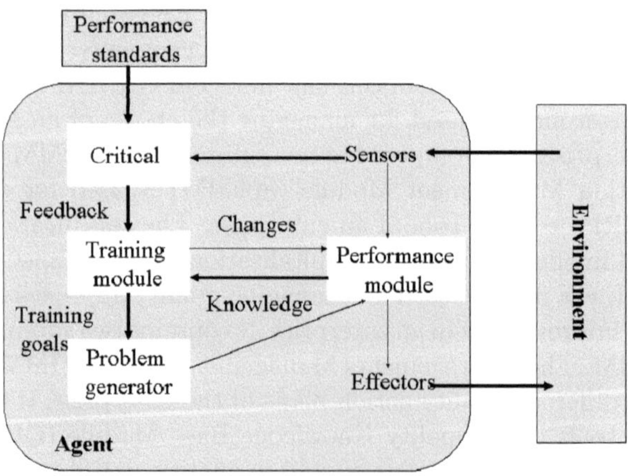

Fig. 15.14. General model of training agent

of the agent for determining how the performance module could be changed for improving the system in the future [10], [13]. The performance module is seen as the global agent when there is no training.

The problems generation provides suggestions of actions for inciting the agent to do exploration. This could allow to choose some solutions which are non-optimised in the short term but that could be the best in the long term.

15.3.3.2 Combining CBR and Multi-Agent Systems

The architecture of the new tool is based on Case Based Reasoning. The use of multi-agent systems will allow to facilitate the development of GRAMOD.

Then CBR needs to be related to Multi-agents systems in order to satisfy user requirements. The reactive agents are not appropriate to our problem because they react only for the environment changes. The BDI architecture could be used but we chose the training agents because of their learning particularity [1], [14], [16]. This system is used for defining how to improve the enterprise performance and for capitalizing the followed process. According to CBR concepts, the new case could be capitalized, but the parameters would also improve the reference model of the enterprise domain. Java has been chosen for developing GRAIMOD. The Jade platform is used in relation with FIPA-ACL for developing the different modules of GRAIMOD (see Fig 15.15). Some changes could be integrated according to the opinion of Jade specialists.

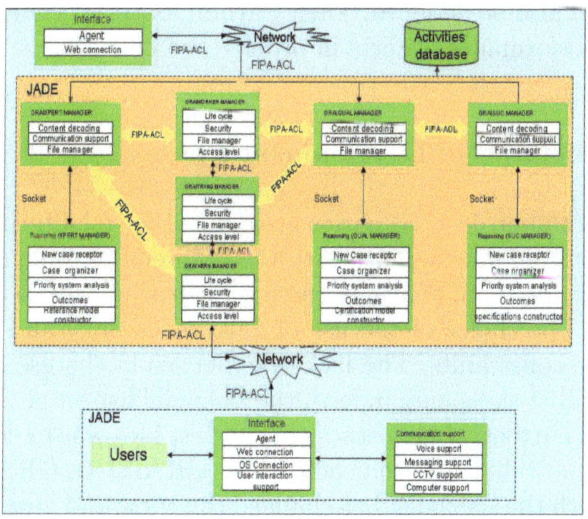

Fig. 15.15. Architecture of GRAIMOD in a Java environment

15.3.3.3 GRAIMOD and Supply Chain Performance

The global objective of this research is to be more efficient in the improvement of enterprises. The supply chain of each enterprise could be reorganized by using the elaborated concepts. The reorganization takes into account both the production typology and the supply chain in the modelling management.

For each enterprise, the supply chain is decomposed into different parts (sourcing, procurement, purchasing, production, distribution, sales, transport and logistics management). For each part GRAIQUAL is used and a quality approach is defined in order to improve this part. Indeed, the optimizing of each part is coherent with the other parts.

The importance of the CBR reasoning is evident because of the use of knowledge capitalization realized for each enterprise domain, for each part of the enterprise supply chain. But the necessity of multi-agent systems is to justify this choice. For instance, with the implementation of SQA in an enterprise, not only will the knowledge relating to this enterprise domain be used, but also the system will evolve during this implementation. The system both provides the new case with data and takes into account the particularity of this new case. The defined multi-agent system is well-suited to this kind of work: use and capitalization of knowledge.

Multi-agent systems architecture also facilitates the communication between each different module of GRAIMOD by defining connections from each module to the others. The improvement of quality in the whole enterprise has a positive impact on the cost and delivery date. The integration of environmental demands and the transformation of these constraints into marketing advantage could also be deduced. Then Carbon footprint could be considered as the fourth performance criteria in order to be efficient.

15.3.3.4 Difficulties of Development Phase

At this stage of the project, scenarios are defined for facilitating the development of the new tool in coherence with the different factors presented above.

The first difficulty is to translate these concepts into terms appropriate for IT developers. The choice of the use of Java technology and Jade Platform integrates some constraints. The lack of expertise in the use of Java implies the use of specialist skills and, indeed, the financial aspect of the employment of a specialist is currently being studied. The first idea was to develop different tools with compatibility between each other (GRAISUC, GRAIXPERT ...). The meeting with the specialist has changed this view and now the objective is to develop only one tool and different sub-modules (in interaction) of this tool. It means the change of the first scenario of development and the elaboration of another one. It seems that multi-agent systems have to be elaborated separately and then integrated and managed by one tool. The knowledge of experts will definitively define our position.

The last difficulty is related to the type of an open tool. By using Java, the objective is to obtain an open tool. But the rights associated to this tool have to be defined. It is difficult to obtain collaboration with a software development company if there is no money for them after the development. The second possibility would be to define and obtain a special research budget for this development.

15.4 Comparing VLProGraph for CIMOSA and GRAIMOD for GRAI

The concepts and formalisms of CIMOSA and GRAI can be compared. Both have advantages and disadvantages but the objective of this chapter is to compare the use of these methodologies and what they could bring to enterprise improvements.

CIMOSA is defined and used from the management level to the shop floor level, so it is well-adapted for improving enterprises internally. The formalisms are organised for studying in detail each part of the system and implementing in this part the defined improvements. The architecture is similar to the "theory of objects". It means the use of decomposition reasoning for elaborating sub-objects by respecting three axes:

- Generic, partial and specific,
- Function, information, resource and organisation,
- Requirements, Design, and Implantation.

In this way CIMOSA is precise and tools supporting this methodology could be used directly in the information system. The main advantage of CIMOSA is the quality of the structure (architecture).

GRAI is defined for finding both internal and external inconsistencies of the system and correcting them in order to define the optimum organisation of the system. It is based on the system theory and could not be implemented in an information system. It is an external methodology for measuring performance and improving the system. An enterprise could be divided into three subsystems and associated to five models:

- Physical,
- Decisional,
- Informational,
- Process,
- Functional

The models correspond to different views of the same system. The main advantage of GRAI is the complete definition of the decisional model, and the ability to detect inconsistencies in order to correct them.

Table 15.3. Comparison between CIMOSA and GRAI

CIMOSA		GRAI	
STRENGTHS	**WEAKNESSES**	**STRENGTHS**	**WEAKNESSES**
CIMOSA supporting tools could be integrated in enterprise information system	Lack of Human dimension	Human dimension more developed	Requirements definition modelling not so clearly defined
Structuring of Reference Architecture into generic and partial modelling levels	Bloc definitions not always clear	Presence of Decisional structure	Success and quality of modelling depends on expert knowledge
Support for function, information, resource, and organisation of modelling views	Requirements for the definition of modelling not so clearly defined	Addition of the decisional view	GRAI Supporting tools could not be integrated in an enterprise software tool
Specifying the processes and exchange of information using domain processes, events, and object views	No reference models to design the system	Reference models have been developed	Process view is less efficient than CIMOSA process view

VLPROGRAPH is developed as a tool for improving enterprise processes. We could improve material flows and informational flows of a manufacturing system. This tool is able to be connected to ERP tools in order to use their data. So, it is a module which could be implemented in an enterprise. The following Porter diagram shows the advantages and weaknesses of VLPROGRAPH (see Fig. 15.16).

GRAIMOD is essentially used for modelling, analysing, designing and improving enterprises. It would not be implemented in enterprises but rather used as a tool for:

- choosing and implementing ERP or SCM tools,
- detecting and correcting inconsistencies in order to define an action plan,
- defining and implementing quality approach,
- integrating social, societal and environmental dimensions in global enterprise improvements.

The following Porter diagram presents the advantages and weaknesses of this tool (see Fig. 15.17).

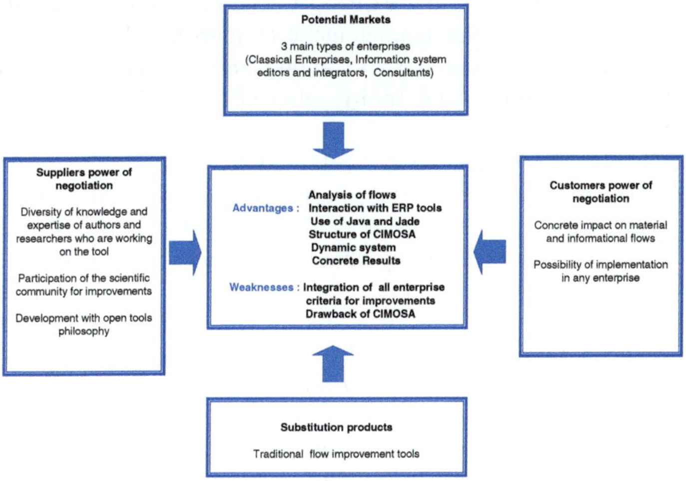

Fig. 15.16. Porter diagram of VLPROGRAPH tool

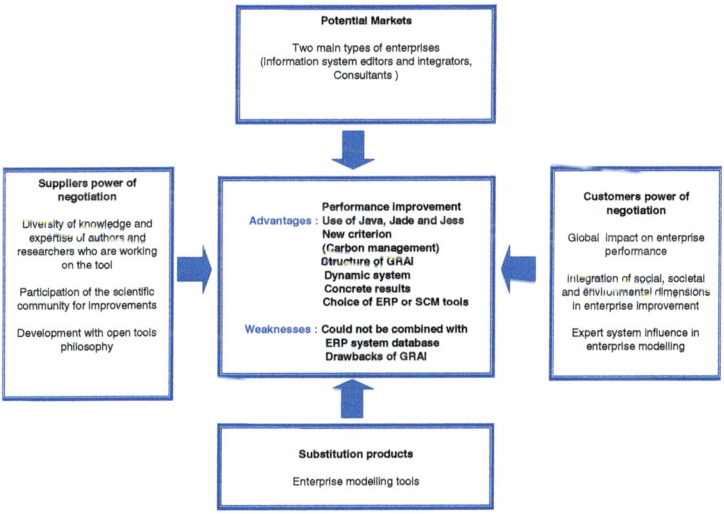

Fig. 15.17. Porter diagram of GRAIMOD

The multi-agent architecture VLPRO-GRAPH based on CIMOSA concepts is well-adapted to the management of material flows in production systems and can be compared to extended actigramme models of GRAI Methodology. For instance, this tool is efficient for modelling a process and defining a new model.

The concepts of GRAIMOD are adapted for managing enterprise performance. The structure of the rule base can be compared to the cube of CIMOSA. For instance, this tool is efficient for adapting (cost, lead time, and carbon management) a supply chain to the actual crisis and improving enterprise performance.

GRAI Methodology and CIMOSA are not the same but some similarities could be found between concepts and tools supporting these methodologies. VLPRO-GRAPH is used at the operational level and GRAIMOD also covers the tactical and the strategic levels.

In addition, both tools are being developed by using multi-agent concepts. It appears that they could be combined for improving globally and in detail the performance of enterprises.

15.5 Perspectives – Conclusion

Enterprises need to reorganise themselves in order to resist the pressures of present economic conditions and then to improve their performance after the crisis. Two methodologies make it possible to aid enterprises with these improvements: CIMOSA and GRAI. VLPRO-GRAPH, a tool for supporting the CIMOSA approach, allows to work in detail on the processes and flows of enterprises and makes them well-organised. GRAIMOD, a tool for supporting GRAI Methodology, is used essentially for improving global enterprise performance.

This chapter presents the extended planning process based on the multi-agent architecture and the special agent called MR-agent which checks the potential problems with material flow in the manufacturing system. The results of numerical experiments are presented as well as the concepts and architecture of GRAIMOD.

The next steps will be performed in two ways: the idea of the first one is to integrate the ROgraph application which can present results with VLPRO-GRAPH architecture; the other way is to do research on results interpretation and prepare the scenarios for responding to problematic situations in the manufacturing system.

For GRAIMOD, the particularity of Icam Vendee (Icam Group), where students spend 50% of their Engineering degree course in industrial apprenticeship, is a great opportunity to obtain applications for improving reference models and the defined case base. GRAIMOD is developed in a Java environment, particularly with Jade platform and FIPA-ACL Language. The possibility of using Jess is studied in order to integrate clearly only Java concepts in the development of the tool.

The development of GRAIMOD associated to GRAI Methodology for managing the different phases of this method and improving enterprises will allow to upgrade the expertise given to SMEs for helping them during the crisis and preparing them for the future.

References

1. Aamodt, A.: Case-Based Reasoning: foundational issues, methodological variations, and system approaches. Artificial Intelligence Communications 7(1), 39–59 (1994)
2. APICS: Dictionary, 12th edn. America Production and inventory Control Society, Inc., Falls Church (2008)
3. Boszko, J.: Factory Organization Structure and Optimization. WNT, Warszawa (1973) (in Polish)
4. Brown, D.C., Chandrasekaran, B.: Expert system for a class of mechanical design activities. In: Gero, J.S. (ed.) Knowledge Engineering in CAD, pp. 259–282. Elsevier, Amsterdam (1985)
5. Burke, E.K., et al.: Structured cases in case-based reasoning reusing and adapting cases for time-tabling problems. The Journal of KBS 13(2-3), 159–165 (2000)
6. Chen, D., Doumeingts, G., Vernadat, F.B.: Architectures for enterprise integration and interoperability. Past, present and future. Computers in Industry 59, 647–659 (2008)
7. Dossou, P.E., Mitchell, P.: Using case based reasoning in GRAIXPERT. In: Proceedings of FAIM 2006, Limerick, Ireland (2006)
8. Dossou, P.E., Mitchell, P.: Implication of Reasoning in GRAIXPERT for modeling Enterprises. In: Omatuet, S., et al. (eds.) Distributed Computing Artificial Intelligence, Bioinformatics, Soft Computing, and Ambient Assisted Living, pp. 374–381. Springer, Heidelberg (2009)
9. Dossou, P.E., Mitchell, P.: How Quality Management could improve the Supply Chain performance of SMES. In: Proceedings of FAIM 2009, Middlesbrough, United Kingdom (2009)
10. Dossou, P.E., Pawlewski, P.: Using Multi-agent system for improving and implementing a new enterprise modeling tool. In: Damazeu, Y. (ed.) Trends in Practical Applications of Agents and Multiagents Systems, pp. 225–232. Springer, Heidelberg (2010)
11. Karageorgos, A., Mehandjiev, N., Weichhart, G., Hammerle, A.: Agent based optimisation of logistics and production planning. Engineering Application of Artificial Intelligencel 16, 335–348 (2003)
12. Pawlewski, P., Goliñska, P., Fertsch, M., Tujillo, J., Pasek, Z.: Multiagent approach for supply chain integration by distributed production planning, scheduling and control system. In: Corchado, J.M., et al. (eds.) International Symposium on Distributed Computing and Artificial Intelligence (DCAI 2008). Advances in Soft Computing, pp. 29–37. Springer (2008)

13. Russels, S.J., Norvig, P.: Artificial Intelligence. A Modern Approach. Prentice-Hall, Englewood Cliffs (1995)
14. Sen, S., Weiss, G.: Learning in Multiagent Systems. In: Weiss, G. (ed.) Multiagent Systems: A Modern Approach to Distributed Artificial Intelligence, ch. 6, pp. 259–298. The MIT Press, Cambridge (1999)
15. Shen, W., Wang, L., Hao, Q.: Agent-Based Distributed Manufacturing Process Planning and Scheduling: A state-of-art survey. IEEE Transactions on Systems, Man and Cybernetics-Part C: Applications and Reviews 36(4), 563–577 (2006)
16. Wooldridge, M.: Intelligent Agents. In: Weiss, G. (ed.) Multiagent Systems: A Modern Approach to Distributed Artificial Intelligence, ch. 1, pp. 27–77. The MIT Press, Cambridge (1999)
17. Xia, Q., et al.: Knowledge architecture and system design for intelligent operation support systems. The Journal Expert Systems with Applications 17(2), 115–127 (1999)
18. Zaborowski, M.: Enterprise Resources The Follow-up Control (in Polish). Wydawnictwo Pracowni Komputerowej Jacka Skalmierskiego, Gliwice (2008)

Author Index

MIX
Papier aus verantwortungsvollen Quellen
Paper from responsible sources
FSC® C105338

If you have any concerns about our products,
you can contact us on
ProductSafety@springernature.com

In case Publisher is established outside the EU,
the EU authorized representative is:
Springer Nature Customer Service Center GmbH
Europaplatz 3, 69115 Heidelberg, Germany

Printed by Libri Plureos GmbH
in Hamburg, Germany